Animal Science: Production and Management of Farm Animals

Animal Science: Production and Management of Farm Animals

Editor: Ashlie Archer

www.callistoreference.com

Callisto Reference,
118-35 Queens Blvd., Suite 400,
Forest Hills, NY 11375, USA

Visit us on the World Wide Web at:
www.callistoreference.com

ISBN: 978-1-64116-244-9 (Hardback)

Cataloging-in-Publication Data

Animal science : production and management of farm animals / edited by Ashlie Archer.
    p. cm.
Includes bibliographical references and index.
ISBN 978-1-64116-244-9
1. Domestic animals. 2. Livestock. 3. Animals. 4. Animal culture. 5. Animal handling.
6. Livestock productivity. I. Archer, Ashlie.
SF41 .A55 2020
636--dc23

Table of Contents

Permissions

List of Contributors

Index

Preface

This book has been an outcome of determined endeavour from a group of educationists in the field. The primary objective was to involve a broad spectrum of professionals from diverse cultural background involved in the field for developing new researches. The book not only targets students but also scholars pursuing higher research for further enhancement of the theoretical and practical applications of the subject.

Animal science is the science concerned with the production and management of animals with potential human benefit. Farm animals like livestock and poultry are raised for milk, meat, eggs and other items. Modern animal production systems rely on animal breeding techniques. Breeding seldom occurs spontaneously. It is initiated to encourage the appearance of desirable traits, such as prolificness, fast growth rates, hardiness, low feed consumption per unit of growth, higher yields, improved fiber qualities, etc. Selective breeding has increased productivity immensely. Purebred breeding is achieved through inbreeding and culling. The advent of transgenic animals and better ways to control gene expression has resulted in significant developments in animal science. This book outlines the processes and applications of biotechnology for agricultural production in detail. It traces the progress in the field of animal science and highlights some of its key concepts and applications. This book is a resource guide for experts as well as students.

It was an honour to edit such a profound book and also a challenging task to compile and examine all the relevant data for accuracy and originality. I wish to acknowledge the efforts of the contributors for submitting such brilliant and diverse chapters in the field and for endlessly working for the completion of the book. Last, but not the least; I thank my family for being a constant source of support in all my research endeavours.

Editor

Effects of *Bacillus subtilis* CSL2 on the composition and functional diversity of the faecal microbiota of broiler chickens challenged with *Salmonella* Gallinarum

Ju Kyoung Oh[1†], Edward Alain B. Pajarillo[1†], Jong Pyo Chae[1], In Ho Kim[1], Dong Soo Yang[2] and Dae-Kyung Kang[1*]

Abstract

Background: The chicken gastrointestinal tract contains a diverse microbiota whose composition and structure play important roles in gut functionality. In this study, microbial shifts resulting from feed supplementation with *Bacillus subtilis* CSL2 were evaluated in broilers challenged and unchallenged with *Salmonella* Gallinarum. To analyse bacterial community composition and functionality, 454 GS-FLX pyrosequencing of 16S rRNA gene amplicons was performed.

Results: The Quantitative Insights into Microbial Ecology (QIIME) pipeline was used to analyse changes in the faecal microbiota over a 24-h period. A total of 718,204 sequences from broiler chickens were recorded and analysed. At the phylum level, Firmicutes, Bacteroidetes, and Proteobacteria were the predominant bacterial taxa. In *Salmonella*-infected chickens (SC), Bacteroidetes were more highly abundant compared to control (NC) and *Bacillus*-treated (BT) chickens. At the genus level, in the NC and BT groups, *Lactobacillus* was present at high abundance, and the abundance of *Turicibacter*, unclassified *Enterobacteriaceae*, and *Bacteroides* increased in SC broilers. Furthermore, taxon-independent analysis showed that the SC and BT groups were compositionally distinct at the end of the 24-h period. Further analysis of functional properties showed that *B. subtilis* CSL2 administration increased gut-associated energy supply mechanisms (i.e. carbohydrate transport and metabolism) to maintain a stable microbiota and protect gut integrity.

Conclusions: This study demonstrated that *S.* Gallinarum infection and *B. subtilis* CSL2 supplementation in the diet of broiler chickens influenced the diversity, composition, and functional diversity of the faecal microbiota. Moreover, the findings offer significant insights to understand potential mechanisms of *Salmonella* infection and the mode of action of probiotics in broiler chickens.

Keywords: *Bacillus subtilis*, Broiler chicken, Microbiome, *Salmonella* Gallinarum, 16S rRNA gene

Background

Poultry is one of the most important meat sources for humans [1]. Due to the increasing demand for food, chicken production has increased tremendously in the past few years [1]. Therefore, husbandry and management are vital for preventing infection and maintaining the health of poultry. Animal health is closely associated with the status of the gastrointestinal tract, whose disruption or dysbiosis leads to detrimental effects [2]. Intestinal homeostasis and functionality are influenced by various factors, such as (1) diet and feed additives, (2) farm conditions and practices, and (3) the resident gut microbiota [3]. The microbiota comprises trillions of microorganisms localised primarily at the distal end of the gastrointestinal tract [4]. These microbial communities mediate digestion of feedstuff, control gut homeostasis, and prevent infection.

Salmonella enterica is an important group of gastrointestinal pathogens that causes food-borne diseases, gastroenteritis, and diarrhoea in animals and humans [2, 5, 6]. Several

* Correspondence: dkkang@dankook.ac.kr
†Equal contributors
1Department of Animal Resources Science, Dankook University, 119 Dandae-ro, Cheonan 31116, Republic of Korea

serovars of *S. enterica* subsp. *enterica*, specifically *S.* Enteritidis and *S.* Gallinarum, are major avian pathogens. These serovars are frequently associated with poultry salmonellosis, which results in severe morbidity and mortality [6]. The pathogenicity and routes of transmission of *Salmonella* spp. have been investigated extensively; however, few studies have addressed *S.* Gallinarum infection and its control in poultry [6].

Probiotics are live microorganisms that confer a wide range of benefits on animals, such as stimulation of immune responses, maintenance of gut barrier function, and prevention of pathogen invasion of the gut [1, 7] The gram-positive bacterium *Bacillus subtilis* has been used as an in-feed probiotic supplement for livestock and poultry [8, 9]. Only recently have studies begun to address the effects of probiotic interventions on the gastrointestinal microbiota. However, previous studies using plating methods, 16S rRNA gene clone libraries, denaturing gradient gel electrophoresis, and so forth have yielded relatively little information [9]. In contrast, high-throughput next-generation sequencing (NGS) methods facilitate rapid quantification and identification of bacterial communities [4, 10]. In addition, simultaneous analysis of multiple samples enables a thorough understanding of microbial communities and therefore prediction of the effect of interventions and infections on microbial functions and community diversity [11]. This study applied 16S rRNA gene sequencing to investigate the composition of the chicken gut microbiota, and the findings suggest a marked effect of *S.* Gallinarum on the overall composition and metabolic functions of the chicken gut microbiota. In addition, the data suggest that *B. subtilis* CSL2 exerts protective effects against *S.* Gallinarum infection by altering the faecal microbiota of broiler chickens.

Methods
Animals and experimental design
A total of 36 Ross-308 broiler chickens were bred and divided into the following three groups: the control group (NC; *n* = 12), *Salmonella*-challenged (SC; *n* = 12) group, and *Bacillus*-treated (BT; *n* = 12) group. From day 1, the NC and SC groups were given the standard basal diet (Additional file 1: Table S1), while the BT group was fed a probiotic-supplemented basal diet containing *Bacillus subtilis* CSL2 (GenBank accession number: KX281166) at 1.0×10^7 colony forming units (CFU)/g feed. Freeze-dried *B. subtilis* CSL2 (1.0×10^{10} CFU/g) diluted with basal diet by mixing for 2 h with a feed mixer (Daedong Tech, Korea), to obtain final concentration of 1.0×10^7 CFU/g feed. Feeding of the respective diets continued until d 31 (before). On d 32, chickens in the SC and BT groups were orally challenged with *S. enterica* subsp. Gallinarum KVCC-BA0700722 (*S.*

Gallinarum) at 1.0×10^8 CFU/mL. The Dankook University Animal Care Committee approved all animal protocols.

All 36 chickens (12 birds per treatment) were tagged and placed in cages randomly, which were equipped with a nipple water dispenser for ad libitum access to water, together with a one-sided self-feeder. The room temperature was maintained at 32 °C for the first week, and then reduced 3 °C weekly until the temperature reached 26 °C. The broiler chickens received no antibiotics or other additives during the study period. No additional chickens were introduced during the duration of the experiment. At 31 and 32 d, 12 chickens per treatment were selected for faecal sampling. Fresh faecal samples were collected aseptically from the rectum of the broiler chickens (on d 31: before; on d 33: after). Finally, faecal contents were placed into sterile tubes and kept on ice until used for microbiota analysis on the same day of the sampling.

Sample preparation and DNA isolation
Genomic DNA isolation from freshly collected faeces was carried out using a technique described previously [11]. Briefly, 0.3 g of faecal extract was placed in a bead-beating tube containing garnet beads. Lysis of host and microbial cells was mediated by both mechanical collisions between beads and chemical disruption of cell membranes. DNA was purified using a Power Faecal® DNA Extraction Kit (MO BIO Laboratories, Inc., USA) as per the manufacturer's instructions. Precipitated DNA was suspended in DNase-free H_2O, and its concentration and purity were assessed by UV/vis spectrophotometry and agarose gel electrophoresis, respectively (Mecasys Co., Ltd, Korea).

454-Pyrosequencing analysis
DNA amplicons from individual broiler chicken samples were amplified using primers for the V1-V3 hypervariable regions of the 16S rRNA gene by polymerase chain reaction (PCR). Forward primers were tagged with 10-bp unique barcode labels at the 5' end along with the adaptor sequence to allow multiple samples to be included in a single 454 GS FLX Titanium sequencing plate, as described previously [11]. Finally, 16S rRNA amplicons were quantified, pooled, and purified for sequencing.

Data processing
The 16S rRNA sequence data generated by the 454 GS FLX Titanium chemistry (Roche) were processed using the Quantitative Insights into Microbial Ecology (QIIME) pipeline. Briefly, sequences that were less than 200 bp or greater than 600 bp in length, were of low quality, contained incorrect primer sequences, and/or

contained more than one ambiguous base were filtered using the split_libraries.py script. After checking for chimeric sequences, sequence data were filtered using the identify_chimeric_seqs.py and filter_fasta.py scripts. Operational taxonomic units (OTU) were picked using the pick_open_reference.py script and the most recent Greengenes reference database (13_8) at a 97% identity threshold. Bacterial composition data from broiler faecal samples were generated using the summarize_taxa_through_plots.py scripts. For alpha diversity measurements, the alpha_diversity.py script was employed to generate Chao1, Shannon, Simpson, and phylogenetic distance (PD) whole-tree values. Rarefaction curves were also generated in the QIIME software. All reads were pooled for each group of broiler chickens. Sequences were rarefied according to sequencing depth to visualise the change in diversity with respect to sampling depth.

Statistical analysis

Statistical and multivariate analyses were performed using the R software (v. 3.1.0; R Core Team, Auckland, New Zealand). The proportions of bacterial taxa (phylum and genus level) were compared between groups before and after the 24 h challenge. To avoid statistical bias and perform valid downstream analysis, the normalisation of OTU table was performed using the base package in R software. For multivariate analysis of bacterial OTU at a 95% identity threshold, the adegenet package in the R software was used to determine the peaks of the bacterial genera that facilitated discrimination according to the defined clustering groups with a user-defined threshold in the canonical loading plot. The NC, SC, and BT groups were labelled accordingly. In discriminant analysis of principal components (DAPC) (dapc {adegenet}), the normalised abundance data of individual samples was employed [12]. The number of principal components was ≥80% of the cumulative variance explained by the eigenvalues of the DAPC plot, and these principal components were subjected to linear discriminant analysis (LDA), resulting in selection of ≥2 linear discriminants for the DAPC plot [12]. The visual outputs of the canonical loading plot and the DAPC plot were then created using (loadingplot {adegenet}) and scatter plot (scatter {ade4}), respectively. Furthermore, functional prediction was carried out using the Phylogenetic Investigation of Communities by Reconstruction of Unobserved States (PICRUSt) based on the Greengenes 16S rRNA database and KEGG Orthologs (KO) [10]. PICRUSt was used to identify differences in the functional potential of bacterial communities among the groups. Using KEGG (level 3) ortholog function predictions, differences among NC, SC, and BT groups were evaluated, and a loading plot was created to identify the most discriminating functions among the

groups after 24 h. Tukey's honestly significant difference (HSD) test was employed to evaluate functional differences among the groups; a P-value <0.05 was considered to indicate significance.

Results

DNA sequence data and quality control

Pyrosequencing analysis generated a total of 718,204 raw sequence reads. The average number of reads per sample was 10,941, and the mean number of sequence reads per group ranged from 7,701 to 14,199 (Table 1). The average number of reads per sample is comparable to previous animal studies that utilised the GS FLX Titanium system [13, 14]. Barcoded primers allowed pooling of samples for individual- and group-based analyses. At a 95% identity cut-off (genus level), 212 unique operational taxonomic units (OTU) were detected in this study using the latest Greengenes database (13_8); these were used for downstream analyses. The recorded mean OTU per group ranged from 421 to 725 (Table 1).

Microbial diversity

Alpha diversity was compared among the NC, SC, and BT groups (Table 1; Additional file 2: Fig. S1). α-diversity parameters were calculated based on the OTU using the phylogenetic diversity (PD) whole tree, Chao1, Shannon, and Simpson methods. Diversity (Shannon and Simpson) values were highest in the NC group and lowest in SC broilers after *Salmonella* challenge (Additional file 2: Fig. S1A). Diversity values are summarised in Table 1. Species richness (Chao1) and bacterial diversity (Shannon and Simpson) exhibited similar trends; i.e. the highest values of both were in the NC group, and lowest in the SC group (Table 1), implying that *Salmonella* infection negatively affects overall microbial diversity. In addition, the rarefaction curves confirmed that *Salmonella* infection decreased the level of microbial diversity (Additional file 2: Fig. S1B).

Effect of *Salmonella* challenge and *B. subtilis* CLS2 administration on microbiota composition

Irrespective of *Salmonella* challenge and *Bacillus* supplementation, the phylum Firmicutes showed the highest abundance (>80%), followed by the phyla Bacteroidetes and Proteobacteria (Fig. 1a). These major bacterial phyla are also major constituents of the gut microbiota of other birds and livestock animals [13]. At the 24-h post-*Salmonella* challenge, the abundance of Proteobacteria was significantly increased, and that of Firmicutes decreased in the SC group, whereas those in the BT group had recovered to levels similar to the NC group. In previous reports, Proteobacteria was highly abundant in *Salmonella*-infected animals [15, 16]. Similarly, dietary supplementation with probiotics altered the microbial

Table 1 Pyrosequencing data and diversity indices of the faecal microbiota of broiler chickens

Group*	Diversity indices (Mean ± standard deviation)					
	No. of reads[a]	OTU	Chao1	Shannon	Simpson	PD
NC (Before)	12,532 ± 6,666	725 ± 420	1,443 ± 752	5.41 ± 1.19	0.92 ± 0.05	48.8 ± 25.75
NC (After)	10824 ± 3,322	712 ± 369	1,555 ± 729	5.79 ± 1.20	0.94 ± 0.04	48.8 ± 21.71
SC (Before)	12,176 ± 8,221	693 ± 447	1,418 ± 938	5.50 ± 0.85	0.93 ± 0.04	45.4 ± 28.49
SC (After)	7701 ± 3,665	421 ± 240	890 ± 482	4.61 ± 1.00	0.85 ± 0.07	30.7 ± 15.04
BT (Before)	8212 ± 4,494	493 ± 274	1,023 ± 574	4.98 ± 1.13	0.88 ± 0.07	36.5 ± 20.07
BT (After)	14,199 ± 5,013	580 ± 254	1,047 ± 360	5.06 ± 1.03	0.90 ± 0.06	36.8 ± 14.35

*Legend: *NC*, negative control; *SC, Salmonella* challenged; *BT, Bacillus* treated; *OTU*, operational taxonomic unit; *PD*, phylogenetic distance (whole tree)
[a]Mean number of raw reads per treatment group. No significant differences ($P < 0.05$) in alpha diversity were detected between groups using compare_alpha_diversity.py script

composition of broiler chickens, resulting in higher abundance of Firmicutes and lower abundance of Bacteroidetes and Proteobacteria [14].

At the genus level, a total of 124 bacterial genera were detected in all broiler samples (Additional file 3: Table S2), which is comparable to other avians [17]. *Lactobacillus*, unclassified *Clostridiaceae, Turicibacter, Bacteroides*, and unclassified *Enterobacteriaceae* were the major bacterial genera in broiler faeces (Fig. 1b). Interestingly, the abundance of *Lactobacillus*, a genus of beneficial bacteria, decreased significantly in the SC group, but increased in the BT group. In addition, the abundance of unclassified *Enterobacteriaceae*, which comprises several pathogenic species, in the SC group increased from 2.9% to 10.9%. A previous study reported an increase in *Lactobacillus* abundance after probiotic administration in chicken

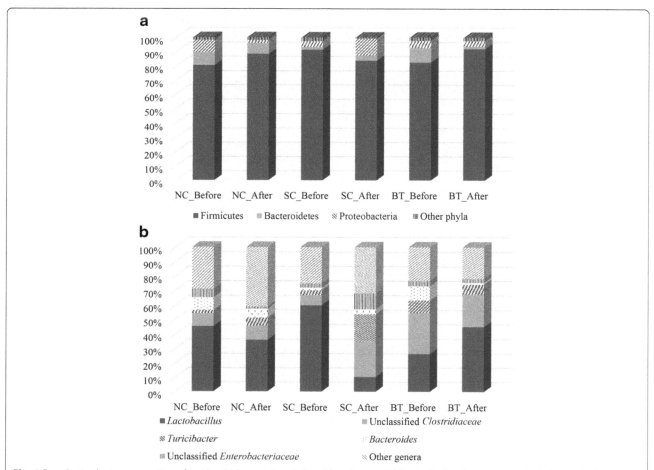

Fig. 1 Faecal microbiota composition of broiler chickens at the phylum (**a**) and genus (**b**) levels. Broiler chickens were divided into the following three groups before and after *Salmonella* challenge: NC, control/basal diet; SC, basal diet challenged with *S*. Gallinarum (SC); and BT, basal diet supplemented with *B. subtilis* CSL2

and pigs [18]. In contrast, SC broilers had a higher abundance of *Turicibacter*, unclassified *Enterobacteriaceae*, and *Bacteroides* than the other groups had. In a previous study, *Salmonella* infection of mice was also found to result in higher *Enterobacteriaceae* and lower *Lactobacillus* abundance [19].

Taxon-independent and functional analyses

Of the 212 bacterial OTU identified in this experiment, the 42 differentially abundant bacterial OTU with >1.0% abundance were used to generate a DAPC plot (Fig. 2a) and canonical loading plot (Fig. 2b). The DAPC plot showed that all groups had similar microbial composition before *Salmonella* challenge. But separate clusters were formed in response to *B. subtilis* CSL2 supplementation and/or *S.* Gallinarum infection, suggestive of distinct microbial communities. These microbial shifts were attributed to subtle changes in the abundance of several bacterial OTU, including unclassified *Neisseriaceae*,

Ruminococcus, and *Candidatus* Arthromitus (Fig. 2b). Unclassified *Neisseriaceae* was the strongest indicator of the presence of distinct microbial clusters; however, the other loading peaks might also exert considerable effects [11].

Functional analysis identified a total of 137 of 265 KEGG functions as differentially abundant (>0.1% mean relative abundance) (Additional file 4: Table S3). These 137 KEGG functions were analysed using a loading plot to identify vital functions among microbial clusters (Fig. 3a). The three most discriminating KEGG functions in broiler microbiota were the phosphotransferase system (PTS), glycan degradation, and replication/recombination/repair proteins (Fig. 3a). The PTS was significantly decreased by *Salmonella* challenge, but recovered to almost normal levels following *Bacillus* administration (NC group) (Fig. 3b). The abundance of genes associated with glycan degradation was significantly decreased in the BT group compared to the NC and ST groups, which might be a unique effect of

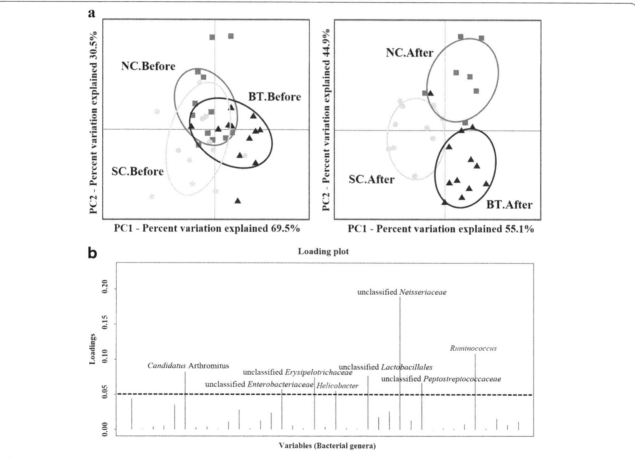

Fig. 2 Taxon-independent multivariate analysis and separation of broiler microbiota. **a** Discriminant analysis of principal components revealed distinct clustering of the control (NC, *grey*), *Salmonella*-challenged (SC, *white*), and *Bacillus*-treated (BT, *black*) groups using OTU at the 97% identity level. Significant differences (P < 0.001) were calculated using compare_categories.py using the PERMANOVA test. **b** Canonical loading plot showing differentially abundant bacterial genera. The individual peaks show the magnitude of the influence of each variable on separation of the NC, SC, and BT groups after challenge of broiler chickens (0.05 threshold level)

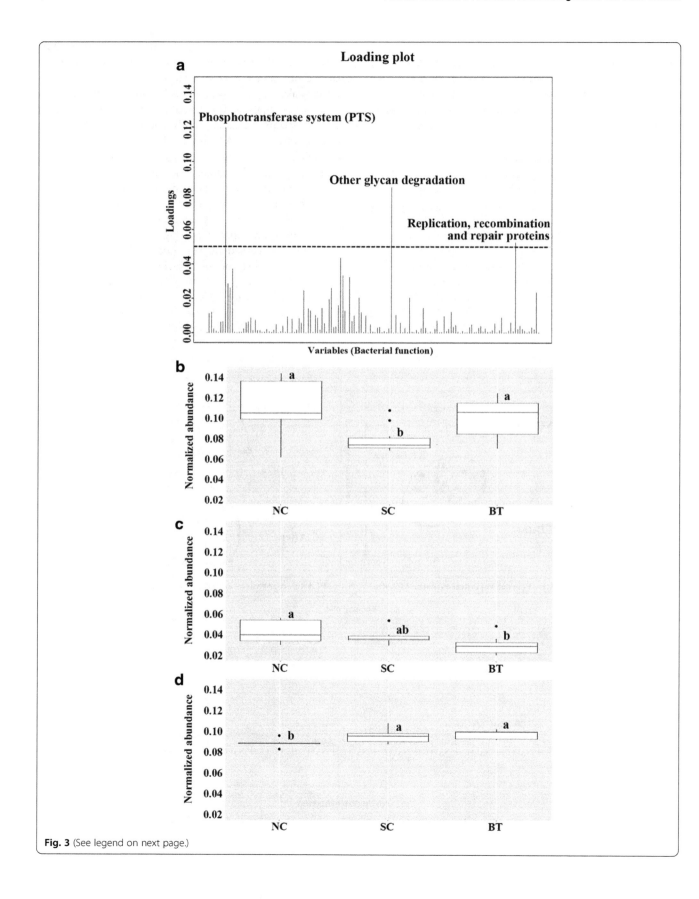

Fig. 3 (See legend on next page.)

supplementation of *B. subtilis* CSL2 (Fig. 3c). In addition, the abundance of genes related to repair and recombination of DNA was highest in the BT group and lowest in the NC group (Fig. 3d).

Discussion

In this study, *B. subtilis* CSL2 administration altered the microbiota of broilers and contributed to protection against *Salmonella* infection. In previous studies, microbial shifts were evident after probiotic administration and pathogenic invasion [14, 19, 20]. In-feed administration of *E. faecium* NCIMB 11181 changed the microbial composition in piglets [14]. Furthermore, pathogen infection led to significant alterations in the chicken gut microbiota [5, 15, 16]. Pathogen infection disrupts gut microbial diversity and overwhelms the microbiota by producing toxins and harmful agents [15, 18, 19]. Lower microbial diversity indicates a reduced ability of the microbiota to maintain gut homeostasis and resist invasion. *B. subtilis* CSL2 might provide protection against pathogen invasion by increasing bacterial diversity and metabolic and cellular functionality. Reduced functional diversity suppresses commensal-microbiota-mediated homeostasis [21].

Firmicutes and Bacteroidetes comprised the majority of the broiler microbiota at the phylum level; these organisms function in energy production and metabolism, particularly starch digestion and microbial fermentation [4, 10, 13]. The increased abundance of Proteobacteria in *Salmonella*-infected broilers suggests gastrointestinal dysbiosis and imbalance. Indeed, Proteobacteria are closely associated with *S. Enteritidis* infection in animals [15, 16]. Moreover, significant changes were detected in the abundance of *Lactobacillus* and *Turicibacter* after 24 h in the SC and BT groups. *Lactobacillus* is a dominant and important gut commensal genus present at frequencies of up to 10^9/g [4]. Increased abundance of *Lactobacillus* in the gastrointestinal tract of chickens is considered beneficial for their health and performance [22]. Certain strains of *Lactobacillus* produce antimicrobial substances [23], exopolysaccharides, and short-chain fatty acids as additional energy sources [24]. *B. subtilis* CSL2 promotes the growth of beneficial lactobacilli, as do other types of in-feed probiotics [9, 14].

Furthermore, *Enterobacteriaceae* abundance was reduced in the BT group after 24 h of *Salmonella* challenge, and increased numbers of *Enterobacteriaceae* in SC broilers might imbalance the microbiota and thus exert harmful effects on the gut [20]. The impact of *S. Gallinarum* and *B. subtilis* CSL2 on the overall composition of the gut microbiota can be detected within 24 h post infection [18]. Surprisingly, in this study *Salmonella* spp. were not detected in the SC group. Previous studies also suggest that *Salmonella* abundance is lower after 72 h, during which time symptoms are manifested [15, 19]. Furthermore, the abundance of the genus *Turicibacter* was significantly increased in *Salmonella*-infected broilers. *Turicibacter* has been reported in mammalian studies to be relevant to infection [25]. Its immunomodulatory and invasive properties resulted in subclinical infection of the gastrointestinal tracts of livestock and poultry animals [25, 26]. However, its ecological role and pathogenic potential remain unclear due to the dearth of studies.

A taxon-independent analysis was performed to prevent taxonomic bias and calculate variation based on all differentially abundant bacterial OTU. It is plausible that *B. subtilis* CSL2 maintains microbial community stability, similar to the case in normal broilers. Given that chicken microbial communities are highly dynamic and delicate, abrupt disturbances might cause greater variations in their microbiota than in other animals [4]. In addition, the loading plot of bacterial OTU (i.e. unclassified *Neisseriaceae*, *Ruminococcus*, and *Candidatus* Arthromitus) abundance in the SC group suggested the effects of *S. Gallinarum* infection in broilers. The identification of bacterial OTU belonging to *Neisseriaceae* (phylum: Proteobacteria) was highly suggestive of microbial clustering, and these organisms include several diarrhoea-causing pathogens [27]. These results imply that establishment of *Salmonella* in the gut requires the support of other pathogens or opportunistic bacteria therein.

In this study, functional prediction was performed to evaluate and compare the metabolic activities of the microbial communities among broiler groups [10]. Three metabolic and cellular functions were significantly affected by the administration of *B. subtilis* CSL2 and *S. Gallinarum* infection, namely the phosphotransferase system, glycan degradation, and DNA repair mechanisms. The greater abundance of genes associated with the PTS system in the BT group might be beneficial to broiler chickens. The PTS system is the major bacterial transport system for carbohydrates, and is involved in the regulation of bacterial fermentation and feed conversion [13, 21, 28]. This result thus indicates the probability of rapid uptake of available simple sugars rather than the digestion of complex carbohydrates. Furthermore, Pérez-Cobas et al. suggested that the PTS system assists bacterial stabilisation of gut-associated stresses in an unstable environment, as well as increasing energy yield, giving the commensal microbiota a competitive advantage over foreign microorganisms [29]. In addition, the reduced abundance of genes related to glycan degradation in broilers fed *B. subtilis* CSL2 implies that glycans are not a primary source of nutrients for the microbiota [30]. Higher glycan degradation is significantly correlated with the presence of pathogens (i.e. *Salmonella* and enterotoxigenic Clostridia) [27, 30]. These results suggest that probiotic supplementation alters the metabolic functions of the microbiota in a way that benefits the host. However, additional studies are required to elucidate the protective effects of *B. subtilis* CSL2 against *S.* Gallinarum infection.

Conclusions

In this study, the characterisation of the chicken faecal microbiota community structure and composition by 16S rRNA gene pyrosequencing revealed that the probiotic strain protected against *Salmonella* infection. The supplementation of *B. subtilis* CSL2 significantly changed the microbial diversity and composition by increasing the abundance of beneficial microorganisms. Conversely, broiler chickens infected with *S.* Gallinarum promoted the growth of potentially harmful bacteria such as *Turicibacter*, *Enterobacteriaceae* and *Neisseriaceae*. Furthermore, the potentially probiotic or pathogenic bacteria influenced the microbial functionality, particularly in the energy transport and metabolism capability of the gut. Overall, *S.* Gallinarum infection and *B. subtilis* CSL2 supplementation in the diet of broiler chickens influenced the diversity, composition, and functional diversity of the faecal microbiota. These results will facilitate prevention of *Salmonella* infection before the onset of symptoms using the potentially probiotic strain *B. subtilis* CSL2. Moreover, the findings offer significant insights to understand potential mechanisms of *Salmonella* infection and the mode of action of probiotics in broiler chickens.

Additional files

Additional file 1: Table S1. Basal diet composition.

Additional file 2: Figure S1. Rarefaction curves measuring bacterial diversity among broiler communities.

Additional file 3: Table S2. Highly abundant genera in broiler groups.

Additional file 4: Table S3. Highly abundant KEGG functions in broiler groups.

Abbreviations
BT: *Bacillus subtilis* CSL2-treated; DAPC: Discriminant analysis of principal components; KEGG: Kyoto encyclopedia of genes and genomes; NC: Negative control; NGS: Next generation sequencing; OTU: Operational taxonomic unit; PICRUSt: Phylogenetic investigation of communities by reconstruction of unobserved states; QIIME: Quantitative insights into microbial ecology; SC: *Salmonella*-challenged/*Salmonella*-infected chickens

Acknowledgements
Not applicable.

Funding
This work was supported by a grant from the Next-Generation BioGreen 21 Program (PJ01115903), Rural Development Administration, Republic of Korea.

Authors' contributions
JKO and EABP performed data collection and drafted the manuscript together. JPC carried out data analysis. IH Kim participated in the animal experiment. DSY isolated *B. subtilis* CSL2 strain. DKK managed the entire experiments and revised the manuscript. All authors read and approved the final manuscript for publication.

Competing interests
The authors declare that they have no competing interests.

Author details
¹Department of Animal Resources Science, Dankook University, 119 Dandae-ro, Cheonan 31116, Republic of Korea. ²Abson BioChem, Inc, 10-1 Yangjimaeul-gil, Sangrok-gu, Ansan 15524, Republic of Korea.

References
1. Leeson S. Future considerations in poultry nutrition. Poult Sci. 2012;91:1281–5.
2. Matulova M, Varmuzova K, Sisak F, Havlickova H, Babak V, Stejskal K, et al. Chicken innate immune response to oral infection with *Salmonella enterica* serovar Enteritidis. BMC Vet Res. 2013;44:37.
3. Stanley D, Hughes RJ, Moore RJ. Microbiota of the chicken gastrointestinal tract: influence on health, productivity and disease. Appl Microbiol Biotechnol. 2014;98:4301–10.
4. Oakley BB, Lillehoj HS, Kogut MH, Kim WK, Maurer JJ, Pedroso A, et al. The chicken gastrointestinal microbiome. FEMS Microbiol Lett. 2014;360:100–12.
5. Videnska P, Sisak F, Havlickova H, Faldynova M, Rychlik I. Influence of *Salmonella enterica* serovar Enteritidis infection on the composition of chicken cecal microbiota. BMC Vet Res. 2013;9:140.
6. Foley SL, Nayak R, Hanning IB, Johnson TJ, Han J, Ricke SC. Population dynamics of *Salmonella enterica* serotypes in commercial egg and poultry production. Appl Environ Microbiol. 2011;77:4273–9.
7. Yeoman CJ, Chia N, Jeraldo P, Sipos M, Goldenfeld ND, White BA. The microbiome of the chicken gastrointestinal tract. Anim Health Res Rev. 2012;13:89–99.
8. Larsen N, Thorsen L, Kpikpi EN, Stuer-Lauridsen B, Cantor MD, Nielsen B, et al. Characterization of *Bacillus* spp. strains for use as probiotic additives in pig feed. Appl Microbiol Biotechnol. 2013;98:1105–18.
9. Park JH, Kim IH. The effects of the supplementation of *Bacillus subtilis* RX7 and B2A strains on the performance, blood profiles, intestinal *Salmonella* concentration, noxious gas emission, organ weight and breast meat quality of broiler challenged with Salmonella typhimurium. J Anim Physiol Anim Nutr (Berl). 2015;99:326–34.
10. Mohd Shaufi MA, Sieo CC, Chong CW, Gan HM, Ho YW. Deciphering chicken gut microbial dynamics based on high-throughput 16S rRNA metagenomics analyses. Gut Pathog. 2015;7:4.
11. Pajarillo EAB, Chae JP, Kim HB, Kim IH, Kang D-K. Barcoded pyrosequencing-based metagenomic analysis of the faecal microbiome of three purebred pig lines after cohabitation. Appl Microbiol Biotechnol. 2015;99:5647–56.
12. Jombart T, Ahmed I. adegenet 1.3-1: new tools for the analysis of genome-wide SNP data. Bioinformatics. 2011;27:3070–1.
13. Lamendella R, Santo Domingo JW, Ghosh S, Martinson J, Oerther DB. Comparative fecal metagenomics unveils unique functional capacity of the swine gut. BMC Microbiol. 2011;11:103.

14. Pajarillo EAB, Chae JP, Balolong MP, Kim HB, Park C-S, Kang D-K. Effects of probiotic *Enterococcus faecium* NCIMB 11181 administration on swine fecal microbiota diversity and composition using barcoded pyrosequencing. Anim Feed Sci Technol. 2015;201:80–8.

15. Stecher B, Robbiani R, Walker AW, Westendorf AM, Barthel M, Kremer M, et al. *Salmonella enterica* serovar Typhimurium exploits inflammation to compete with the intestinal microbiota. PLoS Biol. 2007;5:2177–89.

16. Sekirov I, Gill N, Jogova M, Tam N, Robertson M, de Llanos R, et al. *Salmonella* SPI-1-mediated neutrophil recruitment during enteric colitis is associated with reduction and alteration in intestinal microbiota. Gut Microbes. 2010;1:30–41.

17. Waite DW, Taylor MW. Exploring the avian gut microbiota: current trends and future directions. Front Microbiol. 2015;6:673.

18. Mountzouris KC, Dalaka E, Palamidi I, Paraskeuas V, Demey V, Theodoropoulos G, et al. Evaluation of yeast dietary supplementation in broilers challenged or not with *Salmonella* on growth performance, cecal microbiota composition and *Salmonella* in ceca, cloacae and carcass skin. Poult Sci. 2015;94:2445–55.

19. Barman M, Unold D, Shifley K, Amir E, Hung K, Bos N, et al. Enteric salmonellosis disrupts the microbial ecology of the murine gastrointestinal tract. Infect Immun. 2008;76:907–15.

20. Videnska P, Faldynova M, Juricova H, Babak V, Sisak F, Havlickova H, et al. Chicken faecal microbiota and disturbances induced by single or repeated therapy with tetracycline and streptomycin. BMC Vet Res. 2013;9:30.

21. Sergeant MJ, Constantinidou C, Cogan TA, Bedford MR, Penn CW, Pallen MJ. Extensive microbial and functional diversity within the chicken cecal microbiome. PLoS One. 2014;9:e91941.

22. Nakphaichit M, Thanomwongwattana S, Phraephaisarn C, Sakamoto N, Keawsompong S, Nakayama J, et al. The effect of including *Lactobacillus reuteri* KUB-AC5 during post-hatch feeding on the growth and ileum microbiota of broiler chickens. Poult Sci. 2011;90:2753–65.

23. Rybal'chenko OV, Orlova OG, Bondarenko VM. Antimicrobial peptides of lactobacilli. Zh Mikrobiol Epidemiol Immunobiol. 2013;4:89–100.

24. Pajarillo EAB, Kim SH, Lee JY, Valeriano VDV, Kang, D-K. Quantitative proteogenomics and the reconstruction of the metabolic pathway in *Lactobacillus mucosae* LM1. Korean J Food Sci Anim Resour. 2015;35:692–2.

25. Cuív PÓ, Klaassens ES, Durkin AS, Harkins DM, Foster L, McCorrison J, et al. Draft genome sequence of *Turicibacter sanguinis* PC909, isolated from human feces. J Bacteriol. 2011;193:1288–9.

26. Bosshard PP, Zbinden R, Altwegg M. *Turicibacter sanguinis* gen. nov., sp. nov., a novel anaerobic, Gram-positive bacterium. Int J Syst Evol Microbiol. 2002;52:1263–6.

27. Deatherage Kaiser BL, Li J, Sanford JA, Kim Y-M, Kronewitter SR, Jones MB, et al. A Multi-omic view of host-pathogen-commensal interplay in *Salmonella*-mediated intestinal infection. PLoS One. 2013;8:e67155.

28. Lee JY, Pajarillo EAB, Kim MJ, Chae JP, Kang D-K. Proteomic and transcriptional analysis of *Lactobacillus johnsonii* PF01 during bile salt exposure by iTRAQ Shotgun proteomics and quantitative RT-PCR. J Proteome Res. 2013;12:432–43.

29. Pérez-Cobas AE, Moya A, Gosalbes MJ, Latorre A. Colonization resistance of the gut microbiota against *Clostridium difficile*. Antibiotics (Basel). 2015;4:337–57.

30. Eilam O, Zarecki R, Oberhardt M, Ursell LK, Kupiec M, Knight R, et al. Glycan degradation (GlyDeR) analysis predicts mammalian gut microbiota abundance and host diet-specific adaptations. MBio. 2014;5:e01526–14.

2

Investigation of the immune effects of *Scutellaria baicalensis* on blood leukocytes and selected organs of the chicken's lymphatic system

Bożena Króliczewska[1*†]⬤, Stanisław Graczyk[2†], Jarosław Króliczewski[3*], Aleksandra Pliszczak-Król[2], Dorota Miśta[1] and Wojciech Zawadzki[1]

Abstract

Background: The health of chickens and the welfare of poultry industry are central to the efforts of addressing global food security. Therefore, it is essential to study chicken immunology to maintain and improve its health and to find novel and sustainable solutions. This paper presents a study on investigation of the effect of *Scutellaria baicalensis* root (SBR) on the immune response of broiler chicken, especially on lymphocytes and heterophils reactivity, regarding their contribution to the development of immunity of the chickens.

Methods: The 121-day-old Hubbard Hi-Y male broiler hybrids were randomly assigned to four treatment groups, three SBR supplemented groups (0.5, 1.0, and 1.5% of SBR) and one control group. Each treatment was replicated five times with six birds per replicate pen in a battery brooder. Blood was collected after 3rd and 6th wk of the experiment, and hemoglobin and hematocrit values were determined, as well as total leukocyte count and differential count were performed. Nitroblue tetrazolium test and phagocytosis assay as nonspecific immune parameters and humoral immune responses to the antigenic challenge by sheep red blood cells were performed. Moreover, the ability of peripheral blood lymphocytes to form radial segmentation (RS) of their nuclei was analyzed. Body weight and relative weight of spleen, liver, and bursa of Fabricius were recorded.

Results: Results showed that mean heterophile/lymphocyte ratio increased in the SBR groups compared to the control group and the blood of the chickens showed lymphocytic depletion. The results also demonstrated that the relative weight of bursa of Fabricius and spleen in groups fed with SBR significantly decreased compared to the control group. This study also showed that the addition of SBR significantly inhibited the formation of RS of nuclei compared to some cytotoxic substances.

Conclusion: We found that SBR supplementation should be carefully evaluated when given to poultry. The excess intake of SBR supplementation may cause immunologic inhibition and may negatively affect the development of immune organs. SBR has inhibited the formation of radial segmentation nuclei showing antimetastatic properties and also the phagocytosis of chicken heterophils.

Keywords: Development of immune organs, Leukocyte, Lymphatic system, Radial segmentation, *Scutellaria baicalensis*, Toxic effect

* Correspondence: bozena.kroliczewska@up.wroc.pl; jakrol@windowslive.com
†Equal contributors
[1]Department of Animal Physiology and Biostructure, Faculty of Veterinary Medicine, Wroclaw University of Environmental and Life Sciences, C.K Norwida 31, 50-375 Wrocław, Poland
[3]Department of Chemical Biology, Faculty of Biotechnology, University of Wrocław, Fryderyka Joliot-Curie 14a, 50-383 Wrocław, Poland
Full list of author information is available at the end of the article

Background

In recent years, extensive research has been done on the potential food applications in food products and poultry feeds, for natural antimicrobial agents against foodborne pathogens that improve the health and performance of animals [1]. Novel types of effective and healthy antimicrobial compounds that could protect food and animal against microbiological contamination and the consumer against infection are in high demand. Medicinal herbs, as a new class of additives to animal feeds, can have beneficial properties such as antioxidant, antimicrobial, and antifungal as well as immunomodulatory effects; these properties make them the increasingly used products nowadays [2]. On the other hand, some herbal substances can interact in potentially dangerous ways with the organism systems [3]. Studies have been carried out to investigate the effect of various medicinal plants possessing immunostimulating and antioxidant properties [4], but they focused on the short-term supplementation (<2 wk) to allay the negative effects [5]. Whereas, the nature of commercial feeding programs makes the longer-term efficacy of dietary immune enhancers an important consideration.

Immune response in poultry can be influenced by genetic background, nutrition, environment, and management, or any combination of the above. Chicken heterophils are the first line of defense that can launch a series of intra- and extracellular antimicrobial mechanisms. Improved or reinforced immune response in poultry creates resistance against diseases, may be as the result of preparedness of immune system against pathogenic agents, which is an important factor in improving the homogeneity, long life, growth and the health of a flock. Therefore, greater emphasis has been placed by the researchers on improving the immune response. On this regard, the herb – *Scutellaria baicalensis* (SB) –has received particular attention. However, the effects of other herbs as well as SB on the immune system and selected organs of lymphatic system of chicken are relatively unknown. The lymphoid tissue is involved in the defensive mechanism against microorganisms. The lymphoid system of chicken consists of unique organs and is divided into two morphologically and functionally distinct components: central lymphoid tissue, represented by thymus and bursa of Fabricius and the peripheral lymphoid tissue, represented by the spleen and all mucosa-associated lymphoid tissue [6].

There are practical reasons to study the immune system of poultry, particularly of the chicken. The health and welfare of poultry are central to the efforts of addressing global food security. Over the past two decades, poultry has increased in the world meat production and is still growing. Poultry are also a significant source of zoonotic infections, which can be exemplified by their viruses and bacteria. Antimicrobial compounds (antibiotics) were commonly included in poultry diets for promoting the growth and control the diseases. However, the European Union (EU) banned feed grade antibiotic growth promoters, not only because of the cross-resistance, but also due to the risk of possible multiple drug resistances in human pathogenic bacteria. Thus, the scientific communities have given more attention toward the potential antimicrobial activities of natural products, although using some of them has resulted in decreased body weights, increased feed conversion per kg of weight gain and insignificant effects on carcass yield and carcass fatness. Consequently, from the above reasons, it is therefore essential that we study chicken immunology to maintain and improve poultry health and to find novel and sustainable solutions for the future, because the broilers reach slaughter weight within few weeks. In the EU, the slaughter age ranges from 21 to 170 d, with the average slaughter age of 42 d [7]. This leaves little time to develop immune system. Hence, in the context that the biologically active plant additives could potentially effect or affect the development of mature immune system, we decided to administer chickens with SBR for 42 d with the control in midterm (21 d).

In our previous studies, we showed that the addition of the SBR to fodder at doses of 0.5, 1.0, and 1.5% of SBR did not worsen the quality or chemical composition of the breast and leg muscles of broiler chickens. The addition of 0.5 and 1.0% of SBR in diet had also little effect on performance and measured blood parameters. In the group fed with 1.5% of SBR in diet, the body weight gain, red blood cell count, and hemoglobin level was found to be higher, while the high-density lipoprotein (HDL), low-density lipoprotein (LDL), and total cholesterol levels in blood serum were lower than in the control group [8, 9]. Therefore, in the present study, we chose the same doses of SBR.

The dry root of SB is one of the most widely used Chinese herbal medicines, listed in the Chinese Pharmacopeia [10]. This Asian plant is well acclimatized and cultivated in central European conditions [11]. The dried roots of this plant have a particularly high flavonoid content (over 25%) [10]. The active components of the root of SB have multiple biological properties including anti-inflammatory, antiviral, anticarcinogenic effect, free radical scavenging, and antioxidant effects, as well as antithrombotic and vasoprotective effects [12]. The main flavones of SB include wogonin, wogonoside, baicalin, and baicalein with ratios to the dry material of about 1.3, 3.55, 5.41, and 12.11%, respectively [10], but its content depend on the growing conditions and isolation methods. List of chemical compounds isolated from SB are listed in Dr. Duke's Phytochemical and Ethnobotanical Database [13].

This study was carried out to investigate the effect of SBR on the immune response of chicken especially on the reactivity of lymphocytes and heterophils, regarding their contribution to the development of immunity. Moreover, a phagocytic test was performed to evaluate the ability of heterophils to ingest yeast cells and to furthermore evaluate the ability of peripheral blood lymphocytes to form RS of their nuclei. At the same time, the lymphocyte cytoskeleton status was evaluated indirectly by inducing the RS of lymphocyte nucleus to determine whether SBR affects the leukocyte function of both groups in the same way.

Methods

Plant material

Baikal skullcap, *Scutellaria baicalensis* Georgi, plants were grown in the University's experimental field from authenticated seeds obtained from the botanical garden of the medicinal plant herbarium at the Wroclaw Medical University, Poland. Voucher specimens were deposited at the herbarium. In the spring, the seeds were sowed in light well-drained sandy soil in partial shade. Plants were watered weekly as needed. Fertilizers were supplied along with the water over the growing season via the drip tape to provide 0.012, 0.008, and 0.01 g/m^2 of N, P_2O_5, and K_2O, respectively. The roots were harvested in autumn from 2-year-old plants. The roots were collected, thoroughly, washed in distilled water, and dried under controlled humidity of 25 °C until a moisture content of 5% was reached [14]. The dried roots were then crushed using a laboratory mill and stored at –20 °C until use.

Analytical HPLC analysis of flavonoids

High-performance liquid chromatography (HPLC) was used for the analytical determination of flavonoids in SB. Samples and standards were analyzed using a Waters 600 system coupled to Waters 2487 UV dual wavelengths absorbance detector (Waters Chromatography Canada Inc.), and Empower PDA software (Milford, MA USA). An Agilent Zorbax SB-C18 column (4.6 × 250 mm, 5 μm) was applied for this analysis. Sample preparation and extraction of flavonoids were done using a previously described method with modifications [15]. Before extraction, the herbal powder was soaked in Britton-Robinson buffer, pH 6.5 at 50 °C for 30 min. Flavone standards, baicalin (99%), baicalein (98%), wogonoside (≥95%), wogonin (≥98%) were purchased from Sigma-Aldrich. All HPLC-grade solvents were filtered through a membrane filter (0.2 μm pore size). The content of the constituents was calculated using standard curves acquired for four flavonoids. All measurements were performed in triplicate.

Animals, diets, and experimental design

One hundred and twenty one-day-old Hubbard Hi-Y male broiler hybrids were vaccinated against Newcastle Disease and Infectious Bronchitis. Vaccines were delivered via spraying method. No other vaccination was performed during the experiment. One-day-old broiler chickens with a mean body weight of 39 g (±1.7 g) were randomly assigned in four treatment groups. Each treatment was replicated five times with six birds per replicate pen in a battery brooder. All the pens were equipped with feeders and water. The birds were fed a starter diet for 21 d, followed by a finishing (grower) diet from d 21 to d 42. The basal diets were formulated based on NRC (National Research Council) guidelines and contained 18.50–20.10% crude protein and 12.13–12.55 MJ/kg metabolizable energy [16]. The composition, preparation, and suitability of the experimental diets were followed as described in previous study [8]. Chemical analysis of the principal components in diet was performed by standard methods as described in the Association of Official Analytical Chemists [17]. The chicks were housed in electrically heated battery pens, and the diets and fresh water were provided ad libitum. The birds were fed either a basal diet or a diet supplemented with ground and dried SBR. The experiment included three groups supplemented with SBR (0.5, 1.0, and 1.5% of SBR) and one control group (C) with no supplementation of SBR.

Performance data

Body weight, feed intake and chicken mortality were determined. All the chickens in each pen were weighed in groups at the beginning and at the end of the experiment. Weight gain was obtained from these data. The feed consumed per pen was recorded; average daily feed intake (ADFI) and feed conversion ratio (FCR) from d 0 to d 42 were calculated.

Hematological parameters

Blood was collected after wk 5 and wk 6 of the experiment. Blood was taken from the brachial vein of six chicken chosen randomly from each treatment group. Hemoglobin (HGB) and hematocrit values (HCT) were determined. The HGB was determined spectrophotometrically [18] and HCT was determined by centrifuging the blood in glass capillaries at 10,000 × g for 5 min. One drop of blood from each sample was smeared on a glass slide. The smears were stained by May-Grünwald-Giemsa staining method (MGG) [19]. The smears were used to perform leukogram, by counting up to 200 leukocytes. The percentage of leukocytes (WBC) including heterophils, lymphocytes, basophils, eosinophils, and monocytes were determined by counting up to 200 cells.

The heterophils/lymphocytes ratio (H/L) was calculated as well.

Immunization

At 5 wk of age, six chicks from each treatment group were injected intravenously in the brachial vein with 0.5 mL of 10% suspension containing 1×10^8/mL packed sheep red blood cells (SRBCs) (Sigma-Aldrich) in PBS (phosphate buffered saline). Blood samples were taken from brachial vein 7 d after immunization (at 6 wk of age). Sera were obtained by centrifuging blood at $3500 \times g$ for 15 min. To determine the antibody response to SRBCs, a direct hemagglutination assay was used and the total antibody (IgM and IgY) response to SRBCs in serum was measured. At first, to inactivate the complement proteins, serum samples were incubated for 30 min at 56 °C. To inactivate IgM component and measure anti-SRBC antibodies (IgY), serum samples were mixed with equal volume of 0.2 mol/L 2-mercaptoethanol for 30 min at 37 °C. Then the serum samples were serially diluted with PBS in 2-fold steps (1:1 – 1:1012) in U-bottomed microplates (96 well, Medlab, Poland), 100 μL/well. In the next step, 25 μL of 2% SRBC suspension was added to each well. The plates were incubated for 18 h at 37 °C. The titer of the well containing 50% SRBCs agglutination was recorded as positive result. The "titer" is defined as the reciprocal of the serum dilution that has an optical density (OD) of 0.5. If a serum has an OD value of ≥ 0.5, the reciprocal of the starting serum dilution that is closest to an OD value of 0.5 will be used as the titer. The serum titer is converted to $\log_2$ titer and the $\log_2$ titer is recorded in thousand units. The $\log_2$ titer data may be calculated [20] using the following formula.

$$\log_2 \text{titer} = \frac{\log_{10}(\text{titer})}{0.301}$$

If a serum does not meet the criteria as described above, then the serum may be tested again using a starting dilution, which may be either higher or lower than the previous starting dilution, or the data for that particular serum may be excluded.

Radial segmentation of lymphocyte nuclei

Blood samples were used to test the RS of lymphocyte nuclei with some modification [21]. The collected blood samples were divided into two equal parts of 0.8 mL. Subsequently, 0.2 mL of oxalates mixture (a solution of 0.57% potassium oxalate and 0.85% ammonium oxalate mixed in a ratio of 1:1) was added to one part of the 0.8 mL heparinized blood (RS induced). The second one without oxalate was used as a control (RS spontaneous). Both the samples were incubated for 3 h at 21–23 °C.

Later, the samples were centrifuged ($1500 \times g$, 10 min), the leukocyte layer situated between erythrocytes and plasma was collected, and then three smears from each chick were prepared. The smears were fixed in methanol and stained according to Pappenheim's panoptic method with MGG stain. The steps that followed involved the differentiation between RS positive (RS+) and RS negative (RS−) lymphocytes (the lymphocytes found were counted up to 200), according to the criteria described in previous study [22]. Those cells whose nuclei have slots with the depth of at least one-third of its diameter were considered RS+. The results of the RS test obtained were in a medium range for each group.

Phagocytosis assay

Phagocytosis assay based on the method of Pliszczak-Król et al. [23] was performed. *Saccharomyces cerevisiae* cells were used to evaluate the phagocytic potential of heterophils. The mixture of heat-inactivated yeast cells (100 μL, 8 in McFarland's scale) and heparinized blood (1 mL) was incubated for 15 min at 37 °C. After incubation, two smears were prepared from each blood sample, air dried, and stained using MGG. The smears were analyzed up to 200 granulocytes, differentiating them as positive phage (Fag+) and negative phage (Fag−). The percentage of phagocytosis by heterophils was then calculated.

Nitroblue tetrazolium assay

The bactericidal activity of phagocytic cells was measured based on the method described by Chung and Secombes [24]. Nitroblue tetrazolium (NBT) reduction assay was performed by spectrophotometric method. A total of 0.1 mL of blood sample was added to 0.1 mL of 1% solution of NBT; control sample with 0.1 mL of PBS instead of NBT solution was prepared, and the samples were then incubated for 30 min at 37 °C and again for 30 min at room temperature for formazan to be formed. Then, 0.05 mL of sample was taken and the blue formazan product formed was dissolved by adding 1 mL of dimethyl sulfoxide (DMSO). The sample was centrifuged ($3000 \times g$, 5 min) and the absorbance of the supernatant was measured at a wavelength of 560 nm.

Determination of body weights and relative weights of immune organs

At the end of the experiment, six chicks from every experimental group were randomly chosen to be weighed and slaughtered. The chicks were individually weighed before slaughter. They were sacrificed by decapitation and left to complete bleeding. The liver, bursa of Fabricius, and spleen were removed and weighed. The body weight (BW) of each chick in grams was determined. On this basis, the relative weight (RW) of all organs was calculated, e.g., the RW of immune organ is equal to the weight of immune organ minus the chicken's body weight.

Statistical analysis

The Shapiro-Wilk or D'Agostino-Pearson normality tests were used to analyze the normally distributed population. The Shapiro-Wilk test works very well if every value is unique, whereas it is not as effective when several values are identical. In those cases, the D'Agostino-Pearson test was used. Statistical analyzes of variance were calculated using ANOVA followed by a *post-hoc* test (Bonferroni multiple comparison test). Multiple comparisons were done only when the ANOVA P-values were significant. The P-value was calculated under the null hypothesis that the samples were drawn from the same distribution. The $P < 0.05$ were considered statistically significant. The effects of the SBR were estimated by analyzing the results of the eight lymphoid organs and SRBC antibody titer by polynomial regression model. Pearson correlation coefficient (r) with two-tailed test of significance was conducted to examine the relationship between certain parameters. Analyzes were performed with Statistica version 10.0 (StatSoft Inc., Tulsa, OK, USA).

Results

Growth performance

The results for the production traits of chickens fed with and without SBR in diet are presented in Table 1.

The results show that dietary SBR did not affect body weight. No mortality rate was recorded for broiler chickens fed with SBR diet for the whole experimental period. The addition of SBR to the chicken fodder has affected the final body weight gain of the birds. On the 42 d of breeding, the final weights of the chickens in the experimental groups were higher than the final weight of

the control group chickens: higher by 4.9% for the group with a 1.0% supplement, higher by 2.6% for the 1.5% group and it is lower by about 7.5% for the 0.5% group ($P < 0.05$). We showed that in broiler chickens fed with SBR supplemented diet, the ADFI was increased ($P < 0.05$) in groups supplemented with higher doses of SBR (1.0 and 1.5%). The increase in FCR was up to 7% significantly higher in the experimental group receiving the fodder with the 0.5% concentration of SBR.

Quantitative analysis of the flavonoids

The quantitation of baicalin, baicalein, and wogonin was achieved using peak area ratios of baicalin, baicalein, wogonin, and wogonoside to the internal standard. The amount (mg/g of dry weight) of baicalin, baicalein, and wogonin in the SBR is presented in Table 2.

Total flavonoids content was 214.5 ± 13.3 mg/g of dry weight (DW). The chemical profile of the SBR extract was obtained with a constant ratio of 5:1 for baicalin/wogonoside. Due to poor solubility of baicalin and baicalein in water, its absolute bioavailability after oral administration is 2.2 and 27.8%, respectively [25]. During our study, chickens adsorbed baicalin at doses of 120.42 mg/d, 251.22 mg/d, and 386.14 mg/d in group receiving 0.5, 1, and 1.5% of SBR, respectively.

Blood hematology

The HGB and HCT contents have been significantly ($P < 0.05$) decreased in group receiving diets supplemented with SBR at d 42 of feeding (Fig. 1). A linear decrease in the HGB and HCT was observed with increasing SBR dose. The Pearson's correlation coefficient (r) for these data was −0.995 and −0.980, respectively ($P < 0.001$). The effect of SBR diet at various percentages on particular type of leukocytes is summarized in Table 3.

Total blood WBC, total lymphocyte, heterophils, and eosinophils counts on d 42 differed markedly among treatments ($P < 0.05$). On d 42, a decrease in WBC counts ($P < 0.05$), which included lymphocytes and eosinophils, was found (Table 3). The Pearson's correlation coefficient showed significant ($P < 0.01$) decreasing

Table 1 Growth performance of chickens fed with experimental diets from 1 to 42 d of age[d]

Parameter	Dietary treatments			
	Control	0.5%	1.0%	1.5%
BW, g				
d 1–42	1549 ± 22^a	1432 ± 39^b	1626 ± 24^c	1589 ± 28^c
BWG, g				
d 1–42	1509 ± 29^a	1394 ± 17^b	1585 ± 26^c	1549 ± 32^c
ADFI, g/d				
d 1–42	70.83 ± 1.23^a	71.02 ± 1.30^a	74.54 ± 1.19^b	75.38 ± 1.28^b
FCR, g/g				
d 1–42	1.97^a	2.12^b	1.98^a	2.04^a

[a,b,c]Mean values within the same row not sharing a common uppercase superscript letter differ significantly ($P \leq 0.05$)
[d]Body weight (BW; g), body weight gain (BWG), average daily feed intake (ADFI; g), and feed conversion ratio (FCR; g/g) averaged over treatment period for chicken (from 1 to 42 d of age)

Table 2 The concentration of baicalin, baicalein, and wogonin present in dry roots of *Scutellaria baicalensis*[a]

Flavonoids	Concentration, mg/g DW[a]	Total flavonoids concentration, mg/g DW[b]
Baicalin	153.2 ± 8.1	214.5 ± 13.3
Baicalein	19.3 ± 2.3	
Wogonin	8.2 ± 1.2	
Wogonoside	29.8 ± 1.7	

[a]All values are mean ± SD ($n = 3$)
[b]dry weight

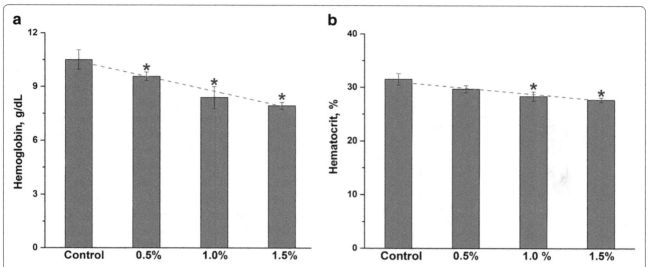

Fig. 1 The effect of SBR diet on RBC. Hemoglobin **a**, and hematocrit **b** were measured in chicken fed SBR containing either 0.5%, 1.0% or 1.5% SBR. The SBR caused a statistically significant (*) reduction in both blood parameters in chicks fed with SBR ($P < 0.05$). Dash line shows the linear relationship between the SBR concentration and HGB or HCT levels. The results are reported as a mean ± SD ($n = 6$)

relationship between SBR concentration and WBC or lymphocytes or eosinophils count.

Significantly higher amounts of heterophils and H/L ratio were determined in chicken fed graded levels of SBR compared to the control group (Table 3). These studies have demonstrated that the Pearson's correlation coefficient showed significant ($P < 0.01$) increasing relationship between SBR concentration and amounts of heterophils, and hence also higher H/L ratio. Monocytes and basophils count did not differ significantly ($P < 0.05$) between the experimental groups, and no correlation was observed.

Phagocytosis assay

The effect of graded levels of dietary SBR on the percentage of heterophils phagocytizing *Saccharomyces cerevisiae* cells is shown in Fig. 2. The inhibition of phagocytosis was measured after d 21 and d 42 of feeding with SBR diet. Statistical analysis has revealed that on d 42 SBR diet of 1.0 and 1.5% decreased ($P < 0.05$) the percentage of phagocytic activity of heterophils significantly (16.7 and 14.5%, respectively). Lower level of SBR diet (0.5%) did not inhibit phagocytosis and there is no difference between d 21 and d 42.

Superoxide anion production/NBT assay

The oxidative radical production in blood was measured by NBT assay. No significant difference ($P > 0.05$) was observed in any other feed-treated groups on d 21 compared to the control group (Fig. 3). There was a significant difference ($P < 0.05$) for the NBT activity between the control and all treatment groups d 42 (Fig. 3). The mean OD values for the heterophils of treatment groups

Table 3 Effect of experimental diets on the percentage of particular types of leukocytes at d 42, ($n = 6$)[d]

Parameter	Dietary treatments				SEM	*P*-value	r[*]
	Control	0.5%	1.0%	1.5%			
WBC, 10^3/μL	3.16[a]	3.07[a]	2.98[b]	2.34[c]	0.49	0.050	−0.884
Lymphocytes, %	63[a]	56.33[b]	53.33[b]	48.83[bc]	5.35	0.048	−0.989
Heterophils, %	30.5[a]	38[b]	39[b]	45.83[b]	6.12	0.032	0.930
Eosinophils, %	3.50[a]	2.33[b]	0.83[c]	1.83[b]	1.19	0.044	−0.757
Basophils, %	2.33	3.17	3.16	2.83	1.12	0.864	0.177
Monocytes, %	0.66	0.16	0.66	0.66	0.38	0.483	0.199
H/L ratio	0.51[a]	0.71[b]	0.73[b]	1.01[b]	0.19	0.043	0.959

[a,b,c]Means in the same row with different superscripts are significantly different ($P < 0.05$)
[d]The results are expressed as the mean for 6 birds and 2 repeats. Control = basal diet; 0.5% = basal diet + 0.5% of SBR, 1.0% = basal diet + 1.0% of SBR, 1.5% = basal diet + 1.5% of SBR
*Pearson's correlation coefficient

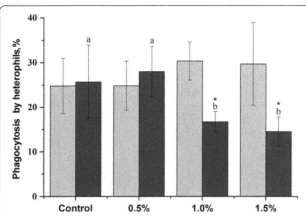

Fig. 2 Percentage of phagocytosis of *Saccharomyces cerevisiae* cells by heterophils. Heterophils were isolated from the blood of broiler chickens fed with graded concentrations of dietary SBR. Different letters indicate the significant difference (*P* < 0.05) among the groups, * indicates significant differences on d 21 and d 42. Light gray—d 21, dark gray—d 42. The results are reported as a mean ± SD (*n* = 6) from duplicate samples

at d 42 (0.5, 1.0, and 1.5%) were found to be 0.117 ± 0.06, 0.098 ± 0.03, and 0.128 ± 0.02, respectively, while for the control group it was 0.068 ± 0.04. Statistical analysis done by NBT assay revealed that at d 42 SBR diet of 1.5% has increased (*P* < 0.05) the superoxide anion production significantly.

Moreover, a positive tendency (*r* = 0.93), reflected by increased immunological indices (NBT reduction), was recorded in the group fed with SBR supplements at the end of the experimental period (d 42).

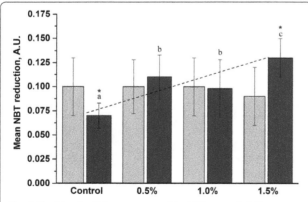

Fig. 3 The blood activity as measured by NBT assay. Activity measured in chicken fed with SBR for 21 d (*light gray*) and 42 d (*dark gray*). All the results were calculated on the estimation of 1000 leukocytes. Different letters indicate significant difference (*P* < 0.05) among the groups at d 21 and d 42. * indicates significant differences between d 21 and d 42 among groups. Dash line shows linear relationship between the SBR concentration and mean NBT on d 42. The results are reported as a mean ± SD (*n* = 6) from duplicate samples

Humoral antibody response to SRBC

We evaluated the effects of SBR on the humoral antibody-mediated responses to SRBCs. SRBCs are natural nonpathogenic antigens, which stimulate a wide range of immune cells with their multiple antigen binding sites after immunization with SRBCs [26]. Immunization with SRBCs resulted in the rise of specific anti-SRBC antibody in serum (Fig. 4). Antibody response against SRBC was detected in all groups. There was a statistically significant (*P* < 0.05) linear decrease (*r* = –0.82) in the antibody titers over immunized groups. The anti-SRBC antibody titers at wk 1 post-immunization in the SBR-treated groups were 6.16, 4.83, and 3.86, respectively. In the control group, the observed level was 5.45. As seen in Fig. 4, there was a consistent trend in SRBCs titer, but the level of anti-SRBC in the group treated with 0.5% of SBR diet was higher (*P* < 0.05) as opposed to the other SBR groups. However, there was not any significant difference between control and 0.05% groups. The antibody response to SRBC was significantly lower (*P* < 0.05) in the group fed with 1.0%, in contrast to control and 0.5% groups. Particularly, we observe a more marked decrease (*P* < 0.01) in the group fed 1.5% SBR.

RS of blood lymphocytes nuclei formation

The influence of SBR on the rate of formation of RS nuclei was studied. The ability to form RS was estimated in parallel to the leukogram analysis [27]. The aim of the experiment was to evaluate the ability of peripheral blood lymphocytes isolated from chicken blood to form the RS of their nuclei. The term radial segmentation (RS) is applied to a characteristic nuclear deformation, occurring in vivo in many neoplastic and leukemic cells, but often also as an in vitro artifact in different conditions [28]. Furthermore, the RS phenomenon is more

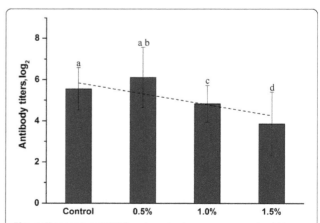

Fig. 4 Changes in anti-SRBC post-vaccine immune response in chicken fed with SBR. Dash line shows the linear relationship between the SBR concentration and antibody titers of immunized groups. The results are reported as a mean ± SD (*n* = 6) from duplicate samples. Different letters indicate statistically significant differences between groups (*P* < 0.05)

frequent in infectious, inflammatory, and necrotizing diseases than in healthy subjects [29]. The normal segmentation of the nucleus is quite different from RS due to the pathologic or toxic environment, for example, storage in a culture medium with in EDTA sodium oxalate.

Radial Segment formation was determined at d 21 and d 42 stage. The levels of RS+ in control group were 14.7 and 15.5, respectively. The percentage of RS+ cells changed nonsignificantly in the chicken fed with 0.5% of SBR. In the groups fed with 1.0 and 1.5% SBR, the percentage of RS+ cells decreased significantly ($P < 0.01$) in both the weeks tested (Fig. 5). However, a two-fold decrease in the percentage of RS+ cells was observed in birds fed with 1.5% of SBR compared to the control group on d 21 and d 42 (3.58 and 7.80, respectively, Fig. 5).

Relative weight of selected organs
Addition of SBR in the diet did not influence ($P > 0.05$) absolute or relative weight (RW) of the liver on d 42; however, there was a linear decrease ($P < 0.05$) in the RW of spleen and bursa of Fabricius when the birds were 42-day-old (Table 4).

The calculated RW of the spleen decreased by 14.22, 19.29, 16.39%, respectively, compared to the control group ($P < 0.05$). Similar results were observed for the RW of bursa of Fabricius. However, RW nonsignificantly increased in group fed with 0.5% SBR. The RW decreased by 17.72 and 23.69% in groups fed with 1.0 and 1.5% of SBR, respectively, compared to control group ($P < 0.05$). A linear decrease in the RW of spleen and bursa of Fabricius was observed with increasing SBR dose ($r = -0.95$ and -0.79 respectively).

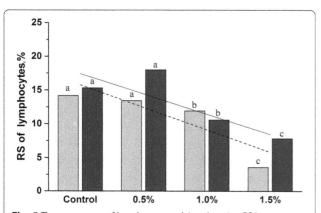

Fig. 5 The percentage of lymphocyte nuclei undergoing RS in experimental groups. Light gray—d 21, dark gray—d 42, showing differences among the groups on d 21 and d 42. Dash line show linear relationship ($r = -0.87$) between the SBR concentration and RS formation on d 21, solid line shows linear relationship ($r = -0.94$) on d 42. Different letters indicate statistically significant differences among groups ($P < 0.05$). Values represent means ± SD ($n = 6$) from duplicate samples

Discussion
Baicalin has been safely used for the treatment in several animal models of diseases and pharmacokinetic properties of baicalin have been well investigated [30]. Both oral administration and IP injection was used for baicalin treatment at a wide range of dosage (10 ~ 800 mg/kg/d) in various animal models [30]. Baicalin and baicalein showed weak absolute bioavailability after oral administration [25]. Similar or lower levels of bioavailability of flavonoids are observed in SBR orally administered animal models [31]. This was the second reason why we chose doses in the range of 0.5–1.5%, which provided average level of flavonoids used in other studies [30].

The presented results concerning chickens fed with supplementary SBR were positive with regard to bird breeding parameters. The data from this study also indicates a trend toward the ADFI of broiler chickens receiving the SBR in their diet ($r = 0.97$). Additional studies are needed to explain growth performance results of the group supplemented with 0.5% SBR, especially concerning body weight (BW).

Evaluation of hematological parameters usually provides significant information on the body's response to injury, they are a good indicator of the physiological and health status of animals and they can be useful to complement the knowledge on unfamiliar effect of feed additives [32].

In this study, HGB and HCT values were significantly lower ($P < 0.05$) in birds that received diets supplemented with SBR. Despite, these levels could be framed within the reference intervals provided by Bounous and Stedman [33]. However, significant ($P < 0.001$) linear decrease for both the parameters was observed during experiment with increasing dose. The values of WBC, heterophils, basophils, and lymphocytes levels were within the physiological limits found in the reference of previous studies, while the percentages of monocytes and eosinophils were lower in all groups [34]. However, even in these cases, we observed a significant ($P < 0.05$) linear change with increasing dose of SBR. In general, the blood of the chickens showed lymphocytic depletion, which may represent a combination of direct effects of SBR components and/or nonspecific stress factors.

The H/L ratio have been widely accepted in many disciplines of avian research as a measure of the chicken's perception of stress in its environment [35] and is useful in assessing the efficiency of the immune system and health condition of birds [36]. The H/L proportion and ratio were found to be significantly increased ($P < 0.01$) in the SBR groups compared to the control group. However, growth of heterophils could be the consequence of mobilization of the immune system to phagocytosis. A number of in vitro and animal studies have shown that bioactive compounds from plants increase immunologic

Table 4 Morphometry of lymphoid organs of broilers fed with diets containing different amounts of SBR, $(n = 6)$[c]

Parameter	Dietary treatments				SEM	P-value	r[*]
	C	0.5%	1.0%	1.5%			
Liver RW, mg/100 g BW	1.97	2.09	2.06	1.78	0.12	0.087	−0.25
Bursa of Fabricius RW, mg/100 g BW	282.70[a]	284.46[a]	232.60[b]	215.72[b]	38.28	0.050	−0.95
Spleen RW, mg/100 g BW	169.63[a]	145.50[b]	136.90[b]	141.82[b]	25.73	0.043	−0.79

[a,b]Means in the same row with different superscripts are significantly different ($P < 0.05$)
[c]The results are expressed as the mean for 6 birds and 2 repeats. Control = basal diet; 0.5% = basal diet + 0.5% of SBR, 1.0% = basal diet + 1.0% of SBR, 1.5% = basal diet + 1.5% of SBR
[*]Pearson's correlation coefficient

activity by increasing phagocytosis [37]. Phagocytosis by granulocytes is the first and major defense mechanism against invasion of bacteria, fungi, and parasites.

Results from the present study demonstrate the inhibition of phagocytosis by heterophils when chickens were fed with medium (1.0%) and high (1.5%) doses of SBR. Similar findings have been recorded in fish by Yin et al. [38] and in mouse by Cai et al. [39]. However, in contrast to our study these researchers have used only baicalin.

Degranulation and production of oxidative burst are closely associated with phagocytosis [40]. In this process, mediated by a multicomponent enzyme complex, nicotinamide adenine dinucleotide phosphate oxidase (NADPH-ox) phagocytic heterophils consume oxygen and produce reactive oxygen species (ROS) [40].

This study demonstrates a significant increase ($P < 0.05$) in the capacity of these cells to reduce NBT on d 42 compared to d 21 (Fig. 2). Therefore, an increased level of active toxic oxygen compounds may suggest a greater efficiency of oxidative response [41]. Flavonoids from SBR, such as baicalin or baicalein, to be a very effective inhibitor of the production of ROS by human leukocytes [42]. In addition to the inhibitory action on the production of ROS, both baicalein and baicalin showed a strong eliminating activity toward $\cdot O_2^-$, but showed no significant effect on scavenging hydroxyl radical ($\cdot OH$) [43].

Many functional and morphological changes are observed during lymphocyte activation in which a major role is ascribed to cytoskeletal microtubules and include the numerous processes occurring in cells. These changes take place in both mature and maturing leukocytes. The similar pattern of functional and morphological changes was observed in lymphocytes activated by mitogens or in neoplastic ones [44].

For the evaluation of cytoskeletal lymphocytes, the indirect test of RS formation was used. Radial segmentation is a fascinating phenomenon, yet to be fully explained, that involves the leucocyte nuclei of various diseases [29]. RS of nuclei may provide a convenient and reliable screening test for metaphase-blocking activities of new substances [45]. Söderström et al. [45] indicated a relation between the ability of lymphocyte nuclei to

deform and the reorganization of cytoskeletal elements, including tubulin, filaments, and actin.

This study has showed that the addition of 1.0 and 1.5% SBR significantly ($P < 0.05$) inhibited the RS nuclei formation. The cytochalasin B possessing cytotoxic activity show such effects [46]. Thus, cytochalasin B converts neutrophils from phagocytic cells into model secretory cell [47]. There is no guarantee that the same mechanism or process is activated by SBR, but the inhibition of phagocytosis and increase in the capacity of these cells to reduce NBT indicate the possibility of similar modes of action. Kumagai et al. [48] showed that SB inhibits the proliferation of lymphocytic leukemia, lymphoma, and myeloma cell lines by the induction of apoptosis.

Birds have central and peripheral lymphoid tissues, which play an important role in the body defense against pathogens [6]. The thymus and bursa of Fabricius are the primary lymphoid organs, whereas secondary lymphoid organs include spleen and all the lymphoid tissues associated to the intestinal mucosa. The spleen is the major site of immune responses to blood-borne antigens and is also a site of hematopoiesis [49].

This study was aimed at elucidating the effects of SBR on the immune responses in chickens. SRBC antigens elicit antibody-mediated responses in chickens [50]. Immunized chicken showed a significantly lower ($P < 0.05$) antibody response to SRBCs, when treated with SBR (1.0 and 1.5%). In the group supplemented with 0.5% of SBR, we observed slight, but not significant, increase in antibody response. Different result from the studies on mice was reported by Jong et al. [51]. The antibody response to T lymphocyte-dependent antigen, (SRBC) was increased at all doses (100, 200, and 400 mg/kg) of SB-treated mouse.

The mechanisms of the weakness of immune responsiveness conferred by SBR in our studies require further investigation.

Considering the RW of organs, only the liver showed similar weight for 42-day-old birds of all the groups receiving SBR diet. In this study, the body weight and RW were used for the evaluation of the state of the development of the spleen, which is responsible for initiating immune reactions. The researchers reported that at 42 d of age, the weight and RW of the spleen and bursa of Fabricius of birds fed with SBR was found to be low and that SBR can inhibit the development of the spleen. Moreover, it was observed that lymphopenia appeared in all SBR-fed groups, which could be a reason for the decreasing weight of spleen, and a possibility that SBR could affect lymphocyte proliferation by changing the production of cyclins and interleukins responsible for their proliferation [52]. Baicalin significantly decreased the production of IL-10. According to the cancer immunoediting theory inhibition of the protective functions of the immune system via overproduction of immunosuppressive cytokines, such as IL-10, may also facilitate tumor escape [53]. Yang et al. [54] reported that baicalin, is responsible for the inhibition of the synthesis of inflammatory mediators such as IL-6 and differentiation of T_H17 cells in spleen. Intraperitoneal administration of baicalein (100 mg/kg) enhanced apoptosis of T and B cells in spleens [55]. Baicalin also could promote T regulatory cell differentiation and upregulate the function of T regulatory cells and may also serve as a natural immunosuppressive compound for treating autoimmune inflammatory diseases [54].

The bursa of Fabricius is responsible for the establishment and maintenance of the B-cell compartment in birds [56]. The most rapid growth of bursa occurs in the first 3–5 wk of life and reaches its maximum size in 4 to 12 wk of age [57]. The RW of the bursa did not decrease in group that was fed with 0.5% of SBR diet. However, the results indicated that excess SBR intake (≥1%) decreased the growth of the bursa in young chicken. The lymphopenia could also be a reason for the decreased weight of the bursa of Fabricius. Atrophy of the bursa due to lymphocytic depletion is also caused by viral diseases, bacterial diseases, mycotoxicosis, nutritional diseases, immunosuppressive drugs such as cyclophosphamide, and chick anemia agents [58].

Conclusion

Antibiotics have played a negative role in the animal industry due to the hidden danger of drug residues to human health. In the past two decades, an intensive amount of research has been focused on the development of alternatives to antibiotics. Chinese herbal medicines like *S. baicalensis* contain the bioactive components, which possess antibacterial and anti-inflammatory properties. Current evidence strongly suggests that SBR supplementation in the diet of chicken should be carefully evaluated, as only a narrow dose range of SBR could be considered safe for the poultry. Dietary SBR supplementation in excess of 1% affected the humoral immunity of chickens and restrained the development of immune organs in birds. The exact mechanism that mediates the immune response observed

in this study is not fully understood, because SBR has a wide variety of effects and requires further investigation. These findings may provide with some information to researchers and feed producers in planning the future diets for poultry. Moreover, the effect of SB is dose-dependent and there is always a potential for overdosing consequently; hence, dosage optimization is strongly recommended. Therefore, these results suggest that further research is required in this area and that the perfect antibiotics alternative does not exist as yet.

Abbreviations

ADFI: Average daily feed intake; BA: Baicalin; DW: Dry weight; FCR: Feed conversion ratio; H/L: Heterophils/lymphocytes ratio; HCT: Haematocrit; HGB: Hemoglobin; HPLC: High-performance liquid chromatography; MGG: May-Grünwald-Giemsa staining method; NBT: Nitroblue tetrazolium; OD: Optical density; ROS: Reactive oxygen species; RS: Radial segmentation; RW: Relative weight; SB: *Scutellaria baicalensis*; SBR: *Scutellaria baicalensis* root; SRBCs: Sheep red blood cells; WBC: Leukocytes

Acknowledgements

We thank Prof Jan Oszmiański, for the kind supply of plant material.

Funding

Publication of this paper was supported by the Wrocław Center for Biotechnology program KNOW (National Scientific Leadership Center) for the 2014–2018 award to BK. The funders had no role in study design, data collection and analysis, decision to publish, or in the preparation of the manuscript.

Authors' contributions

This study was conceived and designed by SG and BK. The experiments were performed by SG, BK, DM and AK-P. The data were analysed by SG, BK, and JK. Reagents/materials/analysis tools were contributed by SG, BK, JK, and WZ. The paper was written by BK and JK, and all authors read and approved the final manuscript.

Competing interests

The authors declare that they have no competing interests.

Author details

[1]Department of Animal Physiology and Biostructure, Faculty of Veterinary Medicine, Wroclaw University of Environmental and Life Sciences, C.K Norwida 31, 50-375 Wrocław, Poland. [2]Department of Immunology, Pathophysiology and Veterinary Preventive Medicine, Faculty of Veterinary Medicine, Wroclaw University of Environmental and Life Sciences, Wrocław, Poland. [3]Department of Chemical Biology, Faculty of Biotechnology, University of Wrocław, Fryderyka Joliot-Curie 14a, 50-383 Wrocław, Poland.

References

1.   Hayek SA, Gyawali R, Ibrahim SA. Antimicrobial natural products. In: Méndez-Vilas A, editor. Microbial pathogens and strategies for combating them: Science, technology and education. Spain: Formatex Research Center; 2013. p. 910–21.

2. Hardy B. The issue of antibiotic use in the livestock industry: what have we learned? Anim Biotechnol. 2002;13:129–47.

3. Asif M. A brief study of toxic effects of some medicinal herbs on kidney. Adv Biomed Res. 2012;1:44.

4. Lojek A, Denev P, Ciz M, Vasicek O, Kratchanova M. The effects of biologically active substances in medicinal plants on the metabolic activity of neutrophils. Phytochem Rev. 2014;13:499–510.

5. Huff GR, Huff WE, Rath NC, Tellez G. Limited Treatment with β-1,3/1,6-Glucan Improves Production Values of Broiler Chickens Challenged with Escherichia Coli. Poultry Sci. 2006;85:613–8.

6. Akter S, Khan M, Jahan M, Karim M, Islam M. Histomorphological study of the lymphoid tissues of broiler chickens. Bangladesh J Vet Med. 2006;4:87–92.

7. Authority TES. Scientific opinion on the influence of genetic parameters on the welfare and the resistance to stress of commercial broilers. EFSA J. 2010;8:1666.

8. Kroliczewska B, Zawadzki W, Skiba T, Kopec W, Kroliczewski J. The influence of baical skullcap root (Scutellaria baicalensis radix) in the diet of broiler chickens on the chemical composition of the muscles, selected performance traits of the animals and the sensory characteristics of the meat. Vet Med Czech. 2008;53:373–80.

9. Króliczewska B, Jankowska P, Zawadzki W, Oszmianski J. Performance and selected blood parameters of broiler chickens fed diets with skullcap (Scutellaria baicalensis Georgi) root. J Anim Feed Sci. 2004;13:35–8.

10. Tang W, Eisenbrand G. baicalensis Georgi. In: Tang W, Eisenbrand G, editors. Chinese Drugs of Plant Origin. Berlin: Springer Berlin Heidelberg; 1992. p. 919–29.

11. Błach-Olszewska Z, Lamer-Zarawska E. Come Back to Root – Therapeutic Activities of Scutellaria baicalensis Root in Aspect of Innate Immunity Regulation – Part I. Adv Clin Exp Med. 2008;17:337–45.

12. Li C, Lin G, Zuo Z. Pharmacological effects and pharmacokinetics properties of Radix Scutellariae and its bioactive flavones. Biopharm Drug Dispos. 2011;32:427–45.

13. Dr. Duke's Phytochemical and Ethnobotanical Databases. 1996. http://phytochem.nal.usda.gov/. Accessed 15 Feb 2017.

14. Kroliczewska B, Mista D, Zawadzki W, Wypchlo A, Kroliczewski J. Effects of a skullcap root supplement on haematology, serum parameters and antioxidant enzymes in rabbits on a high-cholesterol diet. J Anim Physiol Anim Nutr. 2011;95:114–24.

15. Horvath CR, Martos PA, Saxena PK. Identification and quantification of eight flavones in root and shoot tissues of the medicinal plant Huang-qin (Scutellaria baicalensis Georgi) using high-performance liquid chromatography with diode array and mass spectrometric detection. J Chromatogr A. 2005;1062:199–207.

16. NRC. Nutrient Requirements of Poultry. Ninth Revised Edition. Washington, DC: The National Academies Press; 1994.

17. Horwitz W. Official methods of analysis of the Association of Official Analytical Chemists. 17th ed. Gaithersburg: AOAC Int.; 2000.

18. Winterhalter K. Hemoglobins, Porphyrins and Related Compounds. In: Curtius H, Roth M, editors. Clinical Biochemistry; Principles and Methods. Berlin: de Gruyter; 1974. p. 1305–22.

19. Brown BA. Hematology: Principles and Procedures. Philadelphia: Lea and Febiger; 1993.

20. Ladics GS. Use of SRBC antibody responses for immunotoxicity testing. Methods. 2007;41:9–19.

21. Graczyk S, Pliszczak-Król A. Preliminary studies on radial segmentation (RS) of nuclei in hens' blood lymphocytes. Sci Lett Wroclaw, Agric University. 1996;55:15–24.

22. Pliszczak-Król A. The influence of ACTH on the RS of nuclei and acid phosphatase activity in blood lymphocytes of immunized chickens. Med Weter. 2001;57:676–9.

23. Pliszczak-Król A, Szymonowicz M, Król J, Rybak Z, Graczyk S, Haznar D, et al. Influence of Gelatin-Alginian Matrixes on Morphological and Functional Changes of Blood Leukocytes. Polim Med. 2013;43:153–8.

24. Chung S, Secombes CJ. Analysis of events occurring within teleost macrophages during the respiratory burst. Comp Biochem Physiol B. 1988;89:539–44.

25. Xing J, Chen X, Zhong D. Absorption and enterohepatic circulation of baicalin in rats. Life Sci. 2005;78:140–6.

26. Wijga S, Parmentier HK, Nieuwland MGB, Bovenhuis H. Genetic parameters for levels of natural antibodies in chicken lines divergently selected for specific antibody response. Poultry Sci. 2009;88:1805–10.

27. Graczyk S, Wieliczko A, Pliszczak-Krol A, Janaczyk B. Radial Segmentation of Blood Lymphocytes Nuclei in Pheasants Vaccinated against Newcastle Disease and Haemorrhagic Enteritis. Acta Vet Brno. 2008;77:625–30.

28. Norberg B, Soderstrom N. "Radial segmentation" of the nuclei in lymphocytes and other blood cells induced by some anticoagulants. Scand J Haematol. 1967;4:68–76.

29. Core P, Muti S, Cervini C. Radial Segmentation of Leukocyte Nuclei in Some Rheumatic Disorders. Rheumatol Int. 1994;13:247–9.

30. Zhang Y, Li X, Ciric B, Ma C-G, Gran B, Rostami A, et al. Therapeutic effect of baicalin on experimental autoimmune encephalomyelitis is mediated by SOCS3 regulatory pathway. Sci Rep. 2015;5:17407.

31. Zhang Z-Q, Liua W, Zhuang L, Wang J, Zhang S. Comparative Pharmacokinetics of Baicalin, Wogonoside, Baicalein and Wogonin in Plasma after Oral Administration of Pure Baicalin, Radix Scutellariae and Scutellariae-Paeoniae Couple Extracts in Normal and Ulcerative Colitis Rats. Iran J Pharm Res. 2013;12:399–409.

32. Togun VA, Oseni BSA. Effect of low level inclusion of biscuit dust in broiler finisher diet on pre-pubertal growth and some haematological parameters of unsexed broilers. Res Comm Anim Sci. 2005;1:10–4.

33. Bounous DI, Stedman NL. Normal avian hematology: chicken and turkey. In: Feldman BF, Zinkl JG, Jain NC, editors. Schalm's Veterinary Hematology. Baltimore: Lippincott Williams and Wilkins; 2000. p. 1147–54.

34. Trîncă S, Cernea C, Arion A, Ognean L. The Relevance of Mean Blood Samples in Hematological Investigations of Broiler Chickens. Bulletin UASMV, Veterinary Medicine. 2012;69:209–14.

35. Maxwell MH, Robertson GW. The avian heterophil leucocyte: a review. Worlds Poult Sci J. 1998;54:155–78.

36. Czech A, Merska M, Ognik K. Blood Immunological and Biochemical Indicators in Turkey Hens Fed Diets With a Different Content of the Yeast Yarrowia Lipolytica. Ann Anim Sci. 2014;14:935.

37. Geetha RV, Lakshmi T, Roy A. A review on nature's immune boosters. Intl J Pharm Sci Rev Res. 2012;3:43–52.

38. Yin G, Jeney G, Racz T, Xu P, Jun X, Jeney Z . Effect of two Chinese herbs (Astragalus radix and Scutellaria radix) on non-specific immune response of tilapia, Oreochromis niloticus. Aquaculture. 2006;253:39–47.

39. Cai X, Tan J, Wang L, Mu W. Effect of baicalin on the cellular immunity of mice. J Nanjing Railw Med Coll. 1994;13:65–8.

40. Genovese J, He H, Swaggerty CL, Kogut MH. The avian heterophil. Dev Comp Immunol. 2013;41:334–40.

41. Wang R, Luo J, Kong L. Screening of radical scavengers in Scutellaria baicalensis using HPLC with diode array and chemiluminescence detection. J Sep Sci. 2012;35:2223–7.

42. Shen YC, Chiou WF, Chou YC, Chen CF. Mechanisms in mediating the anti-inflammatory effects of baicalin and baicalein in human leukocytes. Eur J Pharmacol. 2003;465:171–81.

43. Shieh DE, Liu LT, Lin CC. Antioxidant and free radical scavenging effects of baicalein, baicalin and wogonin. Anticancer Res. 2000;20:2861–5.

44. Ding M, Robinson JM, Behrens BC, Vandre DD. The microtubule cytoskeleton in human phagocytic leukocytes is a highly dynamic structure. Eur J Cell Biol. 1995;66:234–45.

45. Söderström U-B, Norberg B, Brandt L. The Oxalate-Induced Radial Segmentation of the Nuclei in Peripheral Blood Lymphocytes of Different Size. Scand J Haematol. 1976;17:57–61.

46. Simmingsköld G, Rydgren L, Norberg B, Söderström U-B, Pontén J. Cytochalasin B Partially Inhibits the Oxalate-Induced Radial Segmentation of Mononucleated Blood Cells. Scand J Haematol. 1977;19:33–8.

47. Smolen JE. Characteristics and Mechanisms of Secretion by Neutrophils. In: Hallett MB, editor. The neutrophil: Cellular biochemistry and physiology. Boca Raton: CRC Press; 1989. p. 23–63.

48. Kumagai T, Muller CI, Desmond JC, Imai Y, Heber D, Koeffler HP. Scutellaria baicalensis, a herbal medicine: anti-proliferative and apoptotic activity against acute lymphocytic leukemia, lymphoma and myeloma cell lines. Leuk Res. 2007;31:523–30.

49. Batista FD, Harwood NE. The who, how and where of antigen presentation to B cells. Nat Rev Immunol. 2009;9:15–27.

50. Boa-Amponsem K, Price SEH, Dunnington EA, Siegel PB. Effect of Route of Inoculation on Humoral Immune Response of White Leghorn Chickens Selected for High or Low Antibody Response to Sheep Red Blood Cells. Poultry Sci. 2001;80:1073–8.

51. Jong KL, Ji SS, Gun YL, Jin HK, Juno HE, Kil JH, et al. Immunomodulatory Effect of Scutellaria radix in Balb/c Mice. Lab Anim Res. 2005;21:353–60.

52. Bonham M, O'Connor JM, Hannigan BM, Strain JJ. The immune system as a physiological indicator of marginal copper status? Br J Nutr. 2002;87:393–403.

53. Orzechowska B, Chaber R, Wisniewska A, Pajtasz-Piasecka E, Jatczak B, Siemieniec I, et al. Baicalin from the extract of Scutellaria baicalensis affects the innate immunity and apoptosis in leukocytes of children with acute lymphocytic leukemia. Int Immunopharmacol. 2014;23:558–67.

54. Yang J, Yang X, Li M. Baicalin, a natural compound, promotes regulatory T cell differentiation. BMC Complement Altern Med. 2012;12:64.

55. Zhang Y. Study on Anti-inflammatory and Immune-modulating Effects of Baicalin and Baicalein [Master's thesis]. Shanghai: Second Military Medical University; 2012.

56. Peng X, Cui Y, Cui W, Deng J, Cui H. The Decrease of Relative Weight, Lesions, and Apoptosis of Bursa of Fabricius Induced by Excess Dietary Selenium in Chickens. Biol Trace Elem Res. 2009;131:33–42.

57. Riddell C. Avian histopathology. In: Lawrence KS, editor. Lymphoid system. USA: American Association of Avian Pathologists; 1987. p. 7–17.

58. Nakamura K, Imada Y, Maeda M. Lymphocytic depletion of bursa of Fabricius and thymus in chickens inoculated with Escherichia coli. Vet Pathol. 1986;23:712–7.

3

Mitochondrial DNA T7719G in *tRNA-Lys* gene affects litter size in Small-tailed Han sheep

Xiaoyong Chen[1,2†], Dan Wang[1†], Hai Xiang[1], Weitao Dun[2], Dave O. H. Brahi[1], Tao Yin[1] and Xingbo Zhao[1*]

Abstract

Background: In farm animals, mitochondrial DNA (mtDNA) effect on economic performance remains hot-topic for breeding and genetic selection. Here, 53 maternal lineages of Small-tailed Han sheep were used to investigate the association of mitochondrial DNA variations and the lambing litter size.

Results: Sequence sweeping of the mitochondrial coding regions discovered 31 non-synonymous mutations, and the association study revealed that T7719G in mtDNA *tRNA-Lys* gene was associated with litter size ($P < 0.05$), manifesting 0.29 lambs per litter between the G and T carriers. Furthermore, using the mixed linear model, we assayed the potential association of the ovine litter size and haplogroups and multiple-level mtDNA haplotypes, including general haplotypes, assembled haplotypes of electron transport chain contained sequences (H-ETC), mitochondrial respiratory complex contained sequences (H-MRC) and mitochondrial genes (H-*gene*, including polypeptide-coding genes, rRNA genes and tRNA genes). The strategy for assembled mitochondrial haplotypes was proposed for the first time in mtDNA association analyses on economic traits, although none of the significant relations could be concluded ($P > 0.05$). In addition, the nuclear major gene *BMPR1B* was significantly correlated with litter size in the flock ($P < 0.05$), however, did not interact with mtDNA T7719G mutation ($P > 0.05$).

Conclusions: Our results highlight mutations of ovine mitochondrial coding genes, suggesting T7719G in *tRNA-Lys* gene be a potentially useful marker for selection of sheep litter size.

Keywords: Association, Haplotype, Mitochondria, Non-synonymous mutation, Reproduction, Sheep

Background

Mitochondria are responsible for ATP production in the electron transport chain (ETC) in cells. The ETC consists of five mitochondrial respiratory complexes (MRCs), of which complex I, complex III, complex IV and complex V are constructed by both mitochondrial and nuclear encoded proteins, and complex II is entirely encoded by nuclear genes. Mitochondrial genome codes 13 polypeptides, 2 rRNAs and 22 tRNAs [1]. Since the first report of mitochondrial DNA (mtDNA) effect on milk production traits in dairy cattle [2], mtDNA effects on various economic traits have been widely studied in livestock, including pigs [3, 4], dairy cattle [5, 6], beef

cattle [7, 8], sheep [9, 10] and chickens [11, 12]. Litter size is one of the vital economic traits for animal breeding and production, which has been studied for decades. For sheep, the nuclear gene, *BMPR1B* was identified as one of causative genes for sheep prolificacy [13], and has been widely used in sheep breeding. For the mtDNA effect on litter size, researchers reported the association with ewe litter size among haplogroups in an Afec-Assaf flock, but did not found the interaction with *BMPR1B* effect [10]. However, previous studies reported poor mtDNA effects [14–17], which made it necessary to uncover the genetic contribution of mtDNA for ewe litter size. In this study, Small-tailed Han sheep, a prolific breed of China, were used to investigate the association of litter size with mtDNA coding genes, in which non-synonymous mutations were considered as possible functional SNPs. Besides the non-synonymous mutation and haplogroup, assembled haplotypes of ETC-

* Correspondence: zhxb@cau.edu.cn
†Equal contributors
[1]College of Animal Science and Technology, China Agricultural University, Beijing 100193, China
Full list of author information is available at the end of the article

contained mtDNA sequences, MRC-contained mtDNA sequences, and mitochondrial genes were analyzed, respectively. The strategy for assembled mitochondrial haplotypes was proposed for the first time in association analyses. In addition, *BMPR1B* effect and interaction with mtDNA mutations were analyzed.

Methods

Animals

In total, 117 lambing Small-tailed Han sheep from 53 maternal lineages (families divided by female ancestors) of the same flock were performed blood sample collection, and recorded one or more times of litter size (Additional file 1: Table S1). All sheep were kept indoors year-round, and fed with mixed silage and hay to meet their nutritional requirements. Litter size (number of lambs born) was recorded, and full pedigree information was collected for all animals as reproductive management of the flock where ewes were mated with the selected rams after spontaneous estrus. All the ewes were genotyped for the *BMPR1B* at >2-month age.

Genotyping of *BMPR1B* and sequencing of mitochondrial coding genes

Genomic DNA was extracted using the standard phenol/chloroform method [18]. The *BMPR1B* was genotyped using PCR-RFLP assay with the *Ava* II restriction enzyme [19]. Mitochondrial complete coding sequences were amplified by 17 primer pairs [20], and PCR products were sequenced in the Sanger method.

Haplotype and haplogroup constitution

All non-synonymous mutations were used to constitute the haplotype and haplogroup. Considering the possible action for mitochondrial function, haplotypes were furthermore assembled by mtDNA sequence of ETC, MRCs and genes respectively. Haplotypes were determined by the online software FaBox [21], and haplogroups were constituted based on network analysis by Network 4.6.1.4 [22].

Association analysis

Association analyses were carried out in the following mixed model by MIXED procedure in SAS software version 9.2 (SAS Institute Inc., Cary, North Carolina, USA).

$$ls = ys + parity + ram + BMPR1B + mutations$$
$$+ BMPR1B \times mutations + ID + EP + e$$

In the model, the effects of lambing year-season (*ys*), parity number (*parity*), service ram (*ram*), *BMPR1B* genotype (*BMPR1B*), mtDNA mutations (*mutations*, including the effects of mtDNA non-synonymous mutations, haplotypes and haplogroups), the interaction

between *BMPR1B* and mtDNA mutations (*BMPR1B* × *mutations*), the polygenic effect (*ID*), the permanent environmental effect (*EP*), and the random residual (*e*) were included. The response variable was the ewe litter size (*ls*). Each cell of these effects contained observations. The polygenic effect corrected the genetic background by the additive genetic relationship matrix, *i.e.* the pedigree information. The permanent environmental effect dealt with the repeated measurement data.

Table 1 Non-synonymous mutations in mitochondrial coding genes and corresponding effects on litter size

Gene	Nucleotide[a]	Codon mutation	Amid acid substitution	Significance[b]
ND1	T3543A	UCA → ACA	S → T	ns
ND2	T4208C	AUA → ACA	M → T	ns
COII	C7500A	CCC → CAC	P → H	ns
ATP6	A8039G	AAC → AGC	N → S	ns
	G8264C	GGA → GCA	G → A	ns
COIII	A9375G	AUA → GUA	M → V	ns
ND4L	C9974T	CCU → UCU	P → S	ns
	G10118A	GGU → AGU	G → S	ns
ND4	G10937A	GAC → AAC	D → N	ns
	G11045A	GUU → AUU	V → I	ns
ND5	G12571C	GGC → GCC	G → A	ns
	G13041A	GCA → ACA	A → T	ns
ND6	C13576T	CUC → UUC	L → F	ns
	T13588C	UAC → CAC	Y → H	ns
	C13777T	CAU → UAU	H → Y	ns
	C13789T	CAU → UAU	H → Y	ns
	T13837C	UCA → CCA	S → P	ns
	T13855C	UUC → CUC	F → L	ns
	A13876G	AUA → GUA	M → V	ns
12SrRNA	T281C	-	-	ns
	C291T	-	-	ns
	A538G	-	-	ns
16SrRNA	A1099T	-	-	ns
	T1112C	-	-	ns
	T2199A	-	-	ns
	C2443T	-	-	ns
	T2634C	-	-	ns
tRNA-Tyr	G5295A	-	-	ns
tRNA-Lys	T7719G	-	-	*
tRNA-His	C11606T	-	-	ns
tRNA-Ser	G11668A	-	-	ns

[a]Mutation positions were defined according to the ovine mitochondrial sequence (GenBank Accession nos.: AF010406)
[b]When a set of statistical inferences were simultaneously considered, multiple comparisons were conducted by the FDR using the R project. "ns" represents "not significant", and "*" represents "significant" at the significant level of 0.05

Inference about the interaction effect was made. If non-significant, that effect was dropped from the model and inference was made about the main effects. If significant, the cell means for the interaction became of interest.

When a set of statistical inferences were simultaneously considered, multiple comparisons were conducted by the false discovery rate (FDR) in the R project (R version 3.2.5) [23].

Results

Mutations in mitochondrial coding genes

In total, 95 mutations in mitochondrial coding genes were discovered (Additional file 1: Table S2), including 64 synonymous and 31 non-synonymous mutations (19 missense mutations in protein coding genes, 8 mutations in rRNAs, and 4 mutations in tRNAs respectively), which were illustrated in Table 1.

BMPR1B genotypes

For *BMPR1B* gene, 117 ewes were genotyped, including ++ genotype of 18 ewes, B+ genotype of 87 ewes and BB genotype of 12 ewes.

Effects of mitochondrial haplotype and haplogroup on ovine litter size

The 31 mitochondrial non-synonymous mutations assigned the Small-tailed Han sheep flock to 44 haplotypes (Additional file 1: Table S3), which were clustered into 4 haplogroups (Fig. 1). Using the mixed linear model, no significant effect on any haplotype or haplogroup was associated with litter size of Small-tailed Han sheep ($P > 0.05$), and the interaction between *BMPR1B* and haplotype or haplogroup also did not significantly affect litter size

($P > 0.05$), while the *BMPR1B* was positively associated with litter size ($P < 0.05$).

Effects of missense mutations and haplotypes in mitochondrial protein coding genes on ovine litter size

Totally 35 H-ETCs were constituted by 19 missense mutations in protein coding sequences, while 15 missense mutations in MRCI assembled 33 H-MRCIs, and both 2 mutations in MRCIV and MRCV constituted 3 H-MRCIVs and 3 H-MRCVs respectively. Intensively, H-*genes* were assembled by mutations of individual genes (Table 2 and Additional file 1: Table S4, S5 and S6). With mixed linear model analyses, no significant association was detected between litter size and the 19 non-synonymous mutations ($P > 0.05$) (Table 1), nor was any haplotype ($P > 0.05$) (Table 2). *BMPR1B* was strongly associated with litter size ($P < 0.05$) (Table 3), but the interaction of *BMPR1B* and mtDNA mutations was not remarkable ($P > 0.05$).

Effects of mutations and haplotypes in mitochondrial rRNA genes on ovine litter size

The 3 mutations in *12SrRNA* sorted the sheep flock into 4 haplotypes (H-*12SrRNA*), and the 5 mutations in *16SrRNA* constituted 15 haplotypes (H-*16SrRNA*) (Table 2 and Additional file 1: Table S7). Association analyses revealed that neither mutation nor haplotype in rRNA genes affected litter size ($P > 0.05$) (Tables 1 and 2). The *BMPR1B* genotype was remarkably associated with litter size ($P < 0.05$) (Table 3), however, was inconspicuously correlated to rRNA mutations ($P > 0.05$).

Effects of mutations in mitochondrial tRNA genes on ovine litter size

There were 4 mutations in tRNA genes, and only one variation was observed in each of them (Table 1),

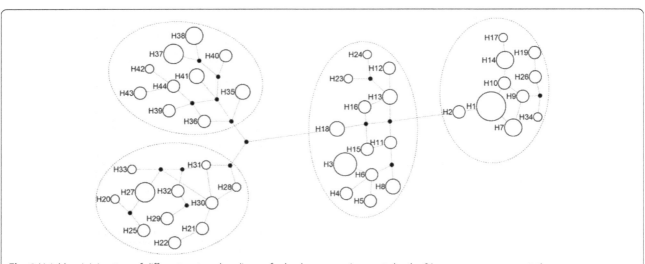

Fig. 1 Neighbor-joining tree of different maternal pedigrees for haplogroup assignments by the 31 non-synonymous mutations

Table 2 Haplotypes constituted by mitochondrial non-synonymous mutations in multiple levels and significant effects on litter size

Functional haplotype[a]	Contained gene	Contained mutation number	Haplotype number	Significance[b]
H	ND1, ND2, ND4L, ND4, ND5, ND6, COII, COIII, ATP6, 12SrRNA, 16SrRNA, tRNA-Tyr, tRNA-Lys, tRNA-His, tRNA-Ser	31	44	ns
H-ETC	ND1, ND2, ND4L, ND4, ND5, ND6, COII, COIII, ATP6	19	35	ns
H-MRCI	ND1, ND2, ND4L, ND4, ND5, ND6	15	33	ns
H-MRCIV	COII, COIII	2	3	ns
H-MRCV	ATP6	2	3	ns
H-ND4L	ND4L	2	4	ns
H-ND4	ND4	2	4	ns
H-ND5	ND5	2	3	ns
H-ND6	ND6	7	10	ns
H-ATP6	ATP6	2	3	ns
H-12SrRNA	12SrRNA	3	4	ns
H-16SrRNA	16SrRNA	5	15	ns

[a]The general haplotype (H) was assembled by all non-synonymous mutations to reflect the integrated characteristics of mtDNA coding regions. Subsequently, the ETC based haplotype (H-ETC) represented the general feature of all ETC-contained mtDNA sequences. The MRC based haplotype (H-MRC) inferred the integrated signal of particular MRC-contained mtDNA sequences. Here, H-MRC included three types, i.e. MRC-I, MRC-IV and MRC-V. The last was the gene-level haplotype (H-gene), which indicated the information of a particular gene-contained mtDNA sequence
[b]When a set of statistical inferences were simultaneously considered, multiple comparisons were conducted by the FDR using the R project. "ns" represents "not significant", and "*" represents "significant" at the significant level of 0.05

therefore no haplotype was constituted. Notably, T7719G in *tRNA-Lys* affected litter size by 0.29 lambs per litter between the G (1.79) and T (1.50) carriers ($P < 0.05$). Even though the *BMPR1B* was in prominent correlation with litter size ($P < 0.05$) (Table 3), there was no interaction with T7719G ($P > 0.05$) (Additional file 1: Table S8).

Discussion

In human diseases, mutations in mitochondrial coding genes led to changes of oxidative phosphorylation enzyme complexes [24]. It is rational to image the possible effects of mtDNA on livestock traits. In this study, the Small-tailed Han sheep were used to explore the correlation of mtDNA and ewe litter size.

In order to investigate the overall mtDNA effects, a novel strategy of assembling mitochondrial haplotypes based on biology functions is proposed. Extensively, there are four levels of mitochondrial haplotypes for mitochondrial biological functions. The general haplotype is assembled by all non-synonymous mutations to reflect the integrated characteristics of mtDNA coding

Table 3 Effects of *BMPR1B* on litter size

Genotype	Ewe number	Parity number	Litter size (Means)[1]
++	18	69	1.4638[a]
B+	87	285	1.7474[b]
BB	12	47	2.1915[c]

[1]"Means" represents the arithmetic average on litter size of sheep with the genotype. The FDR method was used to conduct multiple comparisons, and means with different lowercase letters are different at the significant level of 0.05

regions. Subsequently, the ETC based haplotype (H-ETC) represents the general feature of all ETC-contained mtDNA sequences. The MRC based haplotype infers the integrated signal of particular MRC-contained mtDNA sequences. H-MRC includes four types, i.e. MRC-I, MRC-III, MRC-IV and MRC-V. The last is the gene level, which indicates the information of a particular gene-contained mtDNA sequence.

In the mixed linear model, unnecessary factors interfere with the estimation of interested factors, as the degree of freedom is wasted. Therefore, based on the thoughts of multiple models [10] and variable selection in a mixed linear model [25], a two-step method of environmental variable selection was put forward in the association analyses, in which non-significant environmental factors were ignored to improve accuracy of genetic estimation. In this study, the environmental factors included lambing month, parity of dam, and service ram, and the genetic factors, which we were interested in, included *BMPR1B* genotype, mtDNA mutations and their interaction. We started a full model (Model I) with all the environmental and genetic factors to test if environmental factors were noises. Results revealed that the effects of parity of dam and service ram were not significant on litter size. Subsequently, these two factors were excluded in the aims to construct an optimized model (Model II), which was used to test the significance of genetic factors, especially the effects of mtDNA mutations on litter size.

Many mitochondrial tRNA mutations in human were reported to be associated with wide range of pathological

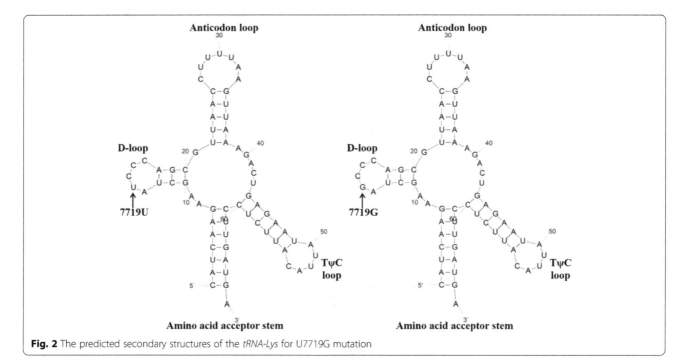

Fig. 2 The predicted secondary structures of the *tRNA-Lys* for U7719G mutation

conditions [26], for example, deafness was correlated with mitochondrial *tRNA-Asp* A7551G mutation [27], hypertension was associated with mitochondrial *tRNA-Ile* A4263G mutation [28], and mitochondrial tRNA mutations might decrease in carcinoma hepatocyte [29]. In sheep, A7755G in *tRNA-Lys* was reported to affect litter size in an Afec-Assaf flock [10], while our study revealed that T7719G in *tRNA-Lys* was significantly associated with ewe litter size in Small-tailed Han sheep, with the means of 1.79 and 1.50 for the G and T carriers, manifesting a difference of 0.29 lambs per litter. Furthermore, the mutation was predicted to produce a transversion of U to G at DHU loop on 2D cloverleaf of the tRNA-Lys structure (Fig. 2).

Conclusions

As a conclusion, we summarize the research procedures for mtDNA effects. Firstly, sweeping mitochondrial variations on maternal lineages; secondly, constituting the haplotype, haplogroup and assembled haplotypes; lastly, analyzing the association between mtDNA mutations (individual mutations, haplotypes and haplogroups) and interested traits. The present study discovered the mtDNA T7719G was linked with ewe litter size. For traits of low heritability, the marker assisted selection could increase the accuracy of breeding and selection. For the further study, the mtDNA T7719G should be put into post-association validation, and may become a genetic marker in the sheep breeding programs.

Additional file

Additional file 1: Table S1. Detail information of used sheep. **Table S2.** Information of variations in mtDNA coding region of Small-tailed Han sheep. **Table S3.** The haplotype constituted by the 31 mitochondrial non-synonymous mutations in mitochondrial coding regions (H). **Table S4.** Assembled haplotypes of mitochondrial electron transport chain contained sequences(H-ETC). **Table S5.** Assembled haplotypes of mitochondrial respiratory complex contained sequences (H-MRCI, H-MRCIV, and H-MRCV). **Table S6.** Assembled haplotypes of protein coding genes (H-*ND4L*, H-*ND4*, H-*ND5*, H-*ND6*, H-*ATP6*). **Table S7.** Assembled haplotypes of rRNA coding genes (H-*12SrRNA*, H-*16SrRNA*). **Table S8.** Statistics for the number of observations in each cell of the interaction between *BMPR1B* and T7719G in *tRNA-Lys* and corresponding effects on litter size.

Abbreviations
mtDNA: Mitochondrial DNA; ETC: Electron transport chain; MRC: Mitochondrial respiratory complex; H: The haplotype constituted by all non-synonymous mutations in mitochondrial coding sequences; HG: The haplogroup classified by H; H-ETC: The assembled haplotypes of mtDNA electron transport chain contained sequences; H-MRC: The assembled haplotypes of mtDNA mitochondrial respiratory complex contained sequences, including H-MRCI, H-MRCIII, H-MRCIV, and H-MRCV, corresponding to complex I, III, IV and V, respectively; H-*gene*: The assembled haplotypes of mitochondrial genes, including polypeptide-coding genes, rRNA genes and tRNA genes

Acknowledgements
The authors thank J. C. Jiang (China Agricultural University) for data analysis. The authors also thank Dr. J. M. Zhong and Ms. D. Miller (Auburn University) for language editing.

Funding
The research was supported by the Project of Science and Technology of Hebei Province (15226308D) and the National Key Basic Research Program of China (2014CB138500). The former collected the sample and fed the experimental sheep, and the latter designed the study, performed the experiments, analyzed the data, interpreted the data, and drafted the manuscript.

Authors' contributions

XZ designed the study. WD and XC collected the sample and fed the experimental sheep. the XC, HX, DB and TY performed the experiments. DW, HX and XC analyzed the data. DW, HX, XC and XZ interpreted the data and drafted the manuscript. All authors read and approved the final manuscript.

Competing interests

The authors declare that they have no competing interests.

Author details

[1]College of Animal Science and Technology, China Agricultural University, Beijing 100193, China. [2]Institute of Animal Science and Veterinary of Hebei Province, Baoding 071000, China.

References

1. Anderson S, Bankier AT, Barrell BG, de Bruijn MH, Coulson AR, Drouin J, et al. Sequence and organization of the human mitochondrial genome. Nature. 1981;290:457–65.
2. Bell BR, Mcdaniel BT, Robison OW. Effects of cytoplasmic inheritance on production traits of dairy-cattle. J Dairy Sci. 1985;68:2038–51.
3. Yen NT, Lin CS, Ju CC, Wang SC, Huang MC. Mitochondrial DNA polymorphism and determination of effects on reproductive trait in pigs. Reprod Domest Anim. 2007;42:387–92.
4. Fernandez AI, Alves E, Fernandez A, de Pedro E, Lopez-Garcia MA, Ovilo C, et al. Mitochondrial genome polymorphisms associated with longissimus muscle composition in Iberian pigs. J Anim Sci. 2008;86:1283–90.
5. Qin YH, Chen SY, Lai SJ. Polymorphisms of mitochondrial Atpase 8/6 genes and association with milk production traits in Holstein cows. Anim Biotechnol. 2012;23:204–12.
6. Paneto JCC, Ferraz JBS, Balieiro JCC, Bittar JFF, Ferreira MBD, Leite MB, et al. Bos indicus or Bos taurus mitochondrial DNA: comparison of productive and reproductive breeding values in a Guzerat dairy herd. Genet Mol Res. 2008;7:592–602.
7. Zhang B, Chen H, Hua LS, Zhang CL, Kang XT, Wang XZ, et al. Novel SNPs of the mtDNA ND5 gene and their associations with several growth traits in the Nanyang cattle breed. Biochem Genet. 2008;46:362–8.
8. Auricélio AM, Bittar JFF, Bassi PB, Ronda JB, Bittar ER, Panetto JCC, et al. Influence of endogamy and mitochondrial DNA on immunological parameters in cattle. BMC Vet Res. 2014;10:79–88.
9. Henry BA, Loughnan R, Hickford J, Young IR, St. John JC, Clarke I. Differences in mitochondrial DNA inheritance and function align with body conformation in genetically lean and fat sheep. J Anim Sci. 2015;93:2083–93.
10. Reicher S, Seroussi E, Weller JI, Rosov A, Gootwine E. Ovine mitochondrial DNA sequence variation and its association with production and reproduction traits within an Afec-Assaf flock. J Anim Sci. 2012;90:2084–91.
11. Li S, Zadworny D, Aggrey SE, Kuhnlein U. Mitochondrial PEPCK: a highly polymorphic gene with alleles co-selected with Marek's disease resistance in chickens. Anim Genet. 1998;29:395–7.
12. Lu WW, Hou LL, Zhang WW, Zhang PF, Chen W, Kang X, et al. Study on heteroplasmic variation and the effect of chicken mitochondrial ND2. Mitochondrial DNA Part A, DNA mapping, sequencing, and analysis. 2016;27:2303-9.
13. Mulsant P, Lecerf F, Fabre S, Schibler L, Monget P, Lanneluc I, et al. Mutation in bone morphogenetic protein receptor-IB is associated with increased ovulation rate in Booroola Merino ewes. Proc Natl Acad Sci U S A. 2001;98:5104–9.
14. Hanford KJ, Snowder GD, Vleck LDV. Models with nuclear, cytoplasmic, and environmental effects for production traits of Columbia sheep. J Anim Sci. 2003;81:1926–32.
15. Van Vleck LD, Snowder GD, Hanford KJ. Models with cytoplasmic effects for birth, weaning, and fleece weights, and litter size, at birth for a population of Targhee sheep. J Anim Sci. 2003;81:61–7.
16. Snowder GD, Hanford KJ, Van Vleck LD. Comparison of models including cytoplasmic effects for traits of Rambouillet sheep. Livest Prod Sci. 2004;90:159–66.
17. Van Vleck LD, Hanford KJ, Snowder GD. Lack of evidence for cytoplasmic effects for four traits of Polypay sheep. J Anim Sci. 2005;83:552–6.
18. Sambrook J, Russell D. Molecular Cloning: A Laboratory Manual. 3rd ed. Cold Spring Harbor, New York: Cold Spring Harbor Laboratory Press; 2001.
19. Liu FL, Liu YB, Wang F, Wang R, Tian CY, Liu MX. Study on the polymorphism of Bone Morphogenetic Protein Receptor IB in China partial sheep. Acta Agr Boreali-Sinica. 2007;22:151–4.
20. Meadows JRS, Hiendleder S, Kijas JW. Haplogroup relationships between domestic and wild sheep resolved using a mitogenome panel. Heredity. 2011;106:700–6.
21. Villesen P. FaBox: an online toolbox for fasta sequences. Mol Ecol Notes. 2007;7:965–8.
22. Network 4.6.1.4. http://www.fluxus-engineering.com. Accessed 5 Sept 2016.
23. R Core Team. R: A language and environment for statistical computing. Vienna: R Foundation for Statistical Computing; 2013. URL http://www.R-project.org/.
24. Taylor RW, Turnbull DM. Mitochondrial DNA mutations in human disease. Nat Rev Genet. 2005;6:389–402.
25. Haapalainen E, Laurinen P, Roning J, Kinnunen H. Estimation of exercise energy expenditure using a Wrist-Worn accelerometer: a linear mixed model approach with fixed-effect variable selection. Seventh International Conference on Machine Learning and Applications, Proceedings. 2008. p. 796–801.
26. A Human Mitochondrial Genome Database. http://www.mitomap.org. Accessed 4 May 2012.
27. Wang M, Peng Y, Zheng J, Zheng B, Jin X, Liu H, et al. A deafness-associated tRNAAsp mutation alters the m1G37 modification, aminoacylation and stability of tRNAAsp and mitochondrial function. Nucleic Acids Res. 2016;44:10974-10985.
28. Chen X, Zhang Y, Xu B, Cai Z, Wang L, Tian J, et al. The mitochondrial calcium uniporter is involved in mitochondrial calcium cycle dysfunction: Underlying mechanism of hypertension associated with mitochondrial tRNA(Ile) A4263G mutation. Int J Biochem Cell Biol. 2016;78:307–14.
29. Li G, Duan YX, Zhang XB, Wu F. Mitochondrial tRNA mutations may be infrequent in hepatocellular carcinoma patients. Genet Mol Res. 2016;15.

4

Innate immune responses induced by lipopolysaccharide and lipoteichoic acid in primary goat mammary epithelial cells

Omar Bulgari[1,2], Xianwen Dong[1,3], Alfred L. Roca[1], Anna M. Caroli[2] and Juan J. Loor[1,4*]

Abstract

Background: Innate immune responses induced by in vitro stimulation of primary mammary epithelial cells (MEC) using Gram-negative lipopolysaccharide (LPS) and Gram-positive lipoteichoic acid (LTA) bacterial cell wall components are well- characterized in bovine species. The objective of the current study was to characterize the downstream regulation of the inflammatory response induced by Toll-like receptors in primary goat MEC (pgMEC). We performed quantitative real-time RT-PCR (qPCR) to measure mRNA levels of 9 genes involved in transcriptional regulation or antibacterial activity: Toll-like receptor 2 (TLR2), Toll-like receptor 4 (TLR4), prostaglandin-endoperoxide synthase 2 (PTGS2), interferon induced protein with tetratricopeptide repeats 3 (IFIT3), interferon regulatory factor 3 (IRF3), myeloid differentiation primary response 88 (MYD88), nuclear factor of kappa light polypeptide gene enhancer in B-cells 1 (NFKB1), Toll interacting protein (TOLLIP), and lactoferrin (LTF). Furthermore, we analyzed 7 cytokines involved in Toll-like receptor signaling pathways: C-C motif chemokine ligand 2 (CCL2), C-C motif chemokine ligand 5 (CCL5), C-X-C motif chemokine ligand 6 (CXCL6), interleukin 8 (CXCL8), interleukin 1 beta (IL1B), interleukin 6 (IL6), and tumor necrosis factor alpha (TNF).

Results: Stimulation of pgMEC with LPS for 3 h led to an increase in expression of CCL2, CXCL6, IL6, CXCL8, PTGS2, IFIT3, MYD88, NFKB1, and TLR4 (P < 0.05). Except for IL6, and PTGS2, the same genes had greater expression than controls at 6 h post-LPS (P < 0.05). Expression of CCL5, PTGS2, IFIT3, NFKB1, TLR4, and TOLLIP was greater than controls after 3 h of incubation with LTA (P < 0.05). Compared to controls, stimulation with LTA for 6 h led to greater expression of PTGS2, IFIT3, NFKB1, and TOLLIP (P < 0.05) whereas the expression of CXCL6, CXCL8, and TLR4 was lower (P < 0.05). At 3 h incubation with both toxins compared to controls a greater expression (P < 0.05) of CCL2, CCL5, CXCL6, CXCL8, IL6, PTGS2, IFIT3, IRF3, MYD88, and NFKB1 was detected. After 6 h of incubation with both toxins, the expression of CCL2, CXCL6, IFIT3, MYD88, NFKB1, and TLR4 was higher than the controls (P < 0.05).

Conclusions: Data indicate that in the goat MEC, LTA induces a weaker inflammatory response than LPS. This may be related to the observation that gram-positive bacteria cause chronic mastitis more often than gram-negative infections.

Keywords: Gene expression, Inflammation, Lactation, Mastitis

* Correspondence: jloor@illinois.edu
[1]Department of Animal Sciences and Division of Nutritional Sciences, University of Illinois at Urbana-Champaign, Urbana, IL 61801, USA
[4]Division of Nutritional Sciences, University of Illinois at Urbana-Champaign, Urbana, IL 61801, USA
Full list of author information is available at the end of the article

Innate immune responses induced by lipopolysaccharide and lipoteichoic acid in primary goat...

29

Background

Mastitis is the most prevalent disease in dairy cattle, causing the largest economic losses to the industry. The economic impact of mastitis on the U.S. dairy industry was estimated at $2 billion in 2009 [1]. The transmission of microorganisms into the mammary gland may involve the transfer of pathogens from other animals directly, from the environment or from the milking process [2]. The most common causal agent of mastitis in goats is *Staphylococcus aureus* followed by *Pasteurella haemolytica*, *Escherichia coli*, *Clostridium perfrigens*, *Streptococcus* sp., *Pseudomonas* sp., and *Nocardia* sp. [3].

Severe clinical mastitis with systemic signs produced by *S. aureus* and *E. coli* may be due to the action of various cytotoxins and endotoxins leading to extensive tissue damage and systemic reactions in the animal [2, 3]. It is well established that mastitis modifies gene expression [4, 5] and decreases animal performance [6, 7]. Toll-like receptors (TLR) play a central role in the innate immune system, and form a first line of defense against infections by recognizing pathogen associated molecular patterns [8]. In the goat, 10 TLRs have been identified, designated TLR1-TLR10 [9]. In particular, TLR2 recognizes lipoteichoic acid (LTA), a major constituent of Gram-positive bacteria, and TLR4 recognizes lipopolysaccharide (LPS) that is common to Gram-negative bacteria [8].

Innate immune responses induced by in vitro stimulation of primary mammary epithelial cells (pMEC) using LPS and LTA bacterial cell wall components are well characterized in bovine species. Numerous studies have demonstrated a potential role for TLR2 and TLR4 in the development of mastitis in dairy cattle [10], resistance to bacteria [11], and ability to affect the level of bacteria in milk [12]. Both LPS and LTA are able to cause an inflammatory response via TLR signaling [13, 14]. Activated TLR2 and TLR4 induce a common signaling pathway known as myeloid differentiation primary response 88 (MYD88)-dependent [15], and leads to the activation of kappa light polypeptide gene enhancer in B-cells 1 (NFKB1) and transcription of several pro-inflammatory genes [16].

Our hypothesis was that primary goat mammary epithelial cells (pgMEC) incubated with LPS or LTA have the capacity to mount innate immune responses that can be evaluated through changes in gene transcription. The objective of the present study was to characterize the downstream regulation of the inflammatory response induced by Toll-like receptors in pgMEC stimulated by LPS or LTA.

Methods

Cell culture and treatments

The pgMEC were isolated according to the method of Ogorevc and Dovč [17]. A cell culture protocol was followed involving the use of growth medium and a lactogenic medium reported in previous studies performed in bovine mammary gland cells [18]. Goat pMEC stored in liquid nitrogen were thawed and cultured in growth medium composed of MEM/EBSS (GE Healthcare, Little Chalfont, United Kingdom) supplemented with 5 mg/L insulin (Thermo Fisher Scientific, Waltham, Massachusetts), 1 mg/L hydrocortisone (Sigma-Aldrich, St. Louis, Missouri), 5 μg/mL transferrin (Sigma-Aldrich), 5 μmol/L ascorbic acid (Sigma-Aldrich), 5 mmol/L sodium acetate (Thermo Fisher Scientific), 10 mL/L penicillin/streptomycin (Sigma-Aldrich), 10% fetal bovine serum (GE Healthcare), 1 mg/L progesterone (Sigma-Aldrich), 0.05% lactalbumin (Sigma-Aldrich), 0.05% α-lactose (Sigma-Aldrich). Media were prepared daily and filtered before use with 0.22 μm Filter Unity Millex MP (EMD Millipore, Billerica, Massachusetts). Thawed cells were seeded in 25 cm^2 flasks (10^6 cells/flask) and cultured until confluence in 5 mL growth medium. At approximately 90% confluence, the cells were washed 3 times with 6 mL PBS (Thermo Fisher Scientific), split following the application of 3 mL 0.25% trypsin (GE Healthcare) and reseeded in new 75 mL flasks at a density of 2.5×10^6 cells/flask (GE Healthcare) in 12 mL fresh growth medium. During growth and treatments the cells were incubated at 37 °C with 5% CO_2 in Incubator KMCC17T0 (Panasonic Healthcare, Tokyo, Japan). After three passages, six 6-well plates were reseeded, 3×10^5 cells/well, in 2.5 mL growth medium.

On the basis of similar studies in bovine pMEC, due to the scarcity of studies on goat cells, agonists inducing an appreciable change in TLR-related genes were selected: LPS from *Escherichia coli* O55:B5 (Sigma-Aldrich) as TLR4 agonist [19, 20] and LTA from *S. aureus* (InvivoGen, San Diego, California) as TLR2 agonist [21, 22]. The use of LPS from *E. coli* 055:B5 strain was also justified by the large number of publications demonstrating its agonist effect on TLR4 receptor in various cell types including mammary cells [20, 23, 24]. The commercial LTA preparation was prepared by the n-butanol extraction method, which preserves its activity while avoiding contamination [25].

After conducting a preliminary study, described in Additional file 1, aimed to select the incubation times and the most suitable concentrations for our purposes, the experiments were performed in 2.5 mL lactogenic medium using 1 μg/mL LPS, 20 μg/mL LTA, and the combination of both (L + L). Lactogenic C medium was composed of Dulbecco's High Glucose Modified Eagle's Medium (GE Healthcare) supplemented with 5 mg/L insulin (Thermo Fisher Scientific), 1 mg/L hydrocortisone (Sigma-Aldrich), 5 μg/mL transferrin (Sigma-Aldrich), 5 μmol/L ascorbic acid (Sigma-Aldrich), 5 mmol/L sodium acetate

(Thermo Fisher Scientific), 10 mL/L penicillin/ streptomycin (Sigma-Aldrich), 1 g/L bovine serum albumin (Sigma-Aldrich), 2.5 mg/L prolactin (Sigma-Aldrich). Triplicate cultures (1 µg/mL LPS; 20 µg/mL LTA; 1 µg/mL LPS + 20 µg/mL LTA) were performed at two incubation times (3 h, 6 h). After incubation, the cell culture supernatant was removed, cells were washed 3 times with PBS 1× and total RNA was extracted from the pgMEC layer. To check cell growth and confluence, a Light Inverted Microscope Primovert (Zeiss, Oberkochen, Germany) integrated with a high definition camera AxioCam ERc 5 s (Zeiss) was used.

RNA extraction, purification, and quality assessment
All these procedures are described in detail in Additional file 1.

Selection of genes, primer design, and quantitative RT-PCR
All these procedures are described in detail in Additional file 1.

Statistical analysis
After normalization with the geometric mean of the internal control genes (*ACTB*, *GAPDH*, and *UXT*), the quantitative PCR data were log$_2$-transformed before statistical analysis to obtain a normal distribution. Statistical analyses were conducted using SAS (v 9.3; SAS Institute Inc., Cary, NC). Data were analyzed using the repeated statement ANOVA with PROC MIXED. The statistical model included time (T; 3 h and 6 h incubation), treatment (TRT; LPS, LTA, LPS + LTA and control), and their interactions (T × TRT) as fixed effects. The Kenward-Roger statement was used for computing the denominator degrees of freedom, whereas spatial power was used as the covariance structure. Data were considered significant at a $P \leq 0.05$ level using the PDIFF statement in SAS. For ease of interpretation, the expression data reported as least squares means were log$_2$ back-transformed.

Results
Microscopy
To verify the aptitude of the cells to develop typical mammary epithelial structure in culture, we carried an overgrowth experiment without harvesting the cells. During cell growth, pgMEC formed a cobblestone-like monolayer (Fig. 1a) that developed into an epithelial island within 3 d (Fig. 1b). By d 8, a central cell cluster within the epithelial islands developed into dense cellular masses (Fig. 1c). Microscopic analysis did not reveal widespread cell death or presence of cellular debris. Our

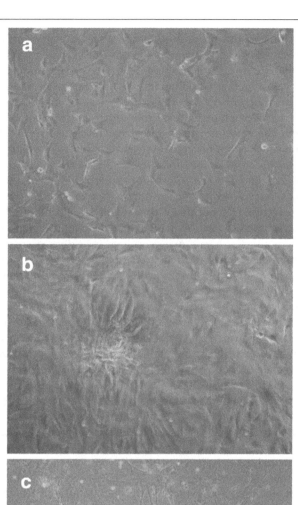

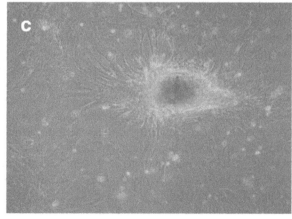

Fig. 1 Establishment of pgMEC in culture on a collagen matrix. **a** Cobblestone-like monolayer. **b** Epithelial island. **c** Dense cellular masses

observations are consistent with previous studies of cellular morphology of pMEC [19, 26, 27].

Gene expression
The quantitative PCR performance results are reported in Table 1. Results of the statistical analyses performed on the expression profiles are in Tables 2 and 3. The expression levels of *IL1B*, *TNF* and *LTF* were deemed undetectable (>30 Ct).

Table 1 Quantitative PCR performance of the measured genes

Gene	Median Ct[a]	Median ΔCt[b]	Slope[c]	(R²)[d]	Efficiency[e]
CCL2	28.62	9.66	−3.29	0.997	2.011
CCL5	28.95	10.05	−3.28	0.991	2.019
CXCL6	24.29	5.23	−3.19	0.999	2.060
CXCL8	29.26	10.34	−3.11	0.994	2.097
IFIT3	24.96	6.04	−3.07	0.993	2.117
IL6	29.11	10.12	−3.34	0.993	1.992
IRF3	24.16	5.27	−3.09	0.991	2.108
MYD88	24.62	5.71	−3.02	0.991	2.143
NFKB1	26.58	7.62	−2.91	0.996	2.204
PTGS2	27.47	8.49	−3.06	0.986	2.120
TLR2	28.51	9.64	−3.31	0.999	2.006
TLR4	30.20	11.28	−2.94	0.999	2.189
TOLLIP	23.59	4.75	−3.35	0.995	1.989

[a]The median is calculated considering all time points and treatments
[b]The median of ΔCt is calculated as [Ct gene - geometrical mean of Ct internal controls] for each time point and treatment
[c]Slope of the standard curve
[d]R² stands for the coefficient of determination of the standard curve
[e]Efficiency is calculated as $[10^{(-1/Slope)}]$

Chemokines and interleukins

We observed a treatment effect for CCL2 ($P < 0.0001$), CCL5 ($P < 0.003$), CXCL6 ($P < 0.0001$), CXCL8 ($P < 0.0001$), and IL6 ($P < 0.001$) (Table 2). Incubation time affected CCL5 ($P < 0.004$), CXCL6 ($P < 0.01$) and CXCL8 genes ($P < 0.0001$) (Table 2). Several significant differences ($P < 0.05$) were found for the interactions between treatment and time (Table 3). Details on these differences are illustrated as follows.

There was an overall increase in most transcript levels in the presence of LPS ($P < 0.0001$), and both toxins ($P < 0.001$) with respects to controls. CCL2 transcription was higher in response to both toxins vs. LTA alone ($P < 0.01$). The combination of both toxins decreased ($P < 0.001$) CCL2 transcription compared to incubation with LPS alone. The highest transcript expression occurred in samples incubated for 3 h in the presence of LPS ($P < 0.0001$). Compared to 3 h, at 6 h incubation the CCL2 transcription was relatively higher in response to LTA ($P < 0.05$) and both toxins ($P < 0.01$), but was lower in the presence of LPS alone ($P < 0.03$).

After 3 h, CCL5 transcript levels increased in samples incubated with both toxins compared to LPS alone ($P < 0.0001$), LTA alone ($P < 0.005$) and control samples ($P < 0.0001$). Incubation for 3 h with LTA alone increased CCL5 transcription with respect to controls ($P < 0.05$). Although no time effect was detected at 3 h for CCL5

Table 2 Log₂ back-transformed LSM of gene transcription for treatment (TRT) and incubation time (T), SEM and P values for TRT and T

| Gene | LSM TRT[d] | | | | LSM T | | SEM | | P-value | |
	Control	LPS	LTA	L + L	3 h	6 h	TRT	T	TRT	T
Cytokines										
CCL2	0.49[c]	1.61[a]	0.54[c]	0.80[b]	0.72	0.80	0.08	0.06	<0.0001	0.0637
CCL5	1.64[b]	1.58[b]	1.65[b]	1.91[a]	1.60[z]	1.78[y]	0.06	0.04	0.0022	0.0034
CXCL6	0.43[c]	1.58[a]	0.40[c]	0.96[b]	0.78[y]	0.66[z]	0.08	0.06	<0.0001	0.0093
CXCL8	0.51[c]	1.64[a]	0.43[c]	0.97[b]	0.97[y]	0.61[z]	0.12	0.09	<0.0001	<0.0001
IL6	1.26[b]	2.65[a]	1.27[b]	1.90[a]	1.66	1.71	0.22	0.18	0.0004	0.8208
Regulatory genes										
IFIT3	1.01[c]	1.13[b]	1.20[ab]	1.24[a]	0.95[z]	1.37[y]	0.04	0.03	<0.0001	<0.0001
IRF3	1.05[b]	1.09	1.13	1.19[a]	1.01[z]	1.22[y]	0.05	0.04	0.0818	<0.0001
MYD88	1.73[b]	2.05[a]	1.81[b]	2.11[a]	1.81[z]	2.03[y]	0.03	0.03	<0.0001	<0.0001
NFKB1	1.13[c]	1.49[a]	1.34[b]	1.51[a]	1.10[z]	1.68[y]	0.05	0.04	<0.0001	<0.0001
PTGS2	1.05[b]	1.30[a]	1.32[a]	1.29[a]	1.03[z]	1.48[y]	0.05	0.04	<0.0001	<0.0001
TLR2	10.39	11.38	10.11	11.24	10.58	10.96	0.09	0.07	0.3747	0.5420
TLR4	1.04[c]	1.49[a]	1.04[c]	1.22[b]	1.18	1.19	0.08	0.06	<0.0001	0.8645
TOLLIP	0.96[b]	0.96[b]	1.04[a]	0.95[b]	0.98	0.97	0.02	0.01	<0.0001	0.3915

[a-c]Different letters represent significant differences between treatments ($P < 0.05$)
The letter a indicates higher transcript levels than b and c. The letter b indicates higher transcript levels than c
[d]Treatments: Control = incubation without toxins; LPS = incubation with 1 µg/mL lipopolysaccharide; LTA = incubation with 20 µg/mL lipoteichoic acid; L + L = incubation with the combination of both toxins
[y-z]Different letters represent significant differences between time points ($P < 0.05$). The letter y indicates higher transcript levels than z

Table 3 Log$_2$ back-transformed LSM of interactions between treatment (TRT) and incubation time (T) on gene transcription, SEM and P values for TRT × T

Gene	T	LSM TRT[d] × T				SEM	P value
		Control	LPS	LTA	L + L	TRT × T	TRT × T
Cytokines							
CCL2	3 h	0.45^c	1.83^{a,y}	0.48^{c,z}	0.69^{b,z}	0.11	0.0040
	6 h	0.53^c	1.41^{a,z}	0.61^{c,y}	0.94^{b,y}		
CCL5	3 h	1.40^{c,z}	1.45^{bc,z}	1.62^b	2.00^a	0.07	0.0018
	6 h	1.91^y	1.72^y	1.68	1.83		
CXCL6	3 h	0.43^c	1.67^a	0.51^{c,y}	0.99^b	0.12	0.0274
	6 h	0.44^c	1.49^a	0.32^{d,z}	0.93^b		
CXCL8	3 h	0.49^c	2.17^{a,y}	0.62^{c,y}	1.34^{b,y}	0.17	0.0085
	6 h	0.52^b	1.24^{a,z}	0.30^{c,z}	0.70^{b,z}		
IL6	3 h	1.01^b	3.01^a	1.15^b	2.17^a	0.28	0.1423
	6 h	1.57	2.33	1.41	1.66		
Regulatory genes							
IFIT3	3 h	0.85^{b,z}	0.96^{a,z}	0.96^{a,z}	1.05^{a,z}	0.05	0.5151
	6 h	1.19^{b,y}	1.34^{a,y}	1.48^{a,y}	1.47^{a,y}		
IRF3	3 h	0.94^{b,z}	1.01^z	0.99^z	1.12^a	0.07	0.3942
	6 h	1.18^y	1.18^y	1.28^y	1.26		
MYD88	3 h	1.64^{b,z}	1.95^{a,z}	1.67^{b,z}	2.02^a	0.05	0.7335
	6 h	1.82^{b,y}	2.16^{a,y}	1.96^{b,y}	2.20^a		
NFKB1	3 h	0.93^{b,z}	1.18^{a,z}	1.12^{a,z}	1.21^{a,z}	0.06	0.5318
	6 h	1.37^{c,y}	1.88^{a,y}	1.62^{b,y}	1.90^{a,y}		
PTGS2	3 h	0.85^{b,z}	1.12^{a,z}	1.07^{a,z}	1.12^{a,z}	0.07	0.2535
	6 h	1.31^{b,y}	1.50^y	1.63^{a,y}	1.48^y		
TLR2	3 h	9.46	10.76	10.80	11.41	0.12	0.2028
	6 h	11.41	12.04	9.47	11.07		
TLR4	3 h	0.99^c	1.42^a	1.28^{ab,y}	1.08^{bc,z}	0.10	<0.0001
	6 h	1.09^b	1.57^a	0.85^{c,z}	1.37^{a,y}		
TOLLIP	3 h	0.96^b	0.97^b	1.03^a	0.97^b	0.03	0.3689
	6 h	0.96^b	0.94^b	1.06^a	0.94^b		

[a-c]Different letters represent significant differences between treatments within the same incubation time ($P < 0.05$). The letter a indicates higher transcript levels than b and c. The letter b indicates higher transcript levels than c
[d]Treatments: LPS = incubation with 1 μg/mL lipopolysaccharide; LTA = incubation with 20 μg/mL lipoteichoic acid; L + L = incubation with the combination of both toxins; Control = incubation without toxins
[y-z]Different letters represent significant differences between time points within the same treatment ($P < 0.05$). The letter y indicates higher transcript levels than z

regardless of treatment, after 6 h the expression of CCL5 increased with LPS alone ($P < 0.02$) and in the controls ($P < 0.0001$).

After 3 and 6 h, treatments with LPS alone or in combination with LTA increased CXCL6 transcription ($P < 0.0001$) when compared to controls and LTA alone. At 3 h ($P < 0.0001$) and 6 h ($P < 0.001$) of incubation, LPS alone increased CXCL6 transcription compared to the incubation with both toxins. A time dependent effect

was detected only in samples incubated with LTA, with a decrease of expression in samples incubated for 6 vs. 3 h ($P < 0.001$). After 3 h, the CXCL8 transcription was higher in LPS samples vs. controls ($P < 0.0001$), LTA alone ($P < 0.0001$) and both toxins ($P < 0.01$). After 6 h, transcription was higher in controls vs. LTA alone ($P < 0.01$) but lower in controls vs. LPS alone ($P < 0.0001$). Furthermore, after 6 h CXCL8 transcription was higher for LPS alone compared to LTA alone ($P < 0.0001$), both toxins vs. LTA alone ($P < 0.0001$), or LPS alone vs. both toxins ($P < 0.002$). Although no time effect was detected at 3 h for CXCL8 regardless of treatment, after 6 h, the expression of CXCL8 decreased with LPS alone ($P < 0.002$), LTA alone ($P < 0.0001$) and both toxins ($P < 0.001$).

Incubation for 3 h with both toxins increased IL6 transcription vs. controls ($P < 0.005$) and LTA alone ($P < 0.02$). After 3 h incubation, LPS alone increased IL6 transcript levels compared to controls and LTA alone ($P < 0.001$).

Other regulatory genes

A treatment effect ($P < 0.0001$) was detected for transcription of IFIT3, MYD88, NFKB1, PTGS2, TLR4 and TOLLIP whereas incubation time affected IFIT3, IRF3, MYD88, NFKB1 and PTGS2 transcription ($P < 0.0001$) (Table 2). Several significant differences ($P < 0.05$) occurred for the interactions between treatment and incubation time (Table 3). Details on these differences are illustrated below.

After 3 h, IFIT3 transcript levels were lower in controls vs. LPS ($P < 0.04$), LTA ($P < 0.03$) and both ($P < 0.001$). The same trend occurred after 6 h when IFIT3 transcription was lower in controls vs. LPS ($P < 0.04$), LTA ($P < 0.001$) and both ($P < 0.001$). Incubation (6 h vs. 3 h) always increased ($P < 0.0001$) IFIT3 transcript levels. We found higher IRF3 transcript levels in samples incubated with both toxins vs. controls ($P < 0.01$) after 3 h incubation. A time dependent increase occurred for LPS ($P < 0.03$), LTA ($P < 0.001$) and controls ($P < 0.002$).

After 3 h, MYD88 transcript levels were lower in controls than LPS ($P < 0.001$) or both toxins ($P < 0.0001$), whereas LTA generated lower transcript levels than LPS alone ($P < 0.003$) or in combination with LTA ($P < 0.001$). After 6 h, MYD88 transcript levels were lower in controls than LPS ($P < 0.001$) or both toxins ($P < 0.001$), whereas LTA generated lower MYD88 transcript levels than LPS alone ($P < 0.05$) or in combination with LTA ($P < 0.02$). Incubation increased MYD88 transcription in samples with LPS ($P < 0.04$), LTA ($P < 0.003$) and controls ($P < 0.04$).

Incubation increased NFKB1 transcription in all samples ($P < 0.0001$). After 3 h, NFKB1 transcript levels were lower in controls than LPS ($P < 0.001$), LTA ($P <

0.002) and both ($P < 0.0001$). After 6 h, *NFKB1* transcription was lower in controls than LPS ($P < 0.0001$), LTA ($P < 0.01$) and both ($P < 0.0001$). Furthermore, at 6 h incubation, transcription was lower in LTA vs. LPS ($P < 0.01$) and both toxins ($P < 0.01$).

After 3 h *PTGS2* transcript levels were lower in controls vs. LPS ($P < 0.001$), LTA ($P < 0.002$) and both toxins ($P < 0.001$). After 6 h only LTA increased *PTGS2* transcript levels vs. controls ($P < 0.004$). Incubation always increased *PTGS2* transcription, i.e. LPS ($P < 0.0001$), LTA ($P < 0.0001$), both toxins ($P < 0.001$) and controls ($P < 0.0001$).

After 3 h, *TLR4* transcript levels were lower in controls than in the presence of LTA ($P < 0.01$) and LPS ($P < 0.001$). Moreover, *TLR4* transcription was higher in samples incubated with LPS vs. both toxins ($P < 0.005$). After 6 h, *TLR4* transcript levels were lower in LTA samples vs. controls ($P < 0.01$), LPS ($P < 0.0001$) and both toxins ($P < 0.0001$), in controls vs. LPS ($P < 0.001$) and both toxins ($P < 0.02$). A time dependent increase was found in samples incubated with both toxins ($P < 0.02$) whereas a time dependent decrease occurred for LTA ($P < 0.0001$).

After 3 h, *TOLLIP* transcript levels were significantly higher in samples incubated with LTA vs. controls ($P < 0.02$), LPS ($P < 0.03$) and both toxins ($P < 0.03$). After 6 h *TOLLIP* transcription was also higher for LTA vs. controls ($P < 0.001$), LPS ($P < 0.001$) and both toxins ($P < 0.0001$). No significant difference was found among treatments and time points in *TLR2* transcription levels.

Discussion

Chemokines and interleukins

Chemokines regulate migration and adhesion of infiltrating cells to an inflamed lesion [28], and inhibition of chemokine expression or secretion significantly reduces cell infiltration [29]. Resident tissue cells such as mesangial cells and inflammatory cells such as monocytes/macrophages stimulate expression and secretion of chemokines [30]. The chemokines *CCL2* and *CCL5*, which belong to the "type I IFN chemokine signature", attract mainly monocytes, natural killer cells and activated lymphocytes [31, 32]. Thus, interferon (IFN) signaling is considered a critical point for host resistance against different pathogens [33], although the end result may be beneficial or detrimental to the host depending on the circumstances [34]. As reported previously in non-ruminants [35], the differential expression of these IFN-regulated chemokines with LPS or LTA could indicate a stronger recruitment of monocytes and lymphocytes in the mammary tissue and milk.

The greater expression of *CCL2* with LPS than LTA was consistent with data from a study with bovine pMEC incubated with LPS purified from *E. coli* strain O55:B5 [19, 20] or heat-inactivated *E. coli* [36], and the lack of effect of LTA isolated from *Streptococcus pyogenes* [19], *S. aureus* [20] or heat-inactivated *S. aureus* [36]. The down-regulation of *CCL2* with L + L than LPS might have been due to an interaction between LPS and LTA. Recent work has led to the speculation that bifidobacteria could induce cross-tolerance in bovine intestinal epithelial cells through their interaction with TLR2 [37]. In addition, it has been speculated that pre-exposure to LTA and lipopeptides which trigger TLR2-mediated signaling led to tolerance to LPS [38]. The lack of LPS effect on *CCL5* is in contradiction to a similar study with bovine MEC using 20 µg/mL LPS from *E. coli* O55:B5 [20]. This discrepancy might be explained by the different concentrations used in the studies.

The chemokines *CCL2* and *CXCL6* have strong chemo-attractant activities [39]. The up-regulation of *CXCL6* with LPS is similar to a previous study where *CCL2* and *CXCL6* increased markedly upon LPS challenge of MEC [19]. Mastitis is strongly associated with increased somatic cell counts in milk, the majority of which is attributable to neutrophils and lymphocytes [40]. Local production of pro-inflammatory cytokines in mammary tissue may have a strong influence on the activation state of the infiltrating neutrophils [41].

The temporal response in *CXCL8* after 3 and 6 h in the presence of LPS is similar to results reported in a previous study incubating bovine MEC with 50 µg/mL LPS or 20 µg/mL LTA, where an initial increase of *CXCL8* transcript levels after 2 h was followed by a decrease after 4 h in the presence of LTA and LPS [19]. In addition, a similar trend has been detected in a study performed with endometrial epithelial cells incubated with LPS where *CXCL8* levels were higher after 3 h incubation vs. 6 h [23].

The cytokine *IL6* is a pleiotropic protein with a strong influence on inflammatory responses, and is a major effector of the acute-phase reaction [42]. Thus, the observation that LPS alone or in combination with LTA up-regulated *IL6* only after 3 h could be explained by its quick mechanism of action, which was also reported previously in bovine MEC [20].

Other regulatory genes

The up-regulation of *IFIT3* with LPS alone compared to controls at 3 and 6 h is consistent with a previous study with bovine MEC using 20 µg/mL LPS from *E. coli* O55:B5 [20]. Activation of TLR4 by LPS induces the MyD88-independent pathway that promotes the internalization of the antigen-receptor LPS-TLR4 complex and activates interferon regulatory factor 3 (*IRF3*) [43]. The observed up-regulation of *IFIT3* with LTA might have been due to the responsiveness of this gene to a large variety of exogenous molecules [44]. The induction of the interferon induced protein with tetratricopeptide repeats (IFIT gene family) by different stimuli is based on the activation of interferon regulatory factors, which recognize the IFN-stimulated response elements (ISRE) in the IFIT promoters and initiate transcription [45].

IRF3 is involved in the MyD88-independent signaling pathway activated by TLR4, which may explain the lack of effect detected in *IRF3* between LTA alone and controls. However, the lack of an increase in *IRF3* transcription with LPS alone was unexpected because *IRF3* should be activated by TLR4 [43]. In a previous study with bovine mammary epithelial cells (MAC-T) [46], no significant *IRF3* increase was detected until 6 h incubation with 1 μg/mL LPS from *E. coli* J5 Rc mutant. The increase in *IRF3* transcription at 3 h incubation with both toxins could be explained by an interaction effect between LPS and LTA on pgMEC.

The published data regarding *MYD88* regulation induced by LPS or LTA are seemingly discordant. For example, a non-significant down-regulation of *MYD88* has been observed after 24 h with 50 μg/mL LPS treatment in immortalized bovine MEC, with no differences detected in primary bovine MEC [19]. In a study performed with immortalized bovine MEC [46], LPS induced the up-regulation of adaptor *MYD88* transcript that increased gradually compared to untreated cells and peaked significantly at 72 h after induction. In endometrial epithelial cells, *MYD88* expression peaks at 6 h after LPS-treatment [23]. Our data were more consistent with a study performed in endometrial stromal cells and whole endometrial cells incubated with LPS and LTA [47]. In that study, LPS stimulation up-regulated *MYD88* expression after 8 h in both cell types, whereas LTA stimulation of whole endometrial cells was associated with a non-significant increase of *MyD88*. Thus, it appears that a positive feedback loop with TLR4-dependent molecular self-regulation of the downstream signaling MyD88 [48] could partly explain our data.

The up-regulation of *NFKB1* with all challenges was consistent with previous studies where bacterial infections up-regulated *NFKB1* transcription in bovine mammary cells, confirming the ability of the mammary gland to mount a robust innate immune response [41, 46, 49]. Furthermore, our data agree with a previous study reporting up-regulation of *NFKB1* in bovine endometrial epithelial cells challenged with LPS [23].

Prostaglandins are one of several inflammatory mediators in the bovine mammary gland with chemotactic activity [50], hence, explaining the up-regulation of *PTGS2* with all challenges after 3 h. The PTGS2 protein is one of the enzymes involved in prostaglandin synthesis that is transiently up-regulated during inflammation [51]. *PTGS2* expression is increased by LTA [52]. The induction of *PTGS2* could have been associated with the action of MyD88 and activation of NFκB as reported previously [53].

The lack of effect on *TLR2* expression in the present study is consistent with a previous study of bovine MEC after 6 h incubation with heat-inactivated *E. coli* or after 30 h incubation with heat-inactivated *S. aureus* [36]. However, both datasets contrast the significant up-regulation of *TLR2* induced by LPS or heat-killed *E. coli* treatment of bovine endometrial cells for 3 and 6 h [23]. It could be possible that LTA inhibited TLR signaling as reported previously in human monocyte-like cells [54].

The greater *TLR4* expression due to LPS when compared to controls is consistent with previous data from a study performed with bovine MEC where *TLR4* was greater than controls in cells incubated for 6 h with 1 μg/mL LPS from *E. coli* [46]. Similar to the decrease that we detected over time for *TLR4* upon LTA challenge, the expression of *TLR4* had decreased in endometrial epithelial cells incubated for 3 and 6 h with 100 μg/mL LPS from *E. coli* after a significant increase at 1 h incubation [23].

The lower *CXCL6* and *CXCL8* expression after 3 and 6 h incubation induced only by LTA coincided with the higher expression of *TOLLIP* (Table 3), which is consistent with its anti-inflammatory role [55–57]. A time-dependent increase in *TOLLIP* has been reported in bovine MEC incubated with 1.0 μg/mL LPS from *E. coli* mutant J5 for 24 h; whereas a time-dependent decrease had occurred between 48 and 72 h of incubation [46]. These data indicate that an up-regulation of *TOLLIP* is necessary to counteract the harmful effects associated with over production of cytokines. In fact, using short hairpin RNA knockdown of TOLLIP in peripheral blood human monocytes, TOLLIP suppresses TNF and IL-6 production after stimulation with TLR2 and TLR4 agonists, and induces secretion of the anti-inflammatory cytokine IL-10 [58].

Conclusions

Consistent with numerous experiments in bovine mammary epithelial cells, our study confirms the capacity of LPS to stimulate inflammatory genes acting as TLR4 agonists in pgMEC. The differences in gene expression responses of goat mammary epithelial cells to LPS and LTA revealed different activation pathways for these components of Gram-negative and Gram-positive bacterial cell walls. Further studies focused on protein expression changes should be carried out to confirm gene transcription variation at the translation level. Furthermore, genes and corresponding proteins involved in cellular apoptosis should be studied in order to investigate potential mechanisms damaging goat mammary tissue in response to inflammatory stimuli. The challenge with LPS compared to LTA generated much stronger and sustained responses that seem to reflect an adaptation to the more acute nature of mastitis caused by coliform bacteria. The lack of response for some pro-inflammatory cytokines during incubation with LTA indicates some degree of tolerance to this agent, consistent with chronic infections of the mammary tissue caused by *Staphylococcal* species.

Abbreviations

ACTB: Actin beta; CCL2: C-C motif chemokine ligand 2; CCL5: C-C motif chemokine ligand 5; CXCL6: C-X-C motif chemokine ligand 6; CXCL8: Interleukin 8; GAPDH: Glyceraldehyde-3-phosphate dehydrogenase; IFIT: Interferon induced protein with tetratricopeptide repeats; IFIT3: Interferon induced protein with tetratricopeptide repeats 3; IFN: Interferon; IL1B: Interleukin 1 beta; IL6: Interleukin 6; IRF3: Interferon regulatory factor 3; ISRE: IFN-stimulated response elements; LPS: Gram-negative lipopolysaccharide; LTA: Gram-positive lipoteichoic acid; LTF: Lactoferrin; MEC: Mammary epithelial cells; MYD88: Myeloid differentiation primary response 88; NFKB1: Nuclear factor of kappa light polypeptide gene enhancer in B-cells 1; pgMEC: Primary goat mammary epithelial cells; pMEC: Primary mammary epithelial cells; PTGS2: Prostaglandin-endoperoxide synthase 2; qPCR: Quantitative real-time PCR; T: Time; TLR: Toll-like receptors; TLR1-TLR10: Toll-like receptors 1–10; TLR2: Toll-like receptor 2; TLR4: Toll-like receptor 4; TNF: Tumor necrosis factor alpha; TOLLIP: Toll interacting protein; TRT: Treatment; UXT: Ubiquitously expressed prefoldin like chaperone

Acknowledgments

We greatly appreciate the support of Prof. Peter Dovč, Department of Animal Science, University of Ljubljana, Slovenia, for providing access to the mammary epithelial cells.

Funding

Funding for this study was provided by the Future Interdisciplinary Research Explorations grant program of the Office of Research, College of ACES, University of Illinois at Urbana-Champaign, through the USDA National Institute of Food and Agriculture Hatch project ILLU-538-395 (Accession Number 0232734) and ILLU-538-914.

Authors' contributions

OB and XD performed the experiments, performed analyses, and analyzed data. ALR, AMC and JJL drafted the manuscript. JJL conceived the experiment and proofread the manuscript. All authors participated in data interpretation. All authors approved the final version of the manuscript.

Authors' information

O. Bulgari is PhD degree candidate at Department of Molecular and Translational Medicine, University of Brescia, Brescia 25123, Italy.
X. Dong is PhD degree candidate at Institute of Animal Nutrition, Sichuan Agricultural University, Chengdu, 611130, China.
A. L. Roca is Associate Professor in the Department of Animal Sciences, University of Illinois at Urbana-Champaign, Urbana, IL, 61801, USA.A. M. Caroli is Professor at Department of Molecular and Translational Medicine, University of Brescia, Brescia 25123, Italy.
J. J. Loor is Associate Professor in the Department of Animal Sciences, University of Illinois at Urbana-Champaign, Urbana, IL, 61801, USA.

Competing interests

The authors declare that they have no competing interests.

Author details

[1]Department of Animal Sciences and Division of Nutritional Sciences, University of Illinois at Urbana-Champaign, Urbana, IL 61801, USA. [2]Department of Molecular and Translational Medicine, University of Brescia, Brescia 25123, Italy. [3]Institute of Animal Nutrition, Sichuan Agricultural University, Chengdu 611130, China. [4]Division of Nutritional Sciences, University of Illinois at Urbana-Champaign, Urbana, IL 61801, USA.

References

1. Viguier C, Arora S, Gilmartin N, Welbeck K, O'Kennedy R. Mastitis detection: current trends and future perspectives. Trends Biotechnol. 2009;27:486–93.
2. Ribeiro MG, Lara GHB, Bicudo SD, Souza AVG, Salerno T, Siqueira AK, et al. An unusual gangrenous goat mastitis caused by *Staphylococcus aureus*, *Clostridium perfringens* and *Escherichia coli* co-infection. Arq Bras Med Vet Zootec. 2007;59:810–2.
3. Radostits OM, Gay C, Hinchcliff K, Constable P. Veterinary medicine - a textbook of the diseases of cattle, horses, sheep, pigs and goats. 10th ed. Edinburgh: Elsevier Saunders; 2007.
4. Moyes KM, Drackley JK, Salak-Johnson JL, Morin DE, Hope JC, Loor JJ. Dietary-induced negative energy balance has minimal effects on innate immunity during a *Streptococcus uberis* mastitis challenge in dairy cows during midlactation. J Dairy Sci. 2009;92:4301–16.
5. Moyes KM, Drackley JK, Morin DE, Loor JJ. Greater expression of *TLR2*, *TLR4*, and *IL6* due to negative energy balance is associated with lower expression of HLA-DRA and HLA-A in bovine blood neutrophils after intramammary mastitis challenge with *Streptococcus uberis*. Funct Integr Genom. 2010;10: 53–61.
6. Loor JJ, Moyes KM, Bionaz M. Functional adaptations of the transcriptome to mastitis-causing pathogens: the mammary gland and beyond. J Mamm Gland Biol Neoplasia. 2011;16:305–22.
7. Huang J, Luo G, Zhang Z, Wang X, Ju Z, Qi C, et al. iTRAQ-proteomics and bioinformatics analyses of mammary tissue from cows with clinical mastitis due to natural infection with *Staphylococci aureus*. BMC Genomics. 2014;15:839.
8. Qian C, Cao X. Regulation of Toll-like receptor signaling pathways in innate immune responses. Ann N Y Acad Sci. 2013;1283:67–74.
9. Tirumurugaan KG, Dhanasekaran S, Dhinakar Raj G, Raja A, Kumanan K, Ramaswamy V. Differential expression of toll-like receptor mRNA in selected tissues of goat (*Capra hircus*). Vet Immunol Immunopathol. 2010;133:296–301.
10. Ma JL, Zhu YH, Zhang L, Zhuge ZY, Liu PQ, Yan XD, et al. Serum concentration and mRNA expression in milk somatic cells of toll-like receptor 2, toll-like receptor 4, and cytokines in dairy cows following intramammary inoculation with *Escherichia coli*. J Dairy Sci. 2011;94:5903–12.
11. Carvajal AM, Huircan P, Lepori A. Single nucleotide polymorphisms in immunity-related genes and their association with mastitis in Chilean dairy cattle. Genet Mol Res. 2013;12:2702–11.
12. Beecher C, Daly M, Ross RP, Flynn J, McCarthy TV, Giblin L. Characterization of the bovine innate immune response in milk somatic cells following intramammary infection with *Streptococcus dysgalactiae* subspecies *dysgalactiae*. J Dairy Sci. 2012;95:5720–9.
13. Poltorak A, He X, Smirnova I, Liu MY, Van Huffel C, Du X, et al. Defective LPS signaling in C3H/HeJ and C57BL/10ScCr mice: mutations in *TLR4* gene. Science. 1998;282:2085–8.
14. Schwandner R, Dziarski R, Wesche H, Rothe M, Kirschning CJ. Peptidoglycan and lipoteichoic acid-induced cell activation is mediated by toll-like receptor 2. J Biol Chem. 1999;274:17406–9.
15. Takeda K, Akira S. Toll-like receptors in innate immunity. Int Immunol. 2005; 17:1–14.

16. Akira S. Pathogen recognition by innate immunity and its signaling. Proc Jpn Acad Ser B Phys Biol Sci. 2009;85:143–56.

17. Ogorevc J, Dovč P. Relative quantification of beta-casein expression in primary goat mammary epithelial cell lines. Genet Mol Res. 2015;14:3481–90.

18. Kadegowda AKG, Bionaz B, Piperova LS, Erdman RA, Loor JJ. Peroxisome proliferator-activated receptor-γ activation and long chain fatty acids alter lipogenic gene networks in bovine mammary epithelial cells to various extents. J Dairy Sci. 2009;92:4276–89.

19. Strandberg Y, Gray C, Vuocolo T, Donaldson L, Broadway M, Tellam R. Lipopolysaccharide and lipoteichoic acid induce different innate immune responses in bovine mammary epithelial cells. Cytokine. 2005; 31:72–86.

20. Gilbert FB, Cunha P, Jensen K, Glass EJ, Foucras G, Robert-Granie C, et al. Differential response of bovine mammary epithelial cells to Staphylococcus aureus or Escherichia coli agonists of the innate immune system. Vet Res. 2013;44:40.

21. Schröder NW, Morath S, Alexander C, Hamann L, Hartung T, Zahringer V, et al. Lipoteichoic acid (LTA) of Streptococcus pneumoniae and Staphylococcus aureus activates immune cells via toll-like receptor (TLR)-2, lipopolysaccharide-binding protein (LBP), and CD14, whereas TLR-4 and MD-2 are not involved. J Biol Chem. 2003;278:15587–94.

22. Bougarn S, Cunha P, Harmache A, Fromageau A, Gilbert FB, Rainard P. Muramyl dipeptide synergizes with Staphylococcus aureus lipoteichoic acid to recruit neutrophils in the mammary gland and to stimulate mammary epithelial cells. Clin Vaccine Immunol. 2010;17:1797–809.

23. Fu Y, Liu B, Feng X, Liu Z, Liang D, Li F, et al. Lipopolysaccharide increases toll-like receptor 4 and downstream toll-like receptor signaling molecules expression in bovine endometrial epithelial cells. Vet Immunol Immunopathol. 2013;151:20–7.

24. Mancek-Keber M, Jerala R. Postulates for validating TLR4 agonists. Eur J Immunol. 2015;45:356–70.

25. Morath S, Geyer A, Hartung T. Structure-function relationship of cytokine induction by lipoteichoic acid from Staphylococcus aureus. J Exp Med. 2001; 193:393–7.

26. Hu H, Wang JQ, Bu DP, Wei HY, Zhou LY, Li F, et al. In vitro culture and characterization of a mammary epithelial cell line from Chinese Holstein dairy cow. PLoS ONE. 2009;4:e7636.

27. Prpar Mihevc S, Ogorevc J, Dovc P. Lineage-specific markers of goat mammary cells in primary culture. In Vitro Cell Dev Biol Anim. 2014;50: 926–36.

28. Nedoszytko B, Sokołowska-Wojdyło M, Ruckemann-Dziurdzińska K, Roszkiewicz J, Nowickiù RJ. Chemokines and cytokines network in the pathogenesis of the inflammatory skin diseases: atopic dermatitis, psoriasis and skin mastocytosis. Postepy Dermatol Alergol. 2014;31:84–91.

29. Haberstroh U, Pocock J, Gómez-Guerrero C, Helmchen U, Hamann A, Gutierrez-Ramos JC, et al. Expression of the chemokines MCP-1/CCL2 and RANTES/CCL5 is differentially regulated by infiltrating inflammatory cells. Kidney Int. 2002;62:1264–76.

30. Rossi D, Zlotnik A. The biology of chemokines and their receptors. Annu Rev Immunol. 2000;18:217–42.

31. Jia T, Leiner I, Dorothee G, Brandl K, Pamer EG. MyD88 and type I interferon receptor-mediated chemokine induction and monocyte recruitment during Listeria monocytogenes infection. J Immunol. 2009;183:1271–8.

32. Lee PY, Li Y, Kumagai Y, Xu Y, Weinstein JS, Kellner ES, et al. Type I interferon modulates monocyte recruitment and maturation in chronic inflammation. Am J Pathol. 2009;175:2023–33.

33. Mancuso G, Midiri A, Biondo C, Beninati C, Zummo S, Galbo R, et al. Type I IFN signaling is crucial for host resistance against different species of pathogenic bacteria. J Immunol. 2007;178:3126–33.

34. Decker T, Muller M, Stockinger S. The yin and yang of type I interferon activity in bacterial infection. Nat Rev Immunol. 2005;5:675–87.

35. Weyrich AS, McIntyre TM, McEver RP, Prescott SM, Zimmerman GA. Monocyte tethering by P-selectin regulates monocyte chemotactic protein-1 and tumor necrosis factor-alpha secretion. Signal integration and NF-kappa B translocation. J Clin Invest. 1995;95:2297–303.

36. Sorg D, Danowski K, Korenkova V, Rusnakova V, Küffner R, Zimmer R, et al. Microfluidic high-throughput RT-qPCR measurements of the immune response of primary bovine mammary epithelial cells cultured from milk to mastitis pathogens. Animal. 2013;7:799–805.

37. Villena J, Aso H, Kitazawa H. Regulation of toll-like receptors-mediated inflammation by immunobiotics in bovine intestinal epitheliocytes: role of signaling pathways and negative regulators. Front Immunol. 2014;5:421.

38. Sato S, Takeuchi O, Fujita T, Tomizawa H, Takeda K, Akira S. A variety of microbial components induce tolerance to lipopolysaccharide by differentially affecting MyD88-dependent and independent pathways. Int Immunol. 2002;14:783–91.

39. Moser B, Wolf M, Walz A, Loetscher P. Chemokines: multiple levels of leukocyte migration control. Trends Immunol. 2004;25:75e84.

40. Schukken YH, Wilson DJ, Welcome F, Garrison-Tikofsky L, Gonzalez RN. Monitoring udder health and milk quality using somatic cell counts. Vet Res. 2003;34:579e96.

41. Bannerman DD, Chockalingam A, Paape MJ, Hope JC. The bovine innate immune response during experimentally-induced Pseudomonas aeruginosa mastitis. Vet Immunol Immunop. 2005;107:201–5.

42. Le JM, Vilcek J. Interleukin 6: a multifunctional cytokine regulating immune reactions and the acute phase protein response. Lab Invest. 1989;61: 588e602.

43. Cao D, Luo J, Chen D, Xu H, Shi H, Jing X, et al. CD36 regulates lipopolysaccharide-induced signaling pathways and mediates the internalization of Escherichia coli in cooperation with TLR4 in goat mammary gland epithelial cells. Sci Rep. 2016;6:23132.

44. Fensterl V, Sen GC. The ISG56/IFIT1 gene family. J Interf Cytok Res. 2011;31: 71–7.

45. Ogawa S, Lozach J, Benner C, Pascual G, Tangirala RK, Westin S, et al. Molecular determinants of crosstalk between nuclear receptors and toll-like receptors. Cell. 2005;122:707–21.

46. Ibeagha-Awemu EM, Lee JW, Ibeagha AE, Bannerman DD, Paape MJ, Zhao X. Bacterial lipopolysaccharide induces increased expression of toll-like receptor (TLR) 4 and downstream TLR signaling molecules in bovine mammary epithelial cells. Vet Res. 2008;39:11.

47. Rashidi N, Mirahmadian M, Jeddi-Tehrani M, Rezania S, Ghasemi J, Kazemnejad S, et al. Lipopolysaccharide and lipoteichoic acid-mediated pro-inflammatory cytokine production and modulation of TLR2, TLR4 and MyD88 expression in human endometrial cells. J Reprod Infertil. 2015;16:72–81.

48. Buchholz BM, Billiar TR, Bauer AJ. Dominant role of the MyD88-dependent signaling pathway in mediating early endotoxin-induced murine ileus. Am J Physiol Gastrointest Liver Physiol. 2010;299:G531–8.

49. Lee JW, Bannerman DD, Paape MJ, Huang MK, Zhao X. Characterization of cytokine expression in milk somatic cells during intramammary infections with Eschericha coli or Staphylococcus aureus by real-time PCR. Vet Res. 2006;37:219–29.

50. Craven N. Chemotactic factors for bovine neutrophils in relation to mastitis. Comp Immunol Microbiol Infect Dis. 1986;9:29–36.

51. Zbinden C, Stephan R, Johler S, Borel N, Bunter J, Bruckmaier RM, et al. The inflammatory response of primary bovine mammary epithelial cells to Staphylococcus aureus strains is linked to the bacterial phenotype. PLoS One. 2014;9:e87374.

52. Lin CH, Kuan IH, Lee HM, Lee WS, Sheu JR, Ho YS, et al. Induction of cyclooxygenase-2 protein by lipoteichoic acid from Staphylococcus aureus in human pulmonary epithelial cells: involvement of a nuclear factor-kB-dependent pathway. Br J Pharmacol. 2001;134:543–52.

53. Carpenter S, Atianand M, Aiello D, Ricci EP, Gandhi P, Hall LL, et al. A long noncoding RNA induced by TLRs mediates both activation and repression of immune response genes. Science. 2013;341:789–92.

54. Kim H, Jung BJ, Jeong J, Chun H, Chung DK. Lipoteichoic acid from Lactobacillus plantarum inhibits the expression of platelet-activating factor receptor induced by Staphylococcus aureus lipoteichoic acid or Escherichia coli lipopolysaccharide in human monocyte-like cells. J Microbiol Biotechnol. 2014;24:1051–8.

55. Zhang G, Ghosh S. Negative regulation of toll-like receptor-mediated signaling by Tollip. J Biol Chem. 2002;277:7059–65.

56. Capelluto DGS. Tollip: a multitasking protein in innate immunity and protein trafficking. Microbes Infect. 2012;14:140–7.

57. Moncayo-Nieto OL, Wilkinson TS, Brittan M, McHugh BJ, Jones RO, Morris AC, et al. Differential response to bacteria, and TOLLIP expression, in the human respiratory tract. BMJ Open Respir Res. 2014;1:e000046.

58. Shah JA, Vary JC, Chau TT, Bang ND, Yen NT, Farrar JJ, et al. Human TOLLIP regulates TLR2 and TLR4 signaling and its polymorphisms are associated with susceptibility to tuberculosis. J Immunol. 2012;189: 1737–46.

5

Association of growth rate with hormone levels and myogenic gene expression profile in broilers

Yingping Xiao[1], Choufei Wu[2], Kaifeng Li[1], Guohong Gui[1], Guolong Zhang[3] and Hua Yang[1*]

Abstract

Background: The growth rate often varies among individual broilers of the same breed under a common management condition. To investigate whether a variation in the growth rate is associated with a difference in hormone levels and myogenic gene expression profile in broilers, a feeding trial was conducted with 10,000 newly hatched Ross 308 chicks in a commercial production facility under standard management. At 38 d of age, 30 fast-, 30 medium-, and 30 slow-growing broilers were selected among 600 healthy male individuals. The levels of insulin-like growth factor-1 (IGF-1), triiodothyronine (T3), thyroxine (T4), and growth hormone in the serum or breast muscle were assayed by ELISA or RIA kits, and the expression levels of several representative pro- and anti-myogenic genes in the breast muscle were also measured by real-time PCR.

Results: Results showed that both absolute and relative weights of the breast muscle were in linear positive correlations with the body weight of broilers ($P < 0.001$). Fast-growing broilers had higher concentrations of IGF-1 than slow-growing broilers ($P < 0.05$) in both the serum and breast muscle. The serum concentration of T3 was significantly higher in fast-growing birds than in slow-growing birds ($P < 0.05$). However, no difference was observed in growth hormone or T4 concentration among three groups of birds. Additionally, a decreased expression of an anti-myogenic gene (*myostatin*) and increased expressions of pro-myogenic genes such as *myogenic differentiation factor 1, myogenin, muscle regulatory factor 4, myogenic factor 5, IGF-1,* and *myocyte enhancer factor 2B, C,* and *D* were observed in fast-growing broilers ($P < 0.05$), relative to slow-growing broilers.

Conclusions: Collectively, these findings suggested that the growth rate is linked to the hormone and myogenic gene expression levels in broiler chickens. Some of these parameters such as serum concentrations of IGF-1 and T3 could be employed to breed for enhanced growth.

Keywords: Breast muscle, Broiler, Growth performance, Hormone, Myogenic gene expression

Background

Muscle growth, also known as myogenesis, is a complicated but precisely regulated developmental process. Myogenic regulatory factors (MRF) and myocyte enhancer factor 2 (MEF2) proteins are key transcription factors that are positively involved in the regulation of skeletal muscle development [1]. The MRF family of transcription factors is comprised of a group of basic helix-loop-helic proteins such as myogenic differentiation factor 1 (MyoD1), myogenic factor 5 (Myf5), myogenin, and MRF4, whereas the

MEF2 family consists of four members denoted as MEF2A, B, C, and D. On the other hand, myostatin (MSTN), a member of the transforming growth factor β (TGF-β) superfamily secreted from skeletal muscle, acts as a potent negative regulator of muscle differentiation and growth [2]. MSTN modifies the muscle fiber-type composition by regulating MyoD and MEF2 expression [3]. Inhibition of MSTN causes myofibre hypertrophy [4], while MSTN over-expression decreases the skeletal muscle mass and fiber size [5].

Additionally, a number of hormones are known to impact on animal growth. The major hormones required to support normal growth in chickens are growth hormone (GH), 3,5,3′-triiodothyronine (T3), thyroxine (T4), and

* Correspondence: yanghua806@hotmail.com
[1]Institute of Quality and Standards for Agro-products, Zhejiang Academy of Agricultural Sciences, Hangzhou 310021, China
Full list of author information is available at the end of the article

insulin-like growth factor-1 (IGF-1) [6]. IGF-1 have been shown to stimulate the growth of the skeletal muscle by enhancing the rate of protein synthesis and therefore, the concentration of IGF-1 is often positively correlated with the body weight (BW) in broiler chickens [6–8].

It is well known that the growth rate of individual broilers of the same breed exhibits a normal Gaussian distribution under a common management condition. However, it remains elusive about whether there is a difference between fast- and slow-growing broilers in growth-related hormone and myogenic gene expression levels. The objective of the present study was to investigate the association of the growth rate with the concentrations of growth-related hormones in the circulation as well as both pro- and anti-myogenic gene expression levels in the breast muscle of broilers.

Methods
Animals
A flock of 10,000 day-of-hatch Ross 308 broiler chicks were raised in a commercial production facility under standard management, with a 24-h photoperiod and 32-34 °C in the first two days, followed by a 16-h photoperiod and a reduction by 2-3 °C per week to a final temperature of 20 °C. The broilers were allowed ad libitum access to water and commercial mash feed. At 38 d of age, 600 healthy male broilers were randomly picked and weighed, from which 30 broilers of the highest, medium, and lowest BW were selected as H, M, and L groups, respectively.

Sampling
Ten birds were randomly selected from each group. Blood (4 mL each) was collected from the wing vein and centrifuged at 3,000 × g and 4 °C for 10 min to obtain serum, which was stored at -80 °C for further analysis. Chickens were then killed by cervical dislocation. The entire breast muscle (including pectoralis major and minor) was weighed after removal. Approximately 3 g of each muscle sample was immediately snap frozen in liquid nitrogen and stored at -80 °C for further analysis as described below.

Tissue preparation and hormone concentration measurement
The breast muscle samples were homogenized (1:19, w/v) in chilled physiological saline. After centrifuging at 4 °C and 5000 × g for 10 min, the supernatants were collected for subsequent assay for various hormones. The IGF-1 concentrations in the serum and muscle were measured with a commercial chicken-specific ELISA kit following the manufacturer's protocols (Jiancheng Bioengineering, Nanjing, China). Serum concentrations of GH, T3 and T4

were measured with the respective commercial RIA kits (Beijing North Institute of Biological Technology, China).

Extraction and quantification of total DNA and proteins
The breast muscle was homogenized in chilled physiological saline (1:10, w/v). DNA and proteins were extracted from the muscle homogenates as described [9]. The concentrations of DNA were measured using a NanoDrop1000 Spectrophotometer (NanoDrop Products, Wilmington, DE), and the protein concentrations were measured with a protein assay kit (Jiancheng Bioengineering, Nanjing, China). The percentages of proteins and nucleic acids in each breast muscle sample were calculated.

RNA extraction and quantitative real-time RT-PCR
Total RNA was extracted from the breast muscle using RNAiso (Takara Bio, Dalian, Liaoning, China). The first-strand cDNA was synthesized using the SuperScript® III Reverse Transcriptase with random primers and an RNase inhibitor (Invitrogen, Carlsbad, CA, USA) as per the manufacturer's instructions. Gene-specific primers for *IGF-1*, *MSTN*, *MyoD1*, *myogenin*, *MRF4*, *Myf5*, *MEF2A*, *MEF2B*, *MEF2C*, and *MEF2D* were designed using Primer Premier 6.0 (Premier, Ontario, Canada) (Table 1). PCR was performed on the ABI 7500 Real-Time PCR Detection System (Applied Biosystems, Foster City, CA, USA) using SYBR Premix Ex Taq II Kit (Takara Bio) and 40 cycles of 95 °C for 15 s and 60 °C for 30 s. All measurements were performed in triplicate. The fold difference was calculated using the $\Delta\Delta$Ct method [10, 11] using the geometric means of 18S rRNA and *glyceraldehyde 3-phosphate dehydrogenase* (*GAPDH*) mRNA for data normalization.

Statistical analysis
All data were presented as means ± standard error of the mean (SEM). Statistics were performed using one-way ANOVA using the SPSS software, version 16.0 (Chicago, IL, USA). The differences were considered to be significant, if $P < 0.05$.

Results
Body weight and the breast muscle weight
The BW of 600 male Ross 308 broilers that were randomly picked exhibited a normal Gaussian distribution ranging from 1,250 to 3,330 g with an average of 2,213 g (Fig. 1). Thirty heaviest (H), 30 medium (M), and 30 lightest (L) broilers were further selected. The average BW was 2,784, 2,221, and 1,493 g for the H, M, and L groups, respectively, and the H group weighed nearly twice as much as the L group (Fig. 2). As expected, the absolute breast muscle weight was also obviously different ($P < 0.001$) among three groups, with the H group doubling the weight of the L group (Fig. 3a). However,

Table 1 Primers Used for Real-Time PCR

Gene	GenBank Accession Number	Primer Sequence (5' to 3')	Product Size, bp
IGF-1	NM_001004384	GAGCTGGTTGATGCTCTTCAGTT CCAGCCTCCTCAGGTCACAACT	148
MSTN	NM_001001461	CGCTACCCGCTGACAGTGGAT CAGGTGAGTGTGCGGGTATTTCT	132
MyoD1	NM_204214	CCGACGGCATGATGGAGTACA GTCGAGGCTGGAAACAACAGAA	131
Myogenin	D90157	GGAGGCTGAAGAAGGTGAACGA CTCTGCAGGCGCTCGATGTACT	127
MRF4	D10599	CAGGCTGGATCAGCAGGACAA GCCGCAGGTGCTCAGGAAGT	106
Myf5	NM_001030363	CAGAGACTCCCCAAAGTGGAGAT GTCCCGGCAGGTGATAGTAGTTC	106
MEF2A	NM_204864	CGGAGGACAGATTCAGCAAACTA GACACTGGAACCGTAACCGACAT	109
MEF2B	XM_430389	CACGCCATCAGCATCAAGTCA GGGGTAGCCCTTGGAGTAGTCAT	156
MEF2C	XM_001231661	GCCGTCTGCCCTCAGTCAACT GGGTGGTGGTACGGTCTCTAGGA	137
MEF2D	NM_001031600	GTGTCTCCCAAGCGACTCACTCT GTGTTGTATGCGGTCGGCAT	109
GAPDH	NM_204305	GCCACACAGAAGACGGTGGAT GTGGACGCTGGGATGATGTTCT	86
18S	AF173612	CCGGACACGGACAGGATTGACA CAGACAAATCGCTCCACCAACTAAG	94

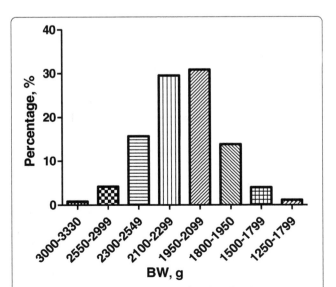

Fig. 1 Body weight (BW) distribution of broilers. The data were based on 600, 38-day-old male broilers that were randomly picked and weighed from a group of 10,000 Ross 308 chickens raised under standard management in a single commercial house

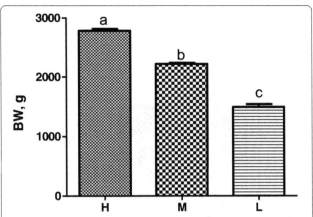

Fig. 2 Average body weight (BW) of three groups of broilers. Each bar represents mean ± standard error (SE) of thirty, 38-day-old male broilers per group. Means with no common letter differ significantly (P < 0.05). H = fast-growing broilers (heavy BW); M = medium-growing broilers (average BW); L = slow-growing broilers (low BW)

the differences among the relative breast muscle weight (breast muscle weight: BW ratio) in the H and M groups were significantly higher than that in the L group (P = 0.026 between H and L; P = 0.045 between M and L) (Fig. 3b).

Concentrations of total DNA, RNA, and proteins in the breast muscle

Total DNA concentration in the breast muscle of the H group was 17% (P = 0.042) and 18% (P = 0.045) lower than that of the M and L groups, whereas no difference was observed between the M and L groups (Table 2). On the other hand, the total RNA concentration between the H and M groups showed no difference (P = 0.884), but was approximately 42% and 39% higher than that in the L group, respectively (Table 2). No difference was seen with the total protein concentration in the breast muscle among three groups of broilers (P = 0.661) (Table 2). The protein:DNA ratio was positively correlated with the BW (P = 0.049), and the RNA:protein ratio showed the same trend (P = 0.086). The RNA:DNA ratio also exhibited an obvious correlation with the BW among the H, M, and L groups (P = 0.006), with the H group being significantly higher than the M group and the M group significantly higher than the L group (Table 2).

Concentrations of hormones in the serum and breast muscle

The serum concentration of IGF-1 showed an obvious linear relationship with the BW (Fig. 4a). The H group had approximately 18% (P = 0.044) and 33% (P = 0.027) higher IGF-1 concentrations than the M and L groups, respectively. The serum concentration of T3 in the L group was approximately 50% (P = 0.017) and 47% (P = 0.029) lower than those in the H and M groups, respectively

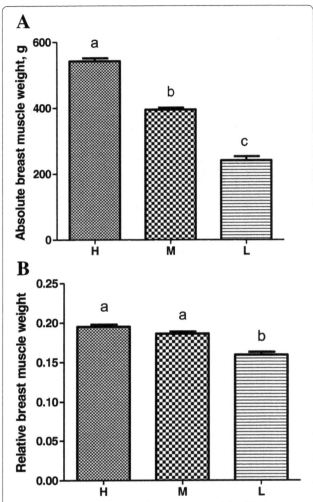

Fig. 3 Absolute (**a**) and relative breast muscle weight (**b**) of three groups of broilers. The relative breast muscle weight is the ratio of absolute breast muscle weight to BW. Each bar represents mean ± standard error (SE) of thirty, 38-day-old male broilers per group. Means with no common letter differ significantly ($P < 0.05$). H = fast-growing broilers (heavy BW); M = medium-growing broilers (average BW); L = slow-growing broilers (low BW)

(Fig. 4b). However, no significant difference was observed in the serum concentrations of GH and T4 among the H, M, and L groups (Fig. 4c and d). Similar to the serum concentration, the IGF-1 concentration in the breast muscle was also linearly correlated with the BW (Fig. 5). Comparing with the L group broilers, the IGF-1 concentration in the breast muscle showed a 18% increase ($P = 0.013$) in the H group.

Expression levels of myogenic genes in the breast muscle

To study the correlation between the expression levels of major myogenic genes and the BW, total RNA was isolated from the breast muscle and subjected to reverse transcription and real-time PCR analysis of the genes for *MyoD1, Myogenin, MRF4, Myf5, MEF2A, MEF2B, MEF2C, MEF2D, IGF-1,* and *MSTN*. As shown in Table 3, the expression levels of all pro-myogenic genes but *MEF2A* showed a significant decrease ($P < 0.05$) in the L group relative to the H group. The pro-myogenic genes, namely *MEF5C*, showed a strong linear correlation ($P < 0.05$) with the BW, with the H group giving the highest expression and the L group the lowest expression. No difference ($P > 0.05$) was observed with *MyoD1, MRF4* or *MEF2B* between the H and M group, whereas *Myogenin* and *MEF2D* showed no difference between the M and L group. As for the expression of *MSTN*, an anti-myogenic gene, the H and M group showed no statistical difference, but both groups were significantly lower than the L group (Table 3). Interestingly, the *IGF-1* mRNA expression level was also significantly reduced ($P = 0.026$), relative to the H or M group (Table 3).

Discussion

Genetic differences among the chicken breeds of different growth rates have been extensively studied and a number of genes and quantitative trait loci have been reported in controlling the growth rate of chickens [12–15]. However, little is known regarding genetic and hormonal variations among individual broilers of the same breed under a common management. This study demonstrated that there is a large variation in the growth rate among individual birds raised under industrial standard management practices, with BW ranging from 1,250 g to 3,330 g. To elucidate the mechanism of the BW variations in a cohort of broiler chickens, the serum concentrations of GH, T3, T4, and IGF-1 were measured in the present study. The result showed that the T3 serum concentration was higher in heavier broilers than lighter ones, with no difference seen with the T4 concentration. This is consistent with an earlier report that the plasma concentration of T3, but not T4, was reduced in the sex-linked dwarf chickens relative to the normal control breed [16]. Broilers with heavier BW had a higher serum concentration of IGF-1 in our study, which is in agreement with an earlier observation

Table 2 The concentrations of DNA, RNA, and proteins in the breast muscle of broilers ($n = 10$)

Item	H	M	L	SEM	P-value
DNA, mg/g	1.68[b]	2.02[a]	2.06[a]	0.07	0.044
RNA, mg/g	3.04[a]	2.97[a]	2.14[b]	0.11	0.036
Protein, mg/g	86.32	88.65	84.97	2.54	0.661
Protein:DNA	49.52[a]	43.87[ab]	39.71[b]	1.61	0.049
RNA:Protein	0.035	0.033	0.025	0.003	0.086
RNA:DNA	1.81[a]	1.47[b]	1.06[c]	0.08	0.006

[a-c]Means in the same row with different superscript letters differ significantly ($P < 0.05$)
H = fast-growing broilers (heavy BW group); M = medium-growing broilers (average BW group); L = slow-growing broilers (low BW group)

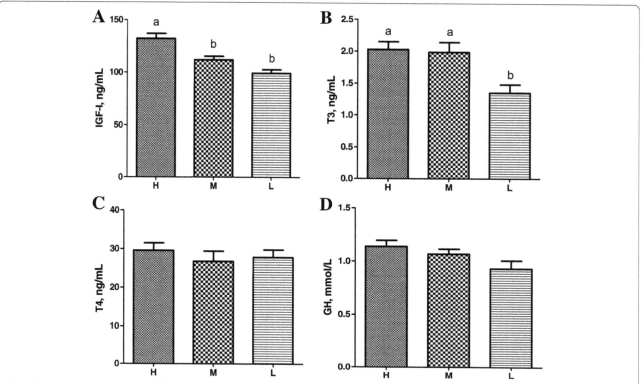

Fig. 4 The serum concentrations of fours hormones including IGF-1 (**a**), T3 (**b**), T4 (**c**), and GH (**d**) in broilers. All hormones were measured by chicken-specific ELISA on 38-day-old male broilers. Each bar represents mean ± standard error (SE) of ten replicates. Means with no common letter differ significantly ($P < 0.05$). H = fast-growing broilers (heavy BW); M = medium-growing broilers (average BW); L = slow-growing broilers (low BW). IGF-1, insulin-like growth factor-1; T3, triiodothyronine; T4, thyroxine; GH, growth hormone

that a reduction in the growth rate was associated with a decrease in circulating concentrations of IGF-1 in chickens [6]. Thus, higher levels of T3 and IGF-1 in the circulation accelerate growth in broilers and can be good predictors of faster growth.

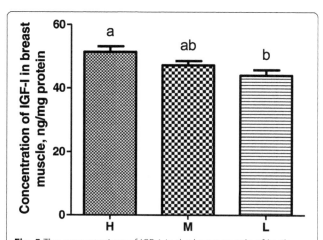

Fig. 5 The concentrations of IGF-1 in the breast muscle of broilers. IGF-1 (insulin-like growth factor-1) was measured by chicken-specific ELISA on 38-day-old male broilers. Each bar represents mean ± standard error (SE) of ten replicates. Means with no common letter differ significantly ($P < 0.05$). H = fast-growing broilers (heavy BW); M = medium-growing broilers (average BW); L = slow-growing broilers (low BW)

Table 3 The relative mRNA levels in the breast muscle of broilers ($n = 10$)

Item	H	M	L	SEM	P-value
MyoD1	0.84[ab]	1.00[a]	0.66[b]	0.065	0.017
Myogenin	2.94[a]	1.00[b]	1.29[b]	0.111	<0.001
MRF4	1.22[a]	1.00[a]	0.78[b]	0.078	0.047
Myf5	0.85[a]	1.00[a]	0.61[b]	0.076	0.044
MEF2A	1.23	1.00	1.14	0.080	0.110
MEF2B	1.06[a]	1.00[a]	0.60[b]	0.042	<0.001
MEF2C	1.22[a]	1.00[b]	0.74[c]	0.062	0.024
MEF2D	1.32[a]	1.00[b]	0.89[b]	0.052	0.015
MSTN	1.15[b]	1.00[b]	1.82[a]	0.101	0.041
IGF-1	0.94[a]	1.00[a]	0.51[b]	0.060	0.031

[a-c]Means in the same row with different superscript letters differ significantly ($P < 0.05$)
H = fast-growing broilers (heavy BW group); M = medium-growing broilers (average BW group); L = slow-growing broilers (low BW group)

Muscle growth is known to be accelerated by activation of the IGF-1 signaling pathway [8, 17], which is consistent with our observation that IGF-1 in the breast muscle were higher at both protein and mRNA levels in fast-growing than slow-growing broilers, which is probably due to increased methylation in the promoter region of the *IGF-1 receptor* gene in slow-growing broilers [18].

Besides *IGF-1*, altered expressions of multiple myogenic genes was also observed between heavy and light broilers in this study. Pro-myogenic genes such as *MyoD1* and *Myf5* are important in the formation of skeletal muscle [19, 20]. Other pro-myogenic genes such as *myogenin* and *MRF4* are directly involved in the differentiation of myotubes [21]. Consistently, we observed a significant increase in the expression of *MyoD1*, *Myf5*, *myogenin*, and *MRF4* in fast-growing birds relative to slow-growing ones. Similarly, three members of the pro-myogenic MEF2 family, namely *MEF2B*, *MEF2C* and *MEF2D*, were also significantly up-regulated in fast-growing broilers. On the other hand, the anti-myogenic *MSTN* mRNA levels are significantly lower in heavy broilers than light ones.

The skeletal muscle is a heterogeneous tissue composed of individual muscle fibers, diversified in size, shape and contractile protein content with different morphology, metabolism, and physiology [22]. Muscle mass can be affected by alterations in satellite cell activities, which dictate mature muscle size [22]. In the present study, fast-growing broilers had a greater breast muscle yield than slow-growing ones, reminiscent of previous studies [8, 23]. It is well known that DNA contents, protein:DNA and RNA:DNA ratios reflect cell population, cell size, and cellular metabolic activities, respectively [24]. An increase in the amount of total cellular RNA is likely due to enhanced synthetic activities [9]. In this study, increased protein:DNA and RNA:DNA ratios and total RNA content, accompanied with a decrease in the DNA content, in the breast muscle of fast-growing broilers suggested that heavier birds has less muscle cells but higher cellular activities. The results coincide with previous studies showing that heavier chickens had greater muscle mass with more satellite cells with enhanced proliferative activities [23, 25].

Hatching weight was shown to be a primary predictor of the market weight in chicks. The average BW of broilers at 42 d of age with heavy (48.3 g) and light hatching weight (41.7 g) became 2,368 g and 2,116 g, respectively [8]. Similarly, broilers on d 41 hatching at 53.1 g was about 1.1-fold heavier that those hatching at 43.5 g [23]. It is important to note that, although the breast muscle weight differs significantly between heavy and light broilers, the abdominal fat content shows no difference at marketing [23], suggesting that the genes involved in adipogenesis may not play a major role in the big variation in BW seen in a commercial broiler production facility.

In this study, we observed a significant difference in both mRNA and protein levels of IGF-1 in the breast muscle and circulation between fast- and slow-growing chickens on d 38. In fact, such a difference in the IGF-1 protein concentration is also evident on day of hatch between the broilers with heavy hatching weight and the ones with light hatching weight [23]. Consistently, IGF-1 levels are positively associated with growth rate in broiler strains divergently selected for high or low growth potential [26, 27], suggesting that IGF-1 can be a good predictor for growth [28]. However, the heritability of IGF-1 is only 0.1 [29] and therefore, its potential might still be limited as a useful parameter for future broiler breeding for enhanced performance. Nevertheless, to our knowledge this is the first study to reveal a positive correlation between IGF-1 and growth rate in a large population of broilers growing under standard commercial management.

It is noted that the serum concentration of T3 was also observed to be significantly different between fast- and slow-growing broilers on d 38 in our study. It will be important to examine a possible difference in its concentration on d 0 in order to explore the growth predictive value of T3.

Conclusions

A variation in the growth rate among individual broilers of the same breed is caused by an alteration in the growth-related hormone and expression of the genes involved in myogenesis. Serum concentration of IGF-1 and T3 is positively correlated with the growth rate of broilers. Most pro-myogenic genes such as *MyoD1*, *Myogenin*, *MRF4*, *Myf5*, *MEF2B*, *MEF2C*, and *MEF2D* are also associated positively with the growth, while *MSTN*, an anti-myogenic gene, shows a negative association. Some of these parameters may have potential to be further explored for the breeding purpose.

Abbreviations

BW: body weight; GH: growth hormone; IGF-1: insulin-like growth factor-1; MEF2: myocyte enhancer factor 2; MRF: myogenic regulatory factors; MSTN: myostatin; Myf5: myogenic factor 5; MyoD1: myogenic differentiation factor 1; T3: triiodothyronine; T4: thyroxine

Acknowledgements

The authors gratefully thank all of the staff of the Wanming Poultry Farm (Hangzhou, China) for their assistance in feeding and care of the animals. We also acknowledge the members of the Institute of Quality and Standards for Agro-products, Zhejiang Academy of Agricultural Sciences for their assistance with sample collections.

Funding

This research was supported by the Special Fund for Public Projects of Zhejiang Province (2016C32073) and International Cooperation Program of Zhejiang Academy of Agricultural Sciences. The funding body has not participated in or interfered with the research.

Authors' contributions

YX, GZ, and HY conceived the study. YX, CW, KL, and GG participated in sample collection and performed all the assays. YX and HY drafted and GZ revised the manuscript. All authors read and approved the final manuscript.

Competing interests

The authors declare that they have no competing interests.

Author details

[1]Institute of Quality and Standards for Agro-products, Zhejiang Academy of Agricultural Sciences, Hangzhou 310021, China. [2]College of Life Sciences, Huzhou University, Huzhou 313000, China. [3]Department of Animal Science, Oklahoma State University, Stillwater, Oklahoma 74078, USA.

References

1. Perry RL, Rudnick MA. Molecular mechanisms regulating myogenic determination and differentiation. Front Biosci. 2000;5:D750–67.
2. Jia Y, Gao G, Song H, Cai D, Yang X, Zhao R. Low-protein diet fed to crossbred sows during pregnancy and lactation enhances myostatin gene expression through epigenetic regulation in skeletal muscle of weaning piglets. Eur J Nutr. 2016;55:1307–14.
3. Hennebry A, Berry C, Siriett V, O'Callaghan P, Chau L, Watson T, et al. Myostatin regulates fiber-type composition of skeletal muscle by regulating mef2 and myod gene expression. Am J Physiol Cell Physiol. 2009;296:C525–34.
4. Wang Q, McPherron AC. Myostatin inhibition induces muscle fibre hypertrophy prior to satellite cell activation. J Physiol. 2012;590:2151–65.
5. Reisz-Porszasz S, Bhasin S, Artaza JN, Shen R, Sinha-Hikim I, Hogue A, et al. Lower skeletal muscle mass in male transgenic mice with muscle-specific overexpression of myostatin. Am J Physiol Endocrinol Metab. 2003;285:E876–88.
6. Scanes CG. Perspectives on the endocrinology of poultry growth and metabolism. Gen Comp Endocrinol. 2009;163:24–32.
7. Boschiero C, Jorge EC, Ninov K, Nones K, do Rosario MF, Coutinho LL, et al. Association of igf1 and kdm5a polymorphisms with performance, fatness and carcass traits in chickens. J Appl Genet. 2013;54:103–12.
8. Wen C, Wu P, Chen Y, Wang T, Zhou Y. Methionine improves the performance and breast muscle growth of broilers with lower hatching weight by altering the expression of genes associated with the insulin-like growth factor-i signalling pathway. Br J Nutr. 2014;111:201–6.
9. Johnson LR, Chandler AM. Rna and DNA of gastric and duodenal mucosa in antrectomized and gastrin-treated rats. Am J Physiol. 1973;224:937–40.
10. Xiao YP, Wu TX, Sun JM, Yang L, Hong QH, Chen AG, et al. Response to dietary l-glutamine supplementation in weaned piglets: A serum metabolomic comparison and hepatic metabolic regulation analysis. J Anim Sci. 2012;90:4421–30.
11. Schmittgen TD, Livak KJ. Analyzing real-time pcr data by the comparative c(t) method. Nat Protoc. 2008;3:1101–8.
12. Zheng Q, Zhang Y, Chen Y, Yang N, Wang XJ, Zhu D. Systematic identification of genes involved in divergent skeletal muscle growth rates of broiler and layer chickens. BMC Genomics. 2009;10:87.
13. Buzala M, Janicki B, Czarnecki R. Consequences of different growth rates in broiler breeder and layer hens on embryogenesis, metabolism and metabolic rate: A review. Poult Sci. 2015;94:728–33.
14. Pauwels J, Coopman F, Cools A, Michiels J, Fremaut D, De Smet S, et al. Selection for growth performance in broiler chickens associates with less diet flexibility. PLoS One. 2015;10, e0127819.
15. Davis RV, Lamont SJ, Rothschild MF, Persia ME, Ashwell CM, Schmidt CJ. Transcriptome analysis of post-hatch breast muscle in legacy and modern broiler chickens reveals enrichment of several regulators of myogenic growth. PLoS One. 2015;10, e0122525.
16. Scanes CG, Marsh J, Decuypere E, Rudas P. Abnormalities in the plasma concentrations of thyroxine, tri-iodothyronine and growth hormone in sex-linked dwarf and autosomal dwarf white leghorn domestic fowl (gallus domesticus). J Endocrinol. 1983;97:127–35.
17. Guernec A, Berri C, Chevalier B, Wacrenier-Cere N, Le Bihan-Duval E, Duclos MJ. Muscle development, insulin-like growth factor-i and myostatin mrna levels in chickens selected for increased breast muscle yield. Growth Horm IGF Res. 2003;13:8–18.
18. Hu Y, Xu H, Li Z, Zheng X, Jia X, Nie Q, et al. Comparison of the genome-wide DNA methylation profiles between fast-growing and slow-growing broilers. PLoS One. 2013;8, e56411.
19. Rudnicki MA, Jaenisch R. The myod family of transcription factors and skeletal myogenesis. Bioessays. 1995;17:203–9.
20. Shin J, McFarland DC, Velleman SG. Heparan sulfate proteoglycans, syndecan-4 and glypican-1, differentially regulate myogenic regulatory transcription factors and paired box 7 expression during turkey satellite cell myogenesis: Implications for muscle growth. Poult Sci. 2012;91:201–7.
21. Comai G, Tajbakhsh S. Molecular and cellular regulation of skeletal myogenesis. Curr Top Dev Biol. 2014;110:1–73.
22. Sobolewska A, Elminowska-Wenda G, Bogucka J, Szpinda M, Walasik K, Bednarczyk M, et al. Myogenesis–possibilities of its stimulation in chickens. Folia Biol (Krakow). 2011;59:85–90.
23. Sklan D, Heifetz S, Halevy O. Heavier chicks at hatch improves marketing body weight by enhancing skeletal muscle growth. Poult Sci. 2003;82:1778–86.
24. Uni Z, Noy Y, Sklan D. Development of the small intestine in heavy and light strain chicks before and after hatching. Br Poult Sci. 1996;37:63–71.
25. Wen C, Chen Y, Wu P, Wang T, Zhou Y. Mstn, mtor and foxo4 are involved in the enhancement of breast muscle growth by methionine in broilers with lower hatching weight. PLoS One. 2014;9, e114236.
26. Scanes CG, Dunnington EA, Buonomo FC, Donoghue DJ, Siegel PB. Plasma concentrations of insulin like growth factors (igf-)i and igf-ii in dwarf and normal chickens of high and low weight selected lines. Growth Dev Aging. 1989;53:151–7.
27. Beccavin C, Chevalier B, Cogburn LA, Simon J, Duclos MJ. Insulin-like growth factors and body growth in chickens divergently selected for high or low growth rate. J Endocrinol. 2001;168:297–306.
28. Duclos MJ. Insulin-like growth factor-i (igf-1) mrna levels and chicken muscle growth. J Physiol Pharmacol. 2005;56 Suppl 3:25–35.
29. Pym RA, Johnson RJ, Etse DB, Eason P. Inheritance of plasma insulin-like growth factor-i and growth rate, food intake, food efficiency and abdominal fatness in chickens. Br Poult Sci. 1991;32:285–93.

6

Net protein and metabolizable protein requirements for maintenance and growth of early-weaned Dorper crossbred male lambs

Tao Ma[1] (iD), Kaidong Deng[2], Yan Tu [1], Naifeng Zhang [1], Bingwen Si [1], Guishan Xu[1,3] and Qiyu Diao[1*]

Abstract

Background: Dorper is an important breed for meat purpose and widely used in the livestock industry of the world. However, the protein requirement of Dorper crossbred has not been investigated. The current paper reports the net protein (NP) and metabolizable protein (MP) requirements of Dorper crossbred ram lambs from 20 to 35 kg BW.

Methods: Thirty-five Dorper × thin-tailed Han crossbred lambs weaned at approximately 50 d of age (20.3 ± 2. 15 kg of BW) were used. Seven lambs of 25 kg BW were slaughtered as the baseline animals at the start of the trial. An intermediate group of seven randomly selected lambs fed ad libitum was slaughtered at 28.6 kg BW. The remaining 21 lambs were randomly divided into three levels of dry matter intake: ad libitum or 70% or 40% of ad libitum intake. Those lambs were slaughtered when the lambs fed ad libitum reached 35 kg BW. Total body N and N retention were measured.

Results: The daily NP and MP requirements for maintenance were 1.89 and 4.52 g/kg metabolic shrunk BW ($SBW^{0.75}$). The partial efficiency of MP utilization for maintenance was 0.42. The NP requirement for growth ranged from 12.1 to 43.5 g/d, for the lambs gaining 100 to 350 g/d, and the partial efficiency of MP utilization for growth was 0.86.

Conclusions: The NP and MP requirements for the maintenance and growth of Dorper crossbred male lambs were lower than the recommendations of American and British nutritional systems.

Keywords: Growth, Lamb, Maintenance, Metabolizable protein, Net protein

Background

In the intensive livestock industry, protein is commonly the most expensive feed component and therefore, it is necessary to have a precise understanding of protein requirement of livestock not only to ensure farm profitability, but also to help reduce nitrogen (N) emission to the environments [1]. Modern feeding systems, such as Agricultural and Food Research Council (AFRC) [2], Commonwealth Scientific and Industrial Research Organisation (CSIRO) [3], and National Research Council (NRC) [4], have reported protein and other nutrient requirements of sheep, which are widely adopted for diet formulation in the world.

The Dorper is a popular sire breed for meat production in both South Africa and United States [5]. In recent years, Dorper sheep have been imported to China to improve the growth performance and carcass traits of indigenous breed, among which the thin-tailed Han sheep is one of the most famous native breeds with high prolificacy, as it carries mutations in both the *BMPR-1B* and *BMP15* genes [6]. Thus, the Dorper × thin-tailed Han crossbreed has become one of the dominant breeds for lamb meat production.

Our research team conducted a series of studies on the nutrient requirements (energy, protein, and minerals) of Dorper × thin-tailed Han crossbred sheep [7–9]. In this paper, we reported the protein requirements of

---
* Correspondence: diaoqiyu@caas.cn
[1]Feed Research Institute/Key laboratory of Feed Biotechnology of the Ministry of Agriculture, Chinese Academy of Agricultural Sciences, Beijing, China
Full list of author information is available at the end of the article

male lambs after weaning ranging from 20 to 35 kg BW by the comparative slaughter technique.

Methods

The current research was conducted from June to September 2011 at the Experimental Station of the Chinese Academy of Agricultural Sciences (CAAS), Nankou (40°22′N, 116°1′E), Beijing, China. The animals were kept in an enclosed animal house and the mean minimum and maximum room temperatures observed during the experimental period were 15.5 and 26.5 °C (average 21.0 °C), respectively.

Animals, treatments, and experimental procedure

Thirty-five Dorper × thin-tailed Han crossbred male lambs weaned at approximately 50 d of age with 20.4 ± 2.15 kg BW were used in a completely randomized design to measure protein requirements for maintenance and growth. The experimental diet was formulated according to NRC [4] with a concentrate-to-forage ratio of 44:56 on a dry matter (DM) basis (Table 1). The lambs with ad libitum intake were fed once daily at 0800 h and allowed 10% of refusals. The amount of feed provided to the restricted feed intake groups was adjusted daily based on the average DM intake (DMI) of the ad libitum group from the previous day. Feed and refusals were sampled daily and frozen at −20 °C until analysis.

Table 1 Ingredient and chemical compositions of the pelleted mixture diet

Items	Value, %DM
Ingredients, DM basis	
Milled Chinese wildrye hay	55.0
Cracked corn grain	29.4
Soybean meal	14.0
Dicalcium phosphate	0.86
Salt	0.50
Mineral/vitamin premix[a]	0.24
Chemical composition	
ME, MJ/kg DM	8.89
DM, % as fed	95.5
CP, % of DM	11.9
EE, % of DM	2.71
Ash, % of DM	6.32
NDF, % of DM	40.9
ADF, % of DM	15.2
Calcium, % of DM	0.68
Phosphorus, % of DM	0.33

[a]Manufactured by Precision Animal Nutrition Research Centre, Beijing, China. The premix contained (per kg): 22.1 g Fe, 13.0 g Cu, 30.2 g Mn, 77.2 g Zn, 19.2 g Se, 53.5 g I, 9.10 g Co, 56.0 g vitamin A, 18.0 g vitamin D_3, and 170 g vitamin E

A comparative slaughter trial was conducted as described previously [9]. Briefly, the initial body composition was measured on seven lambs slaughtered at 20 kg BW (baseline group). An intermediate slaughter group with seven randomly selected lambs fed ad libitum were slaughtered at 28.6 kg BW. The remaining 21 lambs were randomly assigned to three levels of DMI: ad libitum or 70% or 40% of ad libitum intake. Thus, the lambs were pair-fed in seven slaughter groups, with each group consisting of one lamb from each level of intake. When the lambs fed ad libitum of each slaughter group reached 35 kg BW, all lambs within a slaughter group were fasted and slaughtered. Seven lambs in the baseline and intermediate group were slaughtered in one day, respectively, while the remaining 21 lambs were slaughtered in three consecutive days. All lambs were slaughtered by exsanguination after stunning by CO_2 inhalation. Blood, carcass, head, feet, hide, wool, viscera, and adipose tissue removed from the internal organs were weighed. The empty body weight (EBW) was calculated by subtracting the weight of the digestive tract contents from the shrunk BW (SBW), which was measured as BW after a 16-h fast of feed and water. Carcasses and heads were split longitudinally into two identical halves and the muscle, bone, and fat were dissected from the right-half carcass, head, and feet, while the whole hide and whole viscera were ground and homogenized separately and frozen at −20 °C until analysis. Wool was clipped with electrical clippers after slaughter, and subsamples were collected and stored at 4 °C.

Chemical analyses

Feed and orts were oven-dried at 55 °C for 72 h, ground to pass through a 1-mm screen for analyzing DM (method 930.15) [10], ash (method 924.05) [10], ether extract (EE) (method 920.85) [10], and nitrogen (N) [11]. The gross energy (GE) of feed was measured with a bomb calorimeter (C200, IKA Works Inc., Staufen, Germany). The neutral detergent fiber (NDF) and acid detergent fiber (ADF) of feed were measured according to Van Soest et al. [12] and Goering and Van Soest [13], respectively. Calcium of feed was analyzed using an atomic absorption spectrophotometer (M9W-700, Perkin-Elmer Corp., Norwalk, CT, USA) (method 968.08) [10]. The total P of feed was analyzed by the molybdovanadate colorimetric method (method 965.17) [10] using a spectrophotometer (UV-6100, Mapada Instruments Co., Ltd., Shanghai, China).

The DM of all body components except wool was determined following lyophilization for 72 h to constant weight. The samples were then analyzed for N as described above. Wool samples were cut into 2-mm pieces with scissors and analyzed for N as described above.

Data calculations

Metabolizable protein supply: The ratio of metabolizable protein (MP) to OM intake reported in a previous in vivo study (i.e., 69.4, 88.9, and 102.4 g MP/kg OM intake for ad libitum, 70% and 50% of the ad libitum intake, respectively) [14] with Dorper × thin-tailed Han crossbred sheep subjected to the same feeding regime as the present study was used to calculate the individual MP intake (MPI) by the ram lambs.

Prediction of the initial body N content and N retention (NR): The NR in the body of the lambs in the comparative slaughter trial was calculated as the difference between the final and initial body N content. The initial body N content of each animal was calculated from its initial EBW using a regression equation developed from the relationship between the body N content and EBW of the baseline animals ($R^2 = 0.92$, root mean square error [RMSE] = 0.011, n = 7, $P < 0.001$): $\log_{10}$ empty body N, kg = 1.536 (±0.174) + [0.957 (±0.144) × $\log_{10}$ EBW, kg]. The initial SBW of each animal was computed from its initial BW ($R^2 = 0.97$, RMSE = 0.280, n = 7): SBW, kg = 0.571 (±1.457) + [0.915 (±0.069) × BW, kg], and the initial EBW of each animal was computed from its initial BW ($R^2 = 0.86$, RMSE = 0.429, n = 7): EBW, kg = 2.336 (±2.227) + [0.695 (±0.105) × BW, kg].

Protein requirements for maintenance: A linear regression of daily NR on daily N intake (NI) was used to calculate the net protein (NP) requirement for maintenance. The intercept of the regression was assumed the endogenous and metabolic losses of N multiplied by the factor 6.25, which is assumed the maintenance requirement for NP (NP_m, g/kg$^{0.75}$ SBW). The MP required for maintenance (MP_m, g/kg$^{0.75}$ SBW) was then estimated by regressing the NR on MPI and extrapolating the linear regression to zero NR. The efficiency of MP use for maintenance (k_{pm}) was computed as NP_m/MP_m.

Protein requirements for growth: The NP requirements for body weight gain (NP_g) were calculated as the difference between body protein content at different intervals. For example, the NP_g of a lamb with 20 kg SBW and 250 g of average daily gain (ADG) was computed as the difference between body protein contents at 20.25 and 20 kg SBW. Body protein contents were predicted from EBW using an allometric equation according to ARC (1980): $\log_{10}$ protein, kg = a + [b × $\log_{10}$ EBW, kg]. To estimate the partial efficiencies of MP use for body weight gain (k_{pg}), a regression model through the origin was used to partition the utilization of MPI above maintenance for body protein retention as follows: $MPI_g = b \times RP$, where MPI_g (g/kg$^{0.75}$ of SBW) is the MPI above maintenance calculated as the difference between the total MPI and MP_m, RP (g/kg$^{0.75}$ of SBW) is the daily retention of body protein, and the estimated parameter b is the amount of MP (g) required to retain 1 g of protein, and its inverse was assumed to be k_{pg}.

Statistical analyses

The data were analyzed as a completely randomized design using the SAS statistical software package (version 9.1; SAS Institute, Inc., Cary, NC). Intake, body composition, and growth rate were analyzed using a one-way ANOVA. Pairwise comparisons of means were performed by Tukey's multiple range tests once the significance of the treatment effect was declared at $P < 0.05$. The statistical model is: $y_{ij} = \mu + \alpha_i + \varepsilon_{ij}$, where y_{ij} = dependent variable, μ = overall mean of Yij, α_i = effect of the diet (i = 1 to 3), and ε_{ij} = error contribution. Linear regressions were conducted with a GLM, and observations with a studentized residual > 2.5 or < −2.5 were considered outliers. The assumptions of the models, in terms of homoscedasticity, independency, and normality of errors, were examined by plotting residuals against the predicted values.

Results

The intake of OM and N increased with feeding level ($P < 0.001$; Table 2). The lambs fed with ad libitum intake retained greater N than those with either level of restricted feed intake ($P < 0.001$). The lambs fed at 40% of the ad libitum intake had a lower MPI than those in the other two groups ($P < 0.001$), and the MPI did not differ between lambs fed at ad libitum and 70% of the ad libitum intake ($P = 0.286$).

Table 2 Daily protein intake of Dorper × thin-tailed Han crossbred ram lambs at ad libitum (AL) or restricted to 70% or 40% of AL intake[1]

Items	Level of feed intake[2]			SEM	P^4
	AL	70%	40%		
ADG, g	324.2[a]	189.1[b]	37.6[c]	19.3	<0.001
Initial SBW, kg	19.3	19.5	18.5	0.22	0.271
Final SBW, kg	29.9[a]	27.8[a]	20.4[b]	0.91	<0.001
OM intake, g/d	107.6[a]	84.2[b]	53.3[c]	4.37	<0.001
N intake, g/(kg SBW$^{0.75}$ · d)	1.83[a]	1.43[b]	1.01[c]	0.08	<0.001
N retention, g/(kg SBW$^{0.75}$ · d)	0.47[a]	0.29[b]	0.11[c]	0.03	<0.001
MP intake, g/(kg SBW$^{0.75}$ · d)[3]	7.47[a]	7.49[a]	5.46[c]	0.18	<0.001

[1]ADG: average daily gain; SBW: shrunk body weight; MP: metabolizable protein
[2]Ad libitum or restricted to 70% or 40% of ad libitum of an identical diet
[3]Calculated from the ratio of MP supply to OM intake reported by Ma et al. [14]
[4] a,b,c Means bearing different superscripts differ ($P < 0.05$)

Figure 1 shows the linear relationship between NR and NI: NR, $g/(kg\ SBW^{0.75} \cdot d) = -0.303\ (\pm 0.046) + [0.422\ (\pm 0.040) \times N\ intake, g/(kg\ SBW^{0.75} \cdot d)]$ $(R^2 = 0.89, RMSE = 0.054, n = 28, P < 0.001)$. The endogenous and metabolic loss of N, estimated as the intercept of the linear regression, was $303 \pm 46\ mg/kg^{0.75}$ SBW, which corresponds to an NP_m of $1.89 \pm 0.29\ g/kg^{0.75}$ SBW.

Figure 2 shows the linear relationship between NR and MPI: NR, $g/(kg\ SBW^{0.75} \cdot d) = -0.628\ (\pm 0.138) + [0.139\ (\pm 0.020) \times MPI, g/(kg\ SBW^{0.75} \cdot d)]$, $R^2 = 0.70$, RMSE = 0.096, n = 28, $P < 0.001$. The MP required for maintenance by extrapolating the linear regression to zero N retention was $4.52\ g/kg^{0.75}$ SBW. Consequently, the k_{pm} was 0.42 for Dorper × thin-tailed Han crossbred male lambs from 20 to 35 kg of BW.

The partial efficiencies of MP use for body weight gain were estimated using a regression model and assuming that the MPI above maintenance (MPI_g) is partially recovered as body protein for growth (PR_g, $g/kg^{0.75}$ of SBW). The regression equation was: PR_g, $g/(kg^{0.75}\ SBW \cdot d) = 0.002(\pm 0.052) + [0.864 \pm 0.123] \times MPI_g$, $g/(kg^{0.75}$ of SBW $\cdot d)]$ $(R^2 = 0.70, n = 28, RMSE = 0.081)$. The slope of the regression equation (0.86) represents k_{pg}.

The allometric equation between body protein and EBW $(R^2 = 0.96; RMSE = 0.012; n = 20)$ of the ram lambs with free consumption of feed was proposed as $\log_{10}$ protein, $g = 2.385\ (\pm 0.072) + [0.922\ (\pm 0.054) \times \log_{10} EBW, kg]$. The NP_g and MP for growth (MP_g) were therefore calculated accordingly (Tables 3 and 4).

Discussion

The NP_m is the quantity of protein to sustain tissue proteins by counterbalancing the inevitable losses of urinary, fecal, and dermal N [3]. AFRC [15] suggested daily endogenous and metabolic N losses of 350 mg/kg $BW^{0.75}$ for lambs nourished by intra-gastric infusions. However, this method might overestimate the endogenous N requirement due to the lack of conservation of

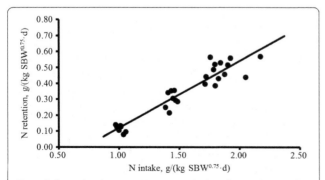

Fig. 1 Relationship between N retention (NR) and N intake (NI) of Dorper × thin-tailed Han crossbred male lambs from 20 to 35 kg of BW. NR, g/kg $SBW^{0.75} = -0.303\ (\pm 0.046) + [0.422\ (\pm 0.040) \times NI$, g/kg of $SBW^{0.75}]$, $R^2 = 0.89$, RMSE = 0.054, n = 28

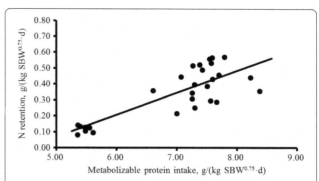

Fig. 2 Relationship between N retention (NR) and metabolizable protein intake (MPI) of Dorper × thin-tailed Han crossbred male lambs from 20 to 35 kg of BW. NR, g/kg $SBW^{0.75} = -0.628\ (\pm 0.138) + [0.139\ (\pm 0.020) \times MPI$, g/kg $SBW^{0.75}]$, $R^2 = 0.70$, RMSE = 0.096, n = 28

protein by the microbial capture of N. CSIRO [3] suggested the NP_m of the sheep is the sum of the endogenous urinary loss $(0.147 \times BW + 3.375)$ and fecal loss (15.2 g/kg DM intake). Thus, the NP_m for a lamb of 28 kg consuming 1.02 kg of DM daily (this is the average DM intake of the 28 lambs used in the current study) is about 23.0 g, which is 10% higher than the current value $(1.72\ g/kg\ BW^{0.75})$. In the current study, the NP_m calculated by partial regression NR on NI $(1.72\ g/kg\ BW^{0.75}$ or $1.89\ g/kg\ SBW^{0.75})$ is slightly higher than that of the Ile de France $(1.56\ g/kg\ SBW^{0.75})$ [16] and Texel crossbreeds $(1.52\ g/kg\ SBW^{0.75})$ [17], but lower than that of Morada Nova $(1.83\ g/kg\ BW^{0.75})$ [18] male lambs measured using the same methods. The lambs fed at the maintenance level had similar NR (around 0.1 to 0.2 g/kg of $SBW^{0.75}$) among all these studies. However, when fed higher than the maintenance level, Dorper crossbred had a relatively higher NR (0.25 to 0.65 g/kg of $SBW^{0.75}$) than the Ile de France [16] and Texel crossbreeds [17], but a lower NR than Morada Nova lambs (0.25 to 1.0 g/kg of $BW^{0.75}$) [18]. Therefore, the variations in NP_m can be explained by the differential utilization efficiency of protein or amino acids by the tissues during the growth of lambs. On the other hand, the above studies were all conducted in a tropical or sub-tropical zone with high humidity, while the current study was conducted in a warm

Table 3 Requirements of net protein (NP) for growth (g/d) of Dorper × thin-tailed Han crossbred male lambs from 20 to 35 kg of body weight (BW)

ADG, g/d	BW, kg			
	20	25	30	35
100	12.4	12.3	12.2	12.1
200	24.9	24.6	24.4	24.2
300	37.3	36.9	36.6	36.3
350	43.5	43.0	42.7	42.3

ADG: average daily gain

Table 4 Requirements of metabolizable protein (MP) for growth (g/d) of Dorper × thin-tailed Han crossbred male lambs from 20 to 35 kg of body weight (BW)

ADG, g/d	BW, kg			
	20	25	30	35
100	14.4	14.3	14.2	14.1
200	29.0	28.6	28.4	28.1
300	43.4	42.9	42.6	42.2
350	50.6	50.0	49.7	49.2

ADG: average daily gain

temperate zone where the weather is always dry during summer. As reviewed by Marai et al. [19], both temperature and humidity could have influence on nutrient digestibility and degradation in the rumen; thus, the experimental condition could be another important factor that contributes to the differences in NP_m apart from methods or animal breeds. We concluded that 1.69 g/kg $BW^{0.75}$ or 1.81 g/kg $SBW^{0.75}$ is suitable for the NP_m of Dorper crossbred male lambs from 20 to 35 kg of BW.

The MP_m obtained in the current study was 4.52 g/kg $SBW^{0.75}$. When scaled to BW, our value (4.37 g/kg $BW^{0.75}$) was close to Liu et al. [20], who found an MP_m of 4.41 g/kg $BW^{0.75}$ from a linear multiple regression of MP requirements against the live weight, live weight gain, and wool growth of sheep (n = 213) in a feeding study. Nevertheless, our result is greater than that suggested by AFRC (2.19 g/kg $BW^{0.75}$) [2], INRA (2.50 g/kg $BW^{0.75}$) [21], or NRC (3.72 g/kg $BW^{0.75}$) [22]. In a more recent study, where a comparative slaughter trial was also used, a MP_m of 2.31 g/kg $SBW^{0.75}$ was observed in Texel crossbred lambs [17]. The methods adopted by UK [2], Australia [3], USA [4], and France [21] were all based on a common overall model, although requirements were expressed in different terms. In the current study, MP was calculated based on the method reported by Ma et al. [14], who conducted an in vivo study and measured MP using sheep with ruminal and duodenal cannula. Therefore, the discrepancy in the calculation of MP is inevitably associated with the methodologies adopted. As there is still a lack of simple and robust methods for calculating MP, this area requires further investigation and examination.

The NP_g values (12.4, 24.9, and 37.3 g/d) determined in the current study are extensively lower than those of early maturing growing lambs (23.5, 30.5, and 50.0 g/d) of 20 kg BW gaining 100, 200, and 300 g/d recommended by NRC [4], assuming a k_{pg} of 0.50. AFRC [2] used two equations proposed by ARC [23], NP_f (g/d) = ADG × (160.4 − 1.22 × BW + 0.0105 × BW^2) and NP_w (g/d) = 3 + 0.1 × NP_f, where NP_f is the NP requirement for fleece-free body growth and NP_w is the NP requirement for wool growth, to predict the protein requirement

for the growth of body and fleece in lambs, respectively. Using those equations, the NP_g (NP_f + NP_w) was approximately 26% to 49% higher (range from 17.4 to 57.0 g/d) than values determined in the current study. Therefore, caution should be taken before applying certain evaluation systems to avoid the overestimation of NP_g of Dorper crossbred lambs. By using the same method, our NP_g results were 26% higher than those of the Texel crossbreed [17] growing from 20 to 35 kg of SBW gaining 100 and 200 g/d, but 20% lower than that of Morada Nova lambs [18] growing from 20 to 30 kg of BW gaining 100, 200, and 300 g/d, respectively. Although the ME (0.76 vs. 0.73 MJ/kg $SBW^{0.75}$), NI (2.12 vs. 2.20 g/kg $SBW^{0.75}$), and N retention/N intake (26.1% vs. 25.9%) were close, the ADG of ad libitum groups were much higher for the Dorper crossbred (324 g) than Texel crossbred (245 g) lambs reported by Galvani et al. [17]. Many factors could be associated with such discrepancies, including breed, physiological stage, and experimental conditions.

The partial efficiency of use of MP_m for NP_m (k_{pm}) was calculated to be 0.42 in the present study. This value is lower than previously adopted 1.0 by AFRC [15] or 0.67 by CSIRO [3], as well as lower than that of Texel crossbred lambs (0.66) [17]. The partial efficiency of use of MP_g for NP_g (k_{pg}) obtained in the current study (0.86) was greater than that adopted by AFRC (0.59) [2], CSIRO (0.70) [3] and by Galvani et al. [17] in Texel crossbred lambs (0.71). Those discrepancies could be attributed to animal factors, including breed, maturity, and physiological status. The method for calculating or determining MP may be another factor contributing to the variability of the efficiency of MP use. NRC [4] suggested a simple equation, MP = CP × (0.64 + 0.16 × % UDP)/100, where UDP is undegraded dietary protein, to convert CP to MP. By using this equation, MP/CP was 0.72 in all treatments, and k_{pm} and k_{pg} were 0.59 (1.89/3.22) and 0.58, respectively. Our previous study showed that a decreased feed intake could increase total-tract N digestibility without affecting ruminal N degradability [24], and the increased duodenal N digestibility could be due to the prolonged gastric empty time. Thus, it could not be expected that MP/CP were identical under a different feeding level. In the current study, MP was calculated from OM intake based on the results of our previous study using 6-month-old Dorper × thin-tailed Han male lambs (41.3 ± 2.8 kg BW) with both ruminal and duodenal cannula fed at three different levels (ad libitum, 70%, and 50% of ad libitum) in which MP supply/OM intake (g/kg) was 69.4, 88.9, and 102.4, respectively. As there is still lack of a simple method for the calculation of MP, this could be a reasonable way to calculate MP in the current study. Nevertheless, considering the difference in animal physiology status (BW, age, and cannulation) between our previous and current studies, further study is still needed to examine the utilization efficiency of MP for both the maintenance and growth for early-weaning lambs.

Conclusions

In conclusion, the current study suggested that the protein requirements for the maintenance and growth of Dorper × thin-tailed Han early-weaned crossbred male lambs were lower than the recommendations of AFRC (1993) and NRC (2007).

Abbreviations

ADF: Acid detergent fiber; ADG: Average daily gain; CP: Crude protein; DM: Dry matter; EBW: Empty body weight; EE: Ether extract; GE: Gross energy; ME: Metabolizable energy; MNI_g: Metabolizable N intake above maintenance; MP: Metabolizable protein; MP_g: MP requirements for body weight gain; MPI_g: MP intake above maintenance; MP_m: Metabolizable protein for maintenance; N: Nitrogen; NDF: Neutral detergent fiber; NP: Net protein; NP_f: Net protein requirements for fleece-free body growth; NP_g: Net protein requirements for body weight gain; NP_m: Net protein for maintenance; NP_w: Net protein requirements for wool growth; NR: Nitrogen retention; OM: Organic matter; RMSE: Root mean square error; RN_g: Body N for growth; SBW: Shrunk body weight; UDP: Undegraded dietary protein

Acknowledgements
We thank the staff (Y.L. Li, Y.F. Zhang, J. Liu, Y.G. Zhao, S.K. Ji, L.T. Zhang, and C. Lou) of Feed Research Institute of Chinese Academy of Agricultural Sciences for their technical assistance.

Funding
This research was supported by the earmarked fund for the China Agriculture Research System (CARS-39). This manuscript was a part of the 1st International Symposium on Young Ruminant (ISYR)-Physiology and Rearing Strategies jointly organised by Laboratory of Ruminant Physiology and Nutrition, Feed Research Institute, Chinese Academy of Agricultural Sciences, China, The Pennsylvania State University, US, Agri-Food and Biosciences Instute, UK, Beijing Dairy Industry Innovation Consortium of Agriculture Research System, Beijing Key Laboratory for Dairy Nutrition.

Authors' contributions
GX carried out the whole animal experiment, including sample collection and determination. TM participated in the statistical analysis and wrote the draft. KD helped to revise the manuscript. YT, NZ, and BS participated in the design and coordination of the study. QD conceived of the study. All authors read and approved the final manuscript.

Competing interest
The authors declare that they have no competing interest.

Author details
^{1}Feed Research Institute/Key laboratory of Feed Biotechnology of the Ministry of Agriculture, Chinese Academy of Agricultural Sciences, Beijing, China. ^{2}College of Animal Science, Jinling Institute of Technology, Nanjing, Jiangsu, China. ^{3}College of Animal Science, Tarim University, Alar, Xinjiang, China.

References
1. Ma T, Xu GS, Deng KD, Ji SK, Tu Y, Zhang NF, et al. Energy requirements of early-weaned Dorper cross-bred female lambs. J Anim Physiol An N. 2016. doi:10.1111/jpn.12480.
2. AFRC. Energy and protein requirements of ruminants. An advisory manual prepared by the Agricultural and Food Research Council Technical Committee on Responses to Nutrients. Wallingford: CAB International; 1993.
3. CSIRO. Nutrient requirements of domesticated ruminants. Collingwood: Commonwealth Scientific and Industrial Research Organisation Publishing; 2007.
4. NRC. Nutrient Requirements of Small Ruminants: Sheep, Goats, Cervids, and New World Camelids. Washington: National Academy Press; 2007.
5. Snowder GD, Duckett SK. Evaluation of the South African Dorper as a terminal sire breed for growth, carcass, and palatability characteristics. J Anim Sci. 2003;81(2):368–75.
6. Chu MX, Liu ZH, Jiao CL, He YQ, Fang L, Ye SC, et al. Mutations in BMPR-IB and BMP-15 genes are associated with litter size in Small Tailed Han sheep (Ovis aries). J Anim Sci. 2007;85:598–603.
7. Deng KD, Diao QY, Jiang CG, Tu Y, Zhang NF, Liu J, et al. Energy requirements for maintenance and growth of Dorper crossbred ram lambs. Livest Sci. 2012;150:102–10.
8. Deng KD, Jiang CG, Tu Y, Zhang NF, Liu J, Ma T. Energy requirements of Dorper crossbred ewe lambs. J Anim Sci. 2014;92:2161–9.
9. Xu GS, Ma T, Ji SK, Deng KD, Tu Y, Jiang CG, et al. Energy requirements for maintenance and growth of early-weane d Dorper crossbred male lambs. Livest Sci. 2015;177:71–8.
10. AOAC. Official methods of analysis. 15th ed. Arlington: AOAC International; 1990.
11. Marshall CM, Walker AF. Comparison of a short method for Kjeldahl digestion using a trace of selenium as catalyst, with other methods. J Sci Food Agri. 1978;29:940–2.
12. Van Soest PJ, Robertson JB, Lewis BA. Methods for dietary fiber, neutral detergent fiber and non-starch polysaccharides in relation to animal nutrition. J Dairy Sci. 1991;74:3583–97.
13. Goering HG, Van Soest JP. Forage fiber analysis. Agricultural Handbook, vol. 379. Washington: UPSDA; 1970.
14. Ma T, Deng KD, Tu Y, Zhang NF, Jiang CG, Liu J, et al. Effect of feed intake on metabolizable protein supply in Dorper × thin-tailed Han crossbred lambs. Small Rumin Res. 2015;132:133–6.
15. AFRC. Technical committee on responses to nutrients, Report 9. Nutritive Requirements of Ruminant Animals: Protein. Nutrition Abstracts and Reviews. Series B. 1992;62:787–835.
16. Silva AMA, Silva Sobrinho AG, Trindade IACM, Resende KT, Bakke OA. Net requirements of protein and energy for maintenance of wool and hair lambs in a tropical region. Small Rumin Res. 2003;49:165–71.
17. Galvani DB, Pires CC, Kozloski GV, Sanchez LMB. Protein requirements of Texel crossbred lambs. Small Rumin Res. 2009;81:55–62.
18. Costa MRGF, Pereira ES, Silva AMA, Paulino PVR, Mizubuti IY, Pimentel PG, et al. Body composition and net energy and protein requirements of Morada Nova lambs. Small Rumin Res. 2013;114:206–13.
19. Marai IFM, El-Darawany AA, Fadiel A, Abdel-Hafez MAM. Physiological traits as affected by heat stress in sheep-a review. Small Rumin Res. 2007;71(1):1–12.
20. Liu SM, Smith TL, Karlsson LJE, Palmer DG, Besier RB. The costs for protein and energy requirements by nematode infection and resistance in Merino sheep. Livest Prod Sci. 2005;97(2):131–9.
21. INRA. Ruminant nutrition: Recommended allowances and feed tables. Paris: John Libbey & Co. Ltd; 1989.
22. NRC. Predicting feed intake of food-producing animals. Washington: National Academy Press; 1987.
23. ARC. The Nutrient Requirements of Ruminant Livestock. Slough: Commonwealth Agricultural Bureaux; 1980.
24. Ma T, Deng KD, Jiang CG, Tu Y, Zhang NF, Liu J, et al. The relationship between microbial N synthesis and urinary excretion of purine derivatives in Dorper × thin-tailed Han crossbred sheep. Small Rumin Res. 2013;112:49–55.

7

Effects of quantitative feed restriction and sex on carcass traits, meat quality and meat lipid profile of Morada Nova lambs

Thiago L. A. C. de Araújo[1], Elzânia S. Pereira[1], Ivone Y. Mizubuti[2], Ana C. N. Campos[1], Marília W. F. Pereira[1], Eduardo L. Heinzen[1], Hilton C. R. Magalhães[3], Leilson R. Bezerra[4], Luciano P da Silva[1] and Ronaldo L. Oliveira[5*]

Abstract

Background: An experiment was conducted to evaluate the effects of feed restriction (FR) and sex on the quantitative and qualitative carcass traits of Morada Nova lambs. Thirty-five animals with an initial body weight of 14.5 ± 0.89 kg and age of 120 d were used in a completely randomized study with a 3×3 factorial scheme consisting of three sexes (11 entire males, 12 castrated males and 12 females) and three levels of feeding (ad libitum – AL and 30% and 60% FR).

Results: Entire males presented greater hot and cold carcass weights ($P < 0.05$), followed by castrated males and females. However, the hot carcass yield was higher for females and castrated males than for entire males. Luminosity values were influenced ($P < 0.05$) by sex, with entire males presenting higher values than castrated males and females. Females showed higher ($P < 0.05$) concentrations of linoleic acid and arachidonic acid in the meat of the *longissimus thoracis* muscle. The meat of animals submitted to AL intake and 30% FR showed similar ($P > 0.05$) concentrations, and the concentrations of palmitic acid, palmitoleic acid, stearic acid, oleic acid and conjugated linoleic acid were higher ($P < 0.05$) than those of animals with 60% FR. The meat of females had a higher ω6/ω3 ratio and lower h/H ratio, and females had greater levels of feeding. The meat of animals on the 60% FR diet had a greater ω6/ω3 ratio, lower h/H ratio and lower concentration of desirable fatty acids in addition to a greater atherogenicity index (AI) and thrombogenicity index (TI).

Conclusion: Lambs of different sexes had carcasses with different quantitative traits without total influence on the chemical and physical meat characteristics. The lipid profile of the meat was less favorable to consumer health when the animals were female or submitted to 60% feed restriction.

Keywords: Dietary restriction, Fatty acid, Hair sheep, Lean meat, Semi-arid condition

Background

Farming ruminants has an unquestionable importance to the economic and food security of many regions of the world, especially for tropical semi-arid regions [1], where sheep are highly relevant [2, 3]. In these regions, the environment interferes intensely with productive management strategies, and making decisions is crucial for the success of animal husbandry.

Production systems in semi-arid regions are based on the use of genetic resources with high adaptability and heat tolerance, which are heavily influenced by the qualitative and quantitative seasonality of food, but these local breeds are now threatened with extinction. Morada Nova is a prominent genetic group of hair sheep in semiarid regions of Brazil. These sheep are an indigenous group used for the production of meat and skin. They are smaller, have lower mortality rates, produce lighter carcasses, and are usually late to slaughter. Adult Morada Nova hair sheep weigh between 45 and 50 kg and can reproduce at approximately 8–9 mons of age with approximately 28 kg of body weight (BW) [4].

* Correspondence: ronaldooliveira@ufba.br
[5]Department of Animal Science, School of Veterinary Medicine and Animal Science/Federal University of Bahia, Salvador City, Bahia State 40.170-110, Brazil
Full list of author information is available at the end of the article

The improvement of animal yield by enhancing sustainable biodiversity may be a pathway toward greater food supplies. Such sustainable increases may be especially important for the 2 billion people reliant on small farms, many of which are undernourished, yet we know little about the efficacy of this approach.

Some productive strategies affect animal performance as well as the chemical and physical quality of the meat produced. The farmer can, for example, opt to obtain production rates and carcasses with different characteristics depending on the sex of the animals [5–7]. In tropical semi-arid regions, animals can be submitted naturally to periods of feed restriction (FR) due to feed supply variations or due to feed management planning, which is commonly used to save resources and reduce costs [8, 9]. Under these conditions, it may be assumed that in addition to lower performance, meat products may have different chemical and physical characteristics if the animals are sold [10, 11].

In the face of global concerns about the safety and nutritional quality of foods, it is necessary to understand the effects of commonly used production strategies not only on productivity but also on aspects related to human health, as is the case with lipid quality parameters. Thus, we simulate the impact of feed restriction in hair sheep of different sexes in semi-arid regions. We then evaluate their effects on carcass characteristics, meat quality and fatty acid profiles in the meat of Morada Nova lambs.

Methods
Animal care and location
This study was conducted at the Department of Animal Science, Federal University of Ceará, located in Fortaleza, CE, Brazil. Protocols (n° 98/2015) were in accordance with the standards established by the Committee of Ethics in Animal Research of the Federal University of Ceará.

Animals, experimental design and management
Experimental lambs were obtained from the Morada Nova sheep breeding facility. The mating season was established with the objective of enabling a selection of animals with little variation in BW. Thirty-five lambs of the Morada Nova breed, including 23 males and 12 females, were selected. Twelve entire males were randomly assigned to the sexual class of castrated males, and the males were castrated using the burdizzo castrating method. Initially, the lambs had 14.5 ± 0.89 kg of BW and 120 d of age. The lambs were distributed in a completely randomized design in a 3×3 factorial scheme. Experimental treatments consisted of three sexes (11 entire males, 12 castrated males and 12 females) and three quantitative feeding levels (ad libitum (AL), 30% and 60% FR). The ration was formulated to

supply the nutritional requirements of late maturity lambs with a gain of 150 g/d as recommended by the National Research Council (NRC) [12]. Before beginning data collection, the animals were randomly assigned to individual boxes provided with feed and water troughs, where they underwent an adaptive period of 15 d. Total mixed rations were provided twice a day (0730 and 1600 h), allowing for up to 10% orts only for animals fed AL. Before each morning feeding, the orts of each animal fed AL were removed and weighed to calculate the intake and feeding level of the lambs submitted to 30% and 60% FR (300 and 600 g/kg of FR). Thus, the restrictions were proportionally based on the intake of animals fed AL of each sex.

The ingredients used in the total ration and their proportions and composition are described in Table 1.

Samples of the roughage, concentrated and feed orts were taken to determine their chemical compositions and dry matter intake (DMI) of the lambs. The lambs were weighed every fifteen days to calculate BW gain (BWG). The trial period lasted 120 d. At the end of the trial period, the animals were weighed to determine total weight gain (TWG) and average daily gain (ADG).

Slaughter, carcass data and meat samples
After 18 h of fasting, the animals were weighed to determine their BW at slaughter (BWS). The animals were then skinned and eviscerated according to the rules established in the Regulation of Brazilian Industrial and Sanitary Inspection of Animal Products. Subsequently, the lambs were stunned with the proper equipment, bled, skinned, and eviscerated. The viscera were weighed when filled, emptied, washed, drained and weighed when empty to determine the contents of the gastrointestinal tract and subsequently the empty BW (EBW) of the animals. The carcasses were identified and weighed to obtain the hot carcass weight (HCW) and yield (HCY) calculated in relation to BWS. After 24 h of cooling at 4 °C, the carcasses were weighed to obtain the cold carcass weight (CCW). Twenty-four hours *post mortem*, the pH was measured using a pH meter (HI-99163, Hanna® instruments, São Paulo, Brazil) by inserting the meter between the 4th and 5th lumbar vertebrae in the *longissimus lumborum* muscle.

The carcasses were sectioned with an electric saw (Ki Junta®, São Paulo, Brazil) along the spine, and the left halves of the carcasses were divided into six commercial cuts (leg, loin, ribs, lower ribs, neck and shoulder), which were individually weighed. A cross-sectional cut was made between the 12th and 13th ribs to expose the *longissimus thoracis (LT)* muscle, which measured the maximum distances between the ends of the muscle in the mediolateral direction (A) and dorsal-ventral (B) to subsequently calculate the rib eye area (REA) according

Table 1 Ingredient proportion and chemical composition of experimental rations

Ingredient	Proportion, % of DM			
Tifton 85 grass hay	60.0			
Ground corn grain	32.72			
Soybean meal	6.30			
Dicalcium phosphate	0.06			
Mineral premix [a]	0.92			
Chemical composition, g/kg of DM	Total ration	Tifton 85 grass hay	Ground corn grain	Soybean meal
Dry matter	907.72	913.40	892.40	910.00
Crude protein	169.32	172.50	102.80	508.80
Ether extract	30.77	25.57	43.18	19.32
Ash	61.93	73.40	13.30	65.90
Neutral detergent fiber	438.65	668.20	112.54	134.63
NDFap [b]	418.32	644.85	97.89	110.41
Acid detergent fiber	201.93	317.54	26.31	102.05
Non-fiber carbohydrate	319.67	58.30	728.19	273.30

[a] Composition, 1 kg of premix: Calcium 225 g to 215 g; Phosphor 40 g; Sulfur 15 g; Sodium 50 g; Magnesium 10 g; Cobalt 11 mg; Iodine 34 mg; Manganese 1,800 mg; Selenium 10 mg; Zinc 2,000 mg; Iron 1,250 mg; Copper 120 mg; Fluor 400 mg; Vitamin A 37.5 mg; Vitamin D$_3$ 0.5 mg and Vitamin E 800 mg
[b] Neutral detergent fiber corrected for ash and protein

to Eq. REA = (A / 2 × B / 2) × π. Subcutaneous fat thickness (SFT) was verified above measure B using a digital caliper. Samples were taken from the *LT* and *longissimus lumborum* (*LL*) muscles, vacuum packed and stored at –20 °C.

Physicochemical meat analyses

The meat color was evaluated using a transverse cut on the back section, which was exposed to atmospheric air for 30 min before reading the oxygen myoglobin, which is the primary element that defines meat color [13]. As described by Miltenburg et al. [14], the coordinates L*, a* and b* were measured at three different points on the muscle, and the triplicates were averaged for each coordinate per animal. These measurements were performed using a Minolta CR-10 colorimeter (Konica® Minolta, Osaka, Japan) that was previously calibrated with the CIELAB system using a blank tile, illuminant D65 and 10° as the standard observation points. L* is related to lightness (L* = 0 black, 100 white); a* (redness) ranges from green (–) to red (+); and b* (yellowness) ranges from blue (–) to yellow (+). Measurements were made from a 2° viewing angle using illuminant C. The color saturation (chroma, C*) was calculated as (a^{*2} + b^{*2})$^{1/2}$ [15].

Meat samples of *LL* muscle were processed in a crusher to determine the water holding capacity (WHC), and cooking weight loss (CWL) was determined according to the American Meat Science Association (AMSA) [16] using *LL* meat samples (triplicate) without visible connective tissue that were previously thawed at 10 °C for 12 h. CWL indicated the difference in the weight of the meat before and after cooking on a preheated grill (George Foreman Jumbo Grill GBZ6BW, Rio de Janeiro, Brazil) at

170 °C. A digital skewer thermometer (Salcasterm 200®, São Paulo, Brazil) was used to monitor the internal temperature of the steak until the center reached 71 °C. Then, each steak was brought to room temperature, removed from the oven after temperature stabilization, and weighed again. The difference between the initial and final weights of a sample was used to determine the CWL, with the value expressed as a percentage.

After cooling at room temperature, the samples were again wrapped in foil and placed in a refrigerator (Consul CHB53C®, Salvador, Brazil) for 12 h at 4 °C. Fillets (5 ± 1) approximately 2 cm long, 1 cm wide and 1 cm high were cut from the meat to be evaluated for Warner-Bratzler shear force (WBSF). The instrumental texture analysis was performed on a TAXT2 texturometer (Stable Micro Systems Ltd., Vienna Court, UK) at 200 mm/min using standard shear blades (1.016 mm thick with a 3.05-mm blade). The instrumental texture analysis was performed according to the Research Center for Meat (US Meat Animal Research Center) and Shackelford et al. [17].

To evaluate lipid oxidation, meat samples of *LL* muscle stored under fast freezing at –20 °C for three months were thawed and crushed. Using the aqueous acid extraction method described by Cherian et al. [18], the 2-thiobarbituric acid reactive substances (TBARS) were measured in mg of malondialdehyde (MDA)/g of tissue.

Meat samples of *LT* muscle were evaluated for moisture, ash and protein contents, following method numbers 930.15, 920.153 and 928.08, respectively [19]. *LT* muscle samples were used to extract and quantify intramuscular fat (IMF). The fat of meat samples was isolated and purified using polar solvents (chloroform and methanol) according to the procedure of Folch et al. [20]. Aliquots of the fat extract were reserved and stored at −20 °C for subsequent use in determining the fatty acid profile.

Fatty acid profile
To determine the fatty acid profile, the fat samples previously extracted from *LT* muscle were converted to fatty acid methyl esters (FAMEs). The FAMEs were prepared using a solution of methanol, ammonium chloride and sulfuric acid, following the procedure described by Hartman and Lago [21].

Samples were analyzed using a chromatograph (GC2010, Shimadzu®, São Paulo, Brazil) equipped with a flame-ionization detector and a biscyanopropyl polydi-methylsiloxane capillary column of stationary phase (SP2560, 100 m × 0.25 mm, d_f 0.20 µm; Supelco®, Bellefonte, PA, USA). The column oven temperature was as follows: the initial temperature was held for 80 °C, increased at 11 °C/min to 180 °C and at 5 °C/min to 220 °C and then maintained for 19 min. Hydrogen was used as a carrier gas at a flow rate of 1.5 mL/min, the split ratio was 1:30, and the injector and detector temperatures were 220 °C. The FAMEs were identified by a comparison of the FAME retention times with those of authentic standards (FAME mix components, Supelco®, Bellefont, PA, USA) following the same injection method. The results were quantified by normalizing the areas of the methyl esters and converted to mg/100 g of meat using a conversion factor of 0.92 for the contribution of fatty acids in lipids [22].

The concentrations of saturated fatty acids (SFAs), unsaturated fatty acids (UFAs), monounsaturated fatty acids (MUFAs), polyunsaturated fatty acids (PUFAs), ω6 and ω3 were calculated based on the fatty acid profile of the meat. Lipid quality indexes were determined using the sum of the desirable fatty acids [23], the thrombo-genicity index (TI), the atherogenicity index (AI) [24], and the ratio between fatty acids hypocholesterolemic acid and hypercholesterolemic acid (h/H) [25]. The activity of enzymes involved in lipid metabolism, such as Δ9 desaturase in C16, Δ9 desaturase in C18 and elon-gase, were calculated according to the methods of Malau-Aduli et al. [26].

Feed chemical analysis
To determine the chemical composition of the feed, tripli-cate samples were dried at 55 °C for 72 h in a forced-air oven, ground with a Willey mill (Tecnal®, São Paulo, Brazil) with a 1-mm sieve, and stored in airtight plastic containers (ASS®, São Paulo, Brazil). The samples were then stored in plastic jars with lids (ASS®, São Paulo, Brazil), labeled, and subjected to further laboratory ana-lysis to measure the contents of dry matter (DM method

967.03), ash (method 942.05), crude protein (CP method 981.10), and ether extract (EE method 920.29) according to the Association of Official Analytical Chemists (AOAC) [27].

The neutral detergent fiber (NDF) content was deter-mined as described by Van Soest et al. [28]. The acid detergent fiber (ADF) contents were determined as described by Robertson and Van Soest [29]. The NDF residue was incinerated in an oven at 600 °C for 4 h to determine the ash content, and the protein concentra-tion was calculated by subtracting the neutral detergent insoluble protein (NDIP). NDF was corrected for the ash and protein contents. The Non-fiber carbohydrate (NFC) content was measured according to Mertens [30] and calculated based on the differences in the equation NFC = 100 − NDF − CP − EE − ash.

Statistical analyses
Variables were subjected to analysis of variance using the GLM procedure of Statistical Analysis System - SAS® software [31] and the following equation: $Y_{ijk} = \mu + S_i + R_j + S_i \times R_j + \varepsilon_{ijk}$, where Y_{ijk} is the dependent or response variable measured in the animal or experimental unit "k" of sexual class "i" at FR "j"; µ is the population mean or global constant; S_i is the effect of sexual class "i"; Rj is the effect of FR "j"; $S_i \times R_j$ is the interaction between effects of sexual class "i" and FR "j"; and ε_{ijk} is unobserved random error. Tukey-Kramer's test was used to compare the means with a significance level of 5% probability ($P < 0.05$), and the same criterion was adopted for interactions between the effects of sex and FR.

Results
Performance and carcass traits
There was an interaction ($P < 0.05$) between sex and FR for ADG, BWS and EBW (Table 2). In sum, females sub-jected to AL intake presented similar ADG, BWS and EBW ($P > 0.05$) to those of entire males and castrated males fed 30% FR (Table 3). Entire males fed AL pre-sented higher ($P < 0.05$) ADG, BWS and EBW due to their higher growth (Table 3, Fig. 1).

Except for SFT, the carcass traits that were analyzed (HCW, HCY, CCW, CCY and REA) were influenced ($P < 0.05$) by sex (Table 4). Entire males showed higher means of HCW and CCW followed by castrated males and females. However, females and castrated males had a higher HCY than did entire males. After cooling and con-sidering the losses caused by this process, only females had the highest yield (CCY), whereas castrated males did not differ from the other sexes. The level of FR did not influence ($P > 0.05$) the HCY and CCY, which indicated that the lower weights due to lower feed intake occurred proportionately throughout the bodies of the animals. HCW, CCW and SFT decreased with increasing FR (60%).

Table 2 Performance of Morada Nova lambs of different sexes subjected to feed restrictions

Variables	Sexes			Feed restrictions			SEM	P-value		
	Ent	Cas	Fem	AL	30%	60%		Sex	Res	Sex × Res
IBW, kg	14.4	14.7	15.5	14.6	14.6	14.4	0.158	0.711	0.860	0.348
DMI, kg	0.66[a]	0.60[b]	0.52[c]	0.80[a]	0.62[b]	0.36[c]	0.009	<.0001	<.0001	0.418
BWS, kg	27.4[a]	23.3[b]	19.9[c]	28.8[a]	24.4[b]	17.3[c]	0.308	<.0001	<.0001	0.047
EBW, kg	21.0[a]	18.1[b]	16.1[c]	23.0[a]	19.0[b]	13.3[c]	0.214	<.0001	<.0001	0.038
ADG, g	106[a]	71.7[b]	42.5[c]	116[a]	80.8[b]	23.3[c]	2.317	<.0001	<.0001	0.029

Ent Entire males, *Cas* Castrated males, *Fem* Females, *AL* ad libitum intake, *30%* 30% feed restriction, *60%* 60% feed restriction, *SEM* standard error of the mean, *Sex* sexes and Res = feed restriction, *IBW* initial body weight, *DMI* dry matter intake, *BWS* body weight at slaughter, *EBW* empty body weight, *TWG* total weight gain, *ADG* average daily gain
[a b c] Means followed by different letters differ between sexes according to a Tukey-Kramer test (P < 0.05)
[a b c] Means followed by different letters differ between feed restrictions according to a Tukey-Kramer test (P < 0.05)

There was an interaction (P < 0.05) between sex and FR in neck weight (Table 4). However, no clear result was evidenced (Table 5). The weights of all commercial cuts were influenced (P < 0.05) by sex and by FR (Table 4). Entire males had heavier cuts, followed by castrated males and females, which reflected the effects observed with the CCW. The weights of the commercial cuts decreased due to the reduction in feeding supply.

Physicochemical meat quality

L* was influenced (P < 0.05) by sex (Table 6). The color parameters a*, b* and C* were not affected by sex or FR. Castrated males and females did not differ (P > 0.05), and they had lower values (P < 0.05) than did entire males. Animals subjected to 60% FR showed a higher 24 h *post-mortem* pH in their meat compared to that of animals subjected to AL intake.

There was no effect (P > 0.05) of FR and sex on the protein content in meat from *LT* muscle (Table 6). However, an interaction (P < 0.05) between sex and FR for this variable was observed. Females fed AL had a higher (P < 0.05) protein content than did females submitted to 30% FR (Table 7). The moisture content in the *LT* muscle of animals with AL intake was lower (P < 0.05) compared to that of animals subjected to 30 and 60% FR (Table 6). Animals submitted to AL intake and 30% FR provided similar (P > 0.05) amounts of IMF, whereas animals subjected to 60% FR had a

reduced (P < 0.05) concentration of IMF. Ash percentage was higher (P < 0.05) in the meat from animals subjected to 60% FR.

Fatty acid profile

It was not possible to separate the peaks of conjugated linoleic acid (CLA) isomers normally identified in meats from ruminants. Thus, the nomenclature used covered all isomers (CLA). There was an interaction (P < 0.05) between sex and FR for elaidic acid (C18:1 t9) and behenic acid (C22:0) in the meat of the *LT* muscle (Table 8). However, after adjusting for multiple comparisons, a clear interaction response was only detected for elaidic acid (Table 9). Probably data set characteristics contributed to the absence of significance after the Tukey-Kramer test, contradicting the initial result of the ANOVA for the behenic acid. Females submitted to 60% FR had higher amounts of elaidic acid than did entire males and castrated males submitted to 60% FR (Table 9).

There was an effect (P < 0.05) of sex only on the concentrations of linoleic acid (C18:2 c9c12) and arachidonic acid (C20:4 c5c8c11c14) (Table 8). Females showed higher (P < 0.05) concentrations of these fatty acids than did entire males and castrated males, which showed no difference (P > 0.05) compared to the other categories. Meat of animals submitted to AL intake and 30% FR showed similar (P > 0.05) and higher (P < 0.05) values compared to those of animals subjected to 60%

Table 3 Interactions between sexes and feed restriction on performance of Morada Nova lambs

Variables	Entire male			Castrated male			Female		
	AL	30%	60%	AL	30%	60%	AL	30%	60%
BWS	35[adA]	27[bdBC]	20[cdEF]	28[aeB]	25[bdBCD]	17[ceFG]	23[afCDE]	21[bdeDEF]	15[bfG]
EBW	27[adA]	21[bdBC]	15[cdEF]	22[aeB]	19[beCD]	13[cdEF]	20[aeBCD]	17[bfDE]	12[ceG]
ADG	164[adA]	110[bdBC]	47[cdDE]	114[aeB]	78[beCD]	21[ceEF]	72[afD]	54[afD]	5[bfF]

AL ad libitum intake, *30%* 30% feeding restriction, *60%* 60% feeding restriction. *BWS* body weight at slaughter (kg), *EBW* empty body weight (kg), *ADG* average daily gain (g)
[abc]; Means followed by different letters in same sexes differ by Tukey-Kramer test (P < 0.05)
[def]; Means followed by different letters in same feeding restriction level differ by Tukey-Kramer test (P < 0.05)
[ABCDEFG]; Means followed by different capital letters in same line differ by Tukey-Kramer test (P < 0.05)

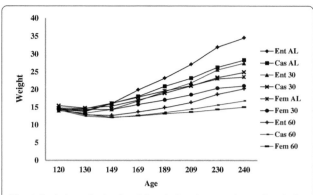

Fig. 1 Evolution of animal weights during the experimental period. Average animal weight (kg) of each treatment in relation to approximate age in days from the beginning (120 d) to the end of the experiment (240 d); Ent AL = Entire males subjected to ad libitum intake; Cas AL = Castrated males subjected to ad libitum intake; Fem AL = Females subjected to ad libitum intake; Ent 30 = Entire males subjected to 30% feed restriction; Cas 30 = Castrated males subjected to 30% feed restriction; Fem 30 = Females subjected to 30% feed restriction; Ent 60 = Entire males subjected to 60% feed restriction; Cas 60 = Castrated males subjected to 60% feed restriction; Fem 60 = Females subjected to 60% feed restriction

FR for concentrations of palmitic acid (C16:0), palmitoleic acid (C16:1c9), stearic acid (C18:0), oleic acid (C18:1c9) and CLA. The concentration of myristic acid (C14:0) was higher ($P < 0.05$) in meat from animals subjected to 30% FR than in meat from animals subjected to 60% FR. Meat from animals subjected to 60% FR showed greater ($P < 0.05$) concentrations of arachidonic acid and eicosapentaenoic acid (C20:5c5c8c11c14c17 - EPA) than did meat from animals subjected to AL intake and 30% FR.

The concentration of PUFAs was greater in the meat of females and lower in the meat of entire males (Table 10). However, the meat of females provided a higher ω6/ω3 ratio and AI in addition to presenting a lower h/H ratio. The activity of the elongase enzyme was higher in the *LT* muscle of castrated males. The sum of SFA, UFA and PUFA was similar ($P > 0.05$) in the meat of animals subjected to AL intake and 30% FR and was higher ($P < 0.05$) compared to that in animals subjected to 60% FR. The meat of animals subjected to 60% FR provided a greater ω6/ω3 ratio, lower h/H ratio and lower concentration of desirable fatty acids, in addition to a greater AI and TI.

Discussion

Performance, carcass traits and physicochemical meat quality

Productions systems in semi-arid regions are based on the use of genetic resources with high adaptability and heat tolerance, which are heavily influenced by qualitative and quantitative seasonality of food [32, 33]. An example is the Morada Nova, an important indigenous breed of hair sheep in northeastern Brazil that is used for meat and skin production and is highly valued on the international market [34]. Weight loss has a strong impact on animal productivity [35], compromising the animal welfare and income of farmers worldwide [33]. In our study, we observed an absence of growth in females with 60% FR and a low ADG in lambs with 30% FR. This effect is a response to a lower amount of nutrients [36] because lambs have higher energy and protein requirements for growth [32, 37], which demand higher intakes.

Table 4 Carcass characteristics and commercial cuts weight of Morada Nova lambs of different sexes subjected to feed restriction

Variables	Sexes			Feed restrictions			SEM	P-value		
	Ent	Cas	Fem	AL	30%	60%		Sex	Res	Sex × Res
HCW, kg	11.5[a]	10.3[b]	8.99[c]	12.7[a]	10.6[b]	7.51[c]	0.139	<.0001	<.0001	0.110
HCY, %	42.1[b]	44.1[a]	45.2[a]	44.4	43.6	43.4	0.325	0.002	0.474	0.967
CCW, kg	11.4[a]	10.18[b]	8.88[c]	12.58[a]	10.51[b]	7.42[c]	0.138	<.0001	<.0001	0.106
CCY, %	41.7[b]	43.6[ab]	44.7[a]	43.9	43.2	42.9	0.315	0.002	0.435	0.965
SFT, mm	1.32	1.10	1.12	1.55[a]	1.25[a]	0.75[b]	0.060	0.283	<.0001	0.185
REA, cm²	11.0[a]	10.3[ab]	9.36[b]	11.7[a]	10.1[b]	8.88[b]	0.205	0.010	<.0001	0.649
Leg, kg	1.84[a]	1.64[b]	1.48[c]	2.02[a]	1.69[b]	1.26[c]	0.022	<.0001	<.0001	0.109
Loin, kg	0.58[a]	0.48[b]	0.42[b]	0.65[a]	0.50[b]	0.32[c]	0.011	<.0001	<.0001	0.102
Neck, kg	0.40[a]	0.35[b]	0.27[c]	0.40[a]	0.33[b]	0.28[c]	0.007	<.0001	<.0001	0.010
Shoulder, kg	1.01[a]	0.93[b]	0.83[c]	1.12[a]	0.93[b]	0.70[c]	0.011	<.0001	<.0001	0.383
Rib, kg	1.07[a]	0.86[b]	0.78[b]	1.11[a]	0.98[a]	0.62[b]	0.024	0.0002	<.0001	0.329
Lower rib, kg	0.85[a]	0.79[a]	0.63[b]	0.96[a]	0.80[b]	0.51[c]	0.019	0.0002	<.0001	0.485

Ent Entire males, *Cas* Castrated males, *Fem* Females, *AL* ad libitum intake, *30%* 30% feed restriction, *60%* 60% feed restriction, *SEM* standard error of the mean, *Sex* sexes and Res = feed restriction, *HCW* hot carcass weight, *HCY* Hot carcass yield, *CCW* cold carcass weight, *CCY* cold carcass yield, *SFT* subcutaneous fat thickness, *REA* Rib eye area
[a b c] Means followed by different letters differ between sexes according to a Tukey-Kramer test ($P < 0.05$)
[a b c] Means followed by different letters differ between feed restrictions according to a Tukey-Kramer test ($P < 0.05$)

Table 5 Interactions between sexes and feed restriction on the weight of commercial cuts of Morada Nova lambs

Variables	Entire male			Castrated male			Female		
	AL	30%	60%	AL	30%	60%	AL	30%	60%
Neck	0.51[adA]	0.36[bBC]	0.33[bdBC]	0.42[adAB]	0.33[bBC]	0.29[bdCD]	0.28[abeCD]	0.32[aBCD]	0.21[beD]

AL ad libitum intake, *30%* 30% feeding restriction, *60%* 60% feeding restriction
[abc]; Means followed by different letters in same sexes differ by a Tukey-Kramer's test ($P < 0.05$)
[def]; Means followed by different letters in same feeding restriction level differ by a Tukey-Kramer test ($P < 0.05$)
[ABCD]; Means followed by different capital letters in same line differ by a Tukey-Kramer test ($P < 0.05$)

The study of van Harten et al. [38] showed that in animals under FR, lipids are mobilized and transformed into energy. This occurs to meet the net energy requirements for maintenance. In addition to genetic and nutritional factors, sex is variable and impacts animal productive responses.

Effects of interactions between sex and FR showed that females performed similarly to entire males and castrated males with 30% FR. Such effects reflect the empty cup weight gain, when females generally present lower rates of protein deposition in the empty body [39].

A sex effect is evident in the regulation of adiposity and muscularity and has been attributed to sexual steroid hormones [40] because testosterone promotes an increase in body mass [41]. These effects were observed in this study, in which entire males had higher carcass weights and commercial cuts. Castrated males [6] and females [42] present higher HCY attributed to increased fat deposit during weight gain [43]. Furthermore, the higher central or intra-abdominal accumulation of fat in male individuals [44] contributed to greater proportions of non-carcass components and consequently had a lower carcass yield.

Dietary restrictions have reduced the accumulation of the body stores of fat and protein, which resulted in lighter carcasses and commercial cuts. Nutritional limitations reduce cell proliferation and differentiation in tissues in response to reduced local production of insulin-like growth factor-1 (IGF-1), as signaled by the state of relative resistance of growth hormone (GH) [45]. In addition, in the post-absorptive state, non-esterified fatty acids, glycerol, alanine and glycine are oxidized, which supplies part of the energy demand [46]. Thus, in situations of lower nutritional intake, some of the body energy reserves are consumed. This situation was observed in animals submitted to the levels of FR used in this research.

The 24 h *post-mortem* pH was higher in the meat from animals subjected to 60% FR, which may be related to the lower content of muscle glycogen caused by lower feed intake and intense mobilization of reserves during the development of these animals. However, the pH remained between 5.5 and 5.8, which is desirable for meat [47]. An increase in the pH of meat can increase

Table 6 Physicochemical quality of Morada Nova lamb meat of different sexes subjected to feed restriction

Variables	Sexes			Feed restrictions			SEM	P-value		
	Ent	Cas	Fem	AL	30%	60%		Sex	Res	Sex × Res
L*	39.7[a]	38.2[b]	37.5[b]	38.2	38.4	38.8	0.244	0.003	0.657	0.659
a*	22.2	21.8	21.9	21.9	22.3	21.7	0.233	0.767	0.541	0.689
b*	9.83	9.43	8.80	9.36	9.93	8.77	0.220	0.197	0.126	0.866
C*	24.3	23.8	23.6	23.8	24.4	23.4	0.320	0.386	0.252	0.771
WHC, %	35.9	34.5	35.0	34.7	32.8	37.9	1.045	0.859	0.154	0.511
CWL, %	38.6	38.5	37.4	37.4	38.5	38.6	0.403	0.443	0.408	0.937
pH	5.61	5.60	5.66	5.57[b]	5.62[ab]	5.69[a]	0.019	0.354	0.044	0.667
Shear force, N	42.9	46.1	39.2	44.0	41.4	42.8	1.623	0.236	0.808	0.702
TBARS	0.72	0.75	0.85	0.82	0.64	0.86	0.041	0.395	0.073	0.534
Moisture, %	76.5	75.9	75.8	75.4[b]	75.9[a]	76.9[a]	0.146	0.167	0.001	0.064
Protein, %	18.6	18.9	19.0	18.9	18.5	19.1	0.117	0.332	0.139	0.018
Fat, %	3.99	4.32	4.18	4.74[a]	4.70[a]	3.05[b]	0.107	0.468	<.0001	0.680
Ash, %	0.92	0.90	0.93	0.92[b]	0.87[b]	0.98[a]	0.009	0.327	<.0001	0.155

Ent Entire males, *Cas* Castrated males, *Fem* Females, *AL* ad libitum intake, *30%* 30% feed restriction, *60%* 60% feed restriction, *SEM* standard error of the mean, *Sex* sexes and Res = feed restriction, *Color parameter L** lightness, 0 black and 100 white, *Color parameter a** redness, ranges from green (−) to red (+), *Color parameter b** yellowness, ranges from blue (−) to yellow (+), *Color parameter C** chroma; C* = (a*2 + b*2)1/2(MacDougall and Taylor, 1975), *WHC* water holding capacity, *CWL* Cooking weight losses, *TBARS* 2-thiobarbituric acid reactive substance, *MDA* malondialdehyde
[a b] Means followed by different letters differ between sexes according to a Tukey-Kramer test ($P < 0.05$)
[a b] Means followed by different letters differ between feed restrictions according to a Tukey-Kramer test ($P < 0.05$)

Table 7 Interactions between sexes and feed restriction on Meat composition of Morada Nova lambs

Variables	Entire male			Castrated male			Female		
	AL	30%	60%	AL	30%	60%	AL	30%	60%
Protein	18.24[AB]	18.70[AB]	18.89[AB]	18.57[AB]	18.66[AB]	19.46[AB]	19.99[A]	18.22[B]	18.93[AB]

AL ad libitum intake, *30%* 30% feeding restriction, *60%* 60% feeding restriction
[abc]; Means followed by different letters in same sex differ by a Tukey-Kramer test (*P* < 0.05)
[def]; Means followed by different letters in same feeding restriction level differ by a Tukey-Kramer test (*P* < 0.05)
[AB]; Means followed by different capital letters in same line differ by a Tukey-Kramer test (*P* < 0.05)

the activity of cytochrome oxidase by reducing the uptake of oxygen by myoglobin, which results in a purplish red color [48]. However, the higher pH in meat from animals subjected to 60% FR was not enough to influence the color and other quality parameters analyzed in this study.

L* was greatly influenced by the pigment contents, especially those of hematin, myoglobin and their forms [49]. Chromophores, such as myoglobin and hemoglobin, absorb visible light by increasing light penetration and consequently decreasing reflectance [50]. Sañudo et al. [51] observed more myoglobin in females (2.90 mg/g of meat) than in entire males

(2.56 mg/g of meat) and reported a similar effect of sex on L*, recording 39.80% for females and 41.26% for entire males. More myoglobin may explain the color variation in the meat of female animals observed in the present study.

Testosterone, besides providing an anabolic effect [41], acts on plasma glucose levels in males but does not alter the phosphorylation of AMP-activated protein kinase in muscle [40], which influences the concentration of IMF in males to a small degree. A similar result was observed in this study because the percentage of meat fat was similar between the sexes and did not reflect the effects of higher HCW in entire males.

Table 8 Fatty acids profile in the meat of the *longissimus thoracis* muscle of Morada Nova lambs of different sexes subjected to feed restriction

Fatty acids, mg/100 g of meat	Sexes			Feed restrictions			SEM	P-value		
	Ent	Cas	Fem	AL	30%	60%		Sex	Res	Sex × Res
C10:0	4.70	6.56	6.05	7.18	5.24	4.89	0.700	0.540	0.392	0.913
C12:0	32.8	59.7	43.6	44.9	35.1	56.1	6.476	0.254	0.432	0.648
C14:0	87.3	89.8	108	96.0[ab]	110[a]	78.5[b]	4.568	0.156	0.027	0.937
C14:1c9	8.51	10.3	9.95	10.9	9.64	8.26	0.610	0.447	0.255	0.436
C15:0	22.6	29.3	33.0	23.6	26.72	34.6	1.984	0.122	0.084	0.572
C16:0	912	914	966	1069[a]	1080[a]	642[b]	30.20	0.711	<.0001	0.816
C16:1c9	41.2	40.3	43.8	49.0[a]	48.0[a]	18.3[b]	1.660	0.685	<.0001	0.551
C17:0	44.4	47.2	48.6	48.9	46. 9	44.4	1.047	0.277	0.245	0.314
C17:1c10	18.9	16.8	17.3	18.9	17.37	16.8	0.901	0.635	0.606	0.300
C18:0	620	687	649	695[a]	701[a]	561[b]	19.03	0.377	0.008	0.770
C18:1c9	1474	1639	1474	186[a]	1814[a]	911[b]	41.65	0.192	<.0001	0.604
C18:1 t9	36.0	36.1	44.5	41.1	38.4	37.1	1.819	0.106	0.666	0.032
C18:2c9c12	70.1[b]	83.1[ab]	91.4[a]	82.8	79.1	82.8	2.388	0.005	0.768	0.582
CLA	3.96	4.92	4.54	5.29[a]	4.99[a]	3.14[b]	0.289	0.410	0.016	0.397
C18:3c9c12c15	18.5	21.6	21.0	22.2	20.5	18.4	0.701	0.201	0.109	0.510
C20:4c5c8c11c14	31.2[b]	37.3[ab]	39.2[a]	33.0	33.2[b]	41.6[a]	1.236	0.045	0.010	0.862
C20:5c5c8c11c14c17	7.57	8.73	8.75	7.74[b]	7.39[b]	9.92[a]	0.216	0.061	<.0001	0.171
C22:0	8.08	9.13	9.02	8.24	8.51	9.48	0.253	0.201	0.143	0.042
Unidentified	550	583	565	618	617	474	-	-	-	-

Ent Entire males, *Cas* Castrated males, *Fem* Females, *AL* ad libitum intake, *30%* 30% feed restriction, *60%* 60% feed restriction, *SEM* standard error of the mean, *Sex* sexes and Res = feed restriction, *C10:0* Capric acid, *C12:0* Lauric acid, *C14:0* Myristic acid, *C14:1c9* Myristoleic acid, *C15:0* Pentadecylic acid, *C16:0* Palmitic acid, *C16:1c9* Palmitoleic acid, *C17:0* Margaric acid, *C17:1c10* Heptadecenoic acid, *C18:0* stearic acid, *C18:1c9* Oleic acid, *C18:1 t9* Elaidic acid, *C18:2c9c12* Linoleic acid, *C18:3c9c12c15* α-Linolenic acid, *C20:4c5c8c11c14* Arachidonic acid, *C20:5c5c8c11c14c17* Eicosapentaenoic acid, *C22:0* Behenic acid, *CLA* Conjugated linoleic acids: rumenic acid and their isomers
[a b] Means followed by different letters differ between feed restrictions according to a Tukey-Kramer test (*P* < 0.05)

Table 9 Interactions between sexes and feed restriction on fatty acids profile in the meat of the *longissimus thoracis* muscle of Morada Nova lambs

Fatty acids, mg/ 100 g of meat	Entire male			Castrated male			Female		
	AL	30%	60%	AL	30%	60%	AL	30%	60%
C18:1 t9	38.6[AB]	41.8[AB]	28.6[B]	45.5[AB]	34.6[AB]	28.1[B]	39.2[AB]	39.9[AB]	54.5[A]
C22:0	8.2	8.6	7.4	9.1	8.0	10.2	7.3	8.9	10.8

AL ad libitum intake, *30%* 30% feeding restriction, *60%* 60% feeding restriction
[AB]; Means followed by different capital letters in same line differ by Tukey-Kramer test ($P < 0.05$)

Variations in meat fat concentration occur mainly due to changes in balance between dietary energy and nutrient requirements [52]; thus, the least amount of energy consumed by animals subjected to 60% FR decreased the deposition of IMF. FR of 60% meets the net energy needs for maintenance and obviously does not prioritize nutrients for the deposition of adipose tissue. However, 30% FR did not cause significant changes to this balance or to the concentration of IMF in *LT* muscle.

Fatty acid profiles

Studies on cattle have indicated increased incorporation of long chain fatty acids into the phospholipids of heifer meat in response to concentrations of plasmalogens [53]. The effects observed on the concentration of linoleic and arachidonic acids in meat of females indicate that this incorporation can also occur in lamb meat but was not observed due to the need for a more detailed analysis. The effect of the interaction between sex and FR on elaidic acid concentration shows that under 60% FR, females accumulate a greater amount of this acid. It becomes important to know the unique physiological functions of specific isomers as well as their origin. A portion of the trans-11 C18:1 isomer produced by ruminal microbes is converted into cis-9, trans-11 C18: 2 by tissue desaturase [54]; however, this cannot occur with

Table 10 Fatty acid classes, ratios, indexes and enzyme activity in meat of *longissimus thoracis* muscle of Morada Nova lambs of different sexes subjected to feed restriction

Index, mg/100 g of meat	Sexes			Feed restrictions			SEM	P-value		
	Ent	Cas	Fem	AL	30%	60%		Sex	Res	Sex × Res
ΣSFA	1731	1837	1853	1988[a]	2007[a]	1425[b]	50.12	0.579	<.0001	0.770
ΣUFA	1707	1896	1748	2131[a]	2069[a]	1151[b]	44.52	0.210	<.0001	0.602
ΣMUFA	1578	1741	1585	1981[a]	1926[a]	997[b]	44.10	0.248	<.0001	0.569
ΣPUFA	129[b]	155[ab]	163[a]	150	142	154	4.222	0.008	0.504	0.716
Σω6	99.0[b]	120[a]	131[a]	116	110	124	3.338	0.003	0.212	0.777
Σω3	26.10	30.3	28.8	29.9	27.9	27.4	0.909	0.188	0.515	0.628
ω6/ω3	3.84[b]	3.99[b]	4.53[a]	3.90[b]	3.96[b]	4.50[a]	0.076	0.003	0.009	0.784
h/H	1.60[ab]	1.77[a]	1.54[b]	1.74[a]	1.65[ab]	1.52[b]	0.027	0.008	0.015	0.661
DFA	2328	2583	2397	2826[a]	2769[a]	1713[b]	56.65	0.185	<.0001	0.661
TI	1.78	1.70	1.81	1.64[b]	1.72[b]	1.93[a]	0.031	0.351	0.003	0.741
AI	0.75[b]	0.72[b]	0.84[a]	0.70[b]	0.74[b]	0.87[a]	0.016	0.020	<.0001	0.706
Δ^9 desaturase C16	4.26	4.17	4.28	4.36	4.28	4.07	0.124	0.925	0.631	0.872
Δ^9 desaturase C18	69.8	69.0	67.9	72.6[a]	72.1[a]	61.9[b]	0.556	0.428	<.0001	0.431
Elongase	68.9[b]	70.9[a]	67.4[b]	69.6	69.1	68.4	0.273	<.0001	0.240	0.179

Ent Entire males, *Cas* Castrated males, *Fem* Females, *AL* ad libitum intake, *30%* 30% feed restriction, *60%* 60% feed restriction, *SEM* standard error of the mean, *Sex* sexes and Res = feed restriction
SFA saturated fatty acids (ΣC10:0, C12:0, C14:0, C15:0, C16:0, C17:0, C18:0, C22:0); *UFA* unsaturated fatty acid (ΣC14:1c9, C16:1c9, C17:1c9, C18:1c9, C18:1 t9, C18:2c9c12, C18:2t9t12, C18:2c9t11, C18:3c6c9c12, C20:4c5c8c11c14, C20:5c5c8c11c14c17), *MUFA* monounsaturated fatty acid (ΣC14:1c9, C16:1c9, C17:1c9, C18:1c9, C18:1 t9), *PUFA* polyunsaturated fatty acid (ΣC18:2c9c12, C18:2t9t12, C18:2c9t11, C18:3c6c9c12, C20:4c5c8c11c14, C20:5c5c8c11c14c17), *ω6* (ΣC18:2c9c12, C18:2t9t12, C20:4c5c8c11c14), *ω3* (C18:3c9c12c15, C20:5c5c8c11c14c17), *ω6:ω3* relation (Σω6/Σω3), *h/H* hypocholesterolemic hypercholesterolemic relation (ΣC18:1c9, C18:2c9c12, C18:3c6c9c12, C20:4c5c8c11c14, C20:5c5c8c11c14c17 / ΣC14:0, C16:0), *DFA* Desirable fatty acids (ΣMUFA, PUFA, C18:0), *TI* Thrombogenicity index [(ΣC14:0, C16:0, C18:0) / Σ(0,5xΣMUFA), (0,5xΣω6), (3xΣω3), (Σω3/Σω6)], *AI* Atherogenicity index {[(ΣC12:0,(4xC14:0), C16:0] / [(Σω3,ω6) + (C18:1c9) + (Σ other MUFA)]}; Enzyme activity Δ^9 desaturase on fatty acid with 16 carbons {100[(C16:1c1) / (ΣC16:1c1, C16:0)]}; Enzyme activity Δ^9 desaturase on fatty acid with 18 carbons {100[(C18:1c1) / (ΣC18:1c1, C18:0)]}; Elongase activity enzyme {100[(ΣC18:0,C18:1c9) / (ΣC16:0,C16:1c9,C18:0,C18:1c9)]}
[a] [b] Means followed by different letters differ between sexes according to a Tukey-Kramer test ($P < 0.05$)
[a] [b] Means followed by different letters differ between feed restrictions according to a Tukey-Kramer test ($P < 0.05$)

Effects of quantitative feed restriction and sex on carcass traits, meat quality and meat lipid profile...

59

elaidic acid (C18:1 t9). The quality lipid indexes showed that the meat of female animals had a lipid profile with less desirable characteristics compared to that of meat from other sexes due to a higher AI and a lower h/H ratio. The higher ω6/ω3 ratio would also indicate lower quality meat fat in females [55]; however, the latest recommendations suggest no rational limit for this ratio if the intake of ω6 and ω3 is within the proper range for human diets [56].

More feeding can explain the higher concentration of palmitic acid, palmitoleic acid, stearic acid and oleic acid in the meat of animals subjected to AL intake and 30% FR. These effects reflect the results observed with the IMF content. As 30% FR did not significantly influence the deposition of IMF, there was no effect on the composition of fatty acids. The kinetics of the feed in the rumen may also have influenced the exposure time of the fatty acids to biohydrogenation [57]. The more severe restriction may have resulted in a lower passage rate, higher biohydrogenation, and higher deposition of SFA. Furthermore, the incorporation of fatty acids synthesized in muscle tissue may have been more effective in animals subjected to AL intake and 30% FR. This is related to a more lipogenic substrate for *de novo* synthesis in muscle adipocytes [58, 59], especially glucose from the propionate originating from the fermentation of carbohydrates in the rumen [60].

In our study, the lowest IMF deposit was in the meat of animals subjected to 60% FR, which justified the lower concentration of CLA in meat, as CLA is preferentially deposited in triglycerides [61, 62]. Similarly, the lowest IMF deposits in the meat of animals subjected to 60% FR may explain the higher concentration of long chain PUFAs (EPA and AA) in meat, which are deposited primarily in phospholipids [63, 64]. The IMF consists of triglycerides deposited in adipocytes and myofibril cytoplasm droplets, structural phospholipids and cholesterol present in membranes [58]. Triglycerides are more mobile, and phospholipids are more stable in muscle [65].

Based on the results of this study, lambs subjected to 60% FR can be expected to produce meat with fatty acid concentrations that are less favorable to consumer health and with a lower amount of desirable fatty acids, a lower h/H ratio, a higher AI and TI and a higher proportion of SFA (55.3%) compared to those of animals submitted to AL intake (48.3%) and 30% FR (49.2%). A lipid profile favorable to the thrombogenicity and atherogenicity in the meat of animals subjected to 60% FR is related to myristic acid, palmitic acid and stearic acid concentrations [24]. Values from 0.9 to 1.94 for TI and 0.59 to 1.15 for AI have been reported in the literature [66–69]; the maximum values of TI (1.93) and AI (0.87) found in this study were within this range.

The activity of the elongase enzyme is related to concentrations of palmitic, palmitoleic and oleic acids [11, 70]. Combined concentrations of these fatty acids resulted in increased activity of elongase in the muscle of castrated male animals. The lower activity of the Δ9 desaturase enzyme C18 in animals subjected to 60% FR could be attributed to lower amounts of oleic acid present in the muscle of these animals [71].

Conclusions

Lambs in different sexes produced carcasses with different characteristics, and except for lightness, sex did not influence meat quality or chemical composition. However, females had a fatty acid profile in their meat that was less favorable to consumer health. FR affected carcass traits without influencing the quality of the meat. IMF content decreased when animals were subjected to 60% FR, but the lipid profile was less favorable to consumer health.

Abbreviations

a*: Redness; ADF: Acid detergent fiber; ADG: Average daily gain; AI: Atherogenicity index; AL: Ad libitum; b*: Yellowness; BW: Body weight; BWS: Body weight at slaughter; C*: Chroma; CCW: Cold carcass weight; CCY: Cold carcass yield; CP: Crude protein; CWL: Cooking weight losses; DFA: Desirable fatty acids; DM: Dry matter; DMI: Dry matter intake; EBW: Empty body weight; EE: Ether extract; FR: Feed restriction; HCW: Hot carcass weight; HCY: Hot carcass yield; IBW: Initial body weight; L*: Lightness; *LL*: *longissimus lumborum*; *LT*: *longissimus thoracis*; MDA: Malondialdehyde; MUFA: Monounsaturated fatty acid; NDF: Neutral detergent fiber; NFC: Non-fiber carbohydrate; NRC: National research council; PUFA: Polyunsaturated fatty acid; REA: Rib eye area; SAS: Statistical analysis software; SFA: Saturated fatty acid; SFT: Subcutaneous fat thickness; TBARS: 2-thiobarbituric acid reactive substances; TI: Thrombogenicity index; TWG: Total weight gain; UFA: Unsaturated fatty acid; WHC: Water holding capacity; ω3: Omega 3; ω6: Omega 6

Acknowledgements

The authors would like to thank Coordination for the Improvement of Higher Education Personnel (CAPES).

Funding

This research was supported by the National Council for Scientific and Technological Development (CNPq-Brazil).

Authors' contributions

TA and ES conceived the study and conducted statistical analysis. TA, ES, AC and MP conducted the laboratory analyses and prepared the manuscript. ES, AC and EH managed the animal model and assisted with manuscript preparation. IM and EH assisted with diet formulation, statistical analysis and manuscript preparation. HM contributed to fatty acid analysis. LB and RO critically revised the manuscript and edited the language. LP contributed to conception and statistical analyses. All authors read and approved the final manuscript.

Competing interests

The authors declare that they have no competing interests.

Author details

[1]Department of Animal Science, Federal University of Ceara, Fortaleza 60356001, Ceara, Brazil. [2]Department of Animal Science, State University of Londrina, Londrina 86051990, Paraná, Brazil. [3]Laboratory of Sensory Analysis, Agency for Agricultural Research (EMBRAPA - Tropical Agroindustry), Fortaleza 60511110, Ceará, Brazil. [4]Department of Animal Science, Campus Professora Cinobelina Elvas, Federal University of Piauí, Bom Jesus 64900000, Piaui, Brazil. [5]Department of Animal Science, School of Veterinary Medicine and Animal Science/Federal University of Bahia, Salvador City, Bahia State 40.170-110, Brazil.

References

1. Mlambo V, Mapiye C. Towards household food and nutrition security in semi-arid areas: What role for condensed tannin-rich ruminant feedstuffs? Food Res. 2015;76:953–61. Int. Elsevier Ltd.

2. Toro-Mujica P, Aguilar C, Vera R, Rivas J, García A. Sheep production systems in the semi-arid zone: Changes and simulated bio-economic performances in a case study in Central Chile. Livest Sci. 2015;180:209–19. Elsevier.

3. Selvaggi M, Laudadio V, Dario C, Tufarelli V. Investigating the genetic polymorphism of sheep milk proteins: a useful tool for dairy production. J Sci Food Agric. 2014;94:3090–9.

4. Pereira ES, Carmo ABR, Costa MRGF, Medeiros AN, Oliveira RL, Pinto AP, et al. Mineral requirements of hair sheep in tropical climates. J Anim Physiol Anim Nutr (Berl). 2016;100:1090–6.

5. Craigie CR, Lambe NR, Richardson RI, Haresign W, Maltin CA, Rehfeldt C, et al. The effect of sex on some carcass and meat quality traits in Texel ewe and ram lambs. Anim Prod Sci. 2012;52:601–7.

6. Sales J. Quantification of the effects of castration on carcass and meat quality of sheep by meta-analysis. Meat Sci. 2014;98:858–68. Elsevier Ltd.

7. Hopkins DL, Mortimer SI. Effect of genotype, gender and age on sheep meat quality and a case study illustrating integration of knowledge. Meat Sci. 2014;98:544–55. Elsevier B.V.

8. Neto SG, Bezerra LR, Medeiros AN, Ferreira MA, Filho ECP, Cândido EP, et al. Feed Restriction and Compensatory Growth in Guzerá Females. Asian Australas J Anim Sci. 2011;24:791–9.

9. Bezerra LR, Neto SG, de Medeiros AN, Mariz TM de A, Oliveira RL, Cândido EP, et al. Feed restriction followed by realimentation in prepubescent Zebu females. Trop Anim Health Prod. 2013;45:1161–9.

10. Madruga MS, Torres TS, Carvalho FF, Queiroga RC, Narain N, Garrutti D, et al. Meat quality of Moxotó and Canindé goats as affected by two levels of feeding. Meat Sci. 2008;80:1019–23.

11. Lopes LS, Martins SR, Chizzotti ML, Busato KC, Oliveira IM, Machado Neto OR, et al. Meat quality and fatty acid profile of Brazilian goats subjected to different nutritional treatments. Meat Sci. 2014;97:602–8. Elsevier Ltd.

12. NRC. Nutrient Requirement of Small Ruminants. Sheep, Goats, Cervids, and New World Camelids. Washington: The National Academies Press; 2007.

13. Cañeque V, Sañudo C. Metodología para el estudio de la calidad de la canal y de la carne en rumiantes. Madrid: Instituto Nacional de Investigación y Tecnologia y Alimenticia; 2001.

14. Miltenburg GA, Wensing T, Smulders FJ, Breukink HJ. Relationship between blood hemoglobin, plasma and tissue iron, muscle heme pigment, and carcass color of veal. J Anim Sci. 1992;70:2766–72.

15. MacDougall DB, Taylor AA. Colour retention in fresh meat stored in oxygen-a commercial scale trial. Int J Food Sci Technol. 2007;10:339–47.

16. AMSA. Guidlines for cooking and sensory evaluation of meat. Chicago: American Meat Science Association. National livestock and meat board; 1978.

17. Shackelford SD, Wheeler TL, Koohmaraie M. Evaluation of slice shear force as an objective method of assessing beef longissimus tenderness. J Anim Sci. 1999;77:2693.

18. Cherian G, Selvaraj RK, Goeger MP, Stitt PA. Muscle fatty acid composition and thiobarbituric acid-reactive substances of broilers fed different cultivars of sorghum. Poult Sci. 2002;81:1415–20.

19. AOAC. Official Analytical Methods of Analysis. 17th ed. Washington: Association of Official Agricultural Chemists; 2002.

20. Folch J, Lees M, Stanley G. A simple method for the isolation and purification of total lipides from animal tissues. J Biol Chem. 1957;226:497–509.

21. Hartman L, Lago RC. Rapid preparation of fatty acid methyl esters from lipids. Lab Pract London. 1973;22:475–6.

22. Weihrauch JL, Posati LP, Anderson BA, Exler J. Lipid conversion factors for calculating fatty acid contents of foods. J Am Oil Chem Soc. 1977;54:36–40.

23. Rhee KS. Fatty acids in meats and meat products. In: Chow CK, editor. Fat. Acids Foods Their Heal. Implic. 2nd ed. New York: Marcel Dekker; 2000.

24. Ulbricht TL, Southgate DA. Coronary heart disease: seven dietary factors. Lancet. 1991;338:985–92.

25. Santos-Silva J, Bessa RJ, Santos-Silva F. Effect of genotype, feeding system and slaughter weight on the quality of light lambs. Livest Prod Sci. 2002;77:187–94.

26. Malau-Aduli AEO, Siebert BD, Bottema CDK, Pitchford WS. A comparison of the fatty acid composition of triacylglycerols in adipose tissue from Limousin and Jersey cattle. Aust J Agric Res. 1997;48:715.

27. AOAC. Official methods of analysis. 15th ed. Washington: Association of Official Agricultural Chemists; 1990.

28. Van Soest PJ, Robertson JB, Lewis BA. Methods for Dietary Fiber, Neutral Detergent Fiber, and Nonstarch Polysaccharides in Relation to Animal Nutrition. J Dairy Sci. 1991;74:3583–97.

29. Robertson JB, Van Soest PJ. The detergent system of analysis and its application to human foods. New York: Marcel Dekker; 1981. p. 123. Anal. Diet. fiber food.

30. Mertens DR. Creating a System for Meeting the Fiber Requirements of Dairy Cows. J Dairy Sci. 1997;80:1463–81.

31. Sas I. SAS User's Guide: Basics. Cary: SAS Inst. Inc.; 2008.

32. Costa MRGF, Pereira ES, Silva AMA, Paulino PVR, Mizubuti IY, Pimentel PG, et al. Body composition and net energy and protein requirements of Morada Nova lambs. Small Ruminant Res. 2013;114:206–13. Elsevier B.V.

33. Cardoso LA, Almeida AM de. Seasonal weight loss-an assessment of losses and implications for animal welfare and production in the tropics: Southern Africa and Western Australia as case studies. Enhancing Anim. Welf. farmer income through Strateg. Anim. Feed. Some case Stud. Rome (Italy): FAO, Rome (Italy), Animal Production and Health Division; 2013. p. 37–44

34. McManus C, Paiva SR, De Araújo RO. Genetics and breeding of sheep in Brazil. Rev Bras Zootec. 2010;39:236–46.

35. Scanlon TT, Almeida AM, van Burgel A, Kilminster T, Milton J, Greeff JC, et al. Live weight parameters and feed intake in Dorper, Damara and Australian Merino lambs exposed to restricted feeding. Small Ruminant Res. 2013;109: 101–6. Elsevier B.V.

36. Almeida AM, Schwalbach LM, De Waal HO, Greyling JPC, Cardoso LA. The effect of supplementation on productive performance of Boer goat bucks fed winter veld hay. Tropl Anim Health Prod. 2006;38:443–9.

37. Pereira ES, Fontenele RM, Silva AMA, Oliveira RL, Ferreira MRG, Mizubuti IY, et al. Body composition and net energy requirements of Brazilian Somali lambs. Ital J Anim Sci. 2014;13:880–6.

38. van Harten S, Kilminster T, Scanlon T, Milton J, Oldham C, Greeff J, et al. Fatty acid composition of the ovine longissimus dorsi muscle: effect of feed restriction in three breeds of different origin. J Sci Food Agric. 2015;96:1777–82.

39. Almeida AK, Resende KT, Tedeschi LO, Fernandes MHMR, Regadas Filho JGL, Teixeira IAMA. Using body composition to determine weight at maturity of male and female Saanen goats. J Anim Sci. 2016;94:2564.

40. Clarke SD, Clarke IJ, Rao A, Cowley MA, Henry BA. Sex Differences in the Metabolic Effects of Testosterone in Sheep. Endocrinology. 2012;153:123–31.

41. Kelly DM, Jones TH. Testosterone: a metabolic hormone in health and disease. J Endocrinol. 2013;217:R25–45.

42. Santos VAC, Cabo A, Raposo P, Silva JA, Azevedo JMT, Silva SR. The effect of carcass weight and sex on carcass composition and meat quality of "Cordeiro Mirandês"—Protected designation of origin lambs. Small Rumin Res. 2015;130:136–40. Elsevier B.V.

43. Nurnberg K, Wegner J, Ender K. Factors influencing fat composition in muscle and adipose tissue of farm animals. Livest Prod Sci. 1998;56:145–56.

44. Shi H, Clegg DJ. Sex differences in the regulation of body weight. Physiol Behav. 2009;97:199–204. Elsevier Inc.

45. Breier BH, Gluckman PD. The regulation of postnatal growth: nutritional influences on endocrine pathways and function of the somatotrophic axis. Livest Prod Sci. 1991;27:77–94.

46. Kozloski GV. Bioquímica dos Ruminantes. Santa Maria: UFSM; 2002.

47. Hedrick HB, Aberle ED, Forrest JC, Judge MD, R.A. M. Principles of Meat Science. 3rd ed. Yowa: Kendall and Hunt; 1994

48. Osório JCDS, Osório MTM, Sañudo C. Características sensoriais da carne ovina. Rev Bras Zootec. 2009;38:292–300.

49. Lindahl G, Lundström K, Tornberg E. Contribution of pigment content, myoglobin forms and internal reflectance to the colour of pork loin and ham from pure breed pigs. Meat Sci. 2001;59:141–51.

50. Krzywicki K. Assessment of relative content of myoglobin, oxymyoglobin

and metmyoglobin at the surface of beef. Meat Sci. 1979;3:1–10.

51. Sañudo C, Sierra I, Olleta JL, Martin L, Campo MM, Santolaria P, et al. Influence of weaning on carcass quality fatty acid composition and meat quality in intensive lamb production systems. Anim Sci. 1998;66:175–88.

52. Geay Y, Bauchart D, Hocquette J-F, Culioli J. Effect of nutritional factors on biochemical, structural and metabolic characteristics of muscles in ruminants, consequences on dietetic value and sensorial qualities of meat. Reprod Nutr Dev. 2001;41:1–26.

53. Bessa RJB, Alves SP, Santos-Silva J. Constraints and potentials for the nutritional modulation of the fatty acid composition of ruminant meat. Eur J Lipid Sci Technol. 2015;117:1325–44.

54. Santora JE, Palmquist DL, Roehrig KL. Trans-vaccenic acid is desaturated to conjugated linoleic acid in mice. J Nutr. 2000;130:208–15.

55. Wood JD, Richardson R, Nute G, Fisher A, Campo M, Kasapidou E, et al. Effects of fatty acids on meat quality: a review. Meat Sci. 2003;66:21–32.

56. FAO. Fats and fatty acids in human nutrition. Report of an expert consultation. Rome: Food and agriculture organization of the united nations; 2010. FAO Food Nutr. Pap.

57. Harvatine KJ, Allen MS. Fat supplements affect fractional rates of ruminal fatty acid biohydrogenation and passage in dairy cows. J Nutr. 2006;136:677–85.

58. Hocquette JF, Gondret F, Baéza E, Médale F, Jurie C, Pethick DW. Intramuscular fat content in meat-producing animals: development, genetic and nutritional control, and identification of putative markers. Animal. 2010;4:303–19.

59. Alves SP, Bessa RJB, Quaresma MAG, Kilminster T, Scanlon T, Oldham C, et al. Does the Fat Tailed Damara Ovine Breed Have a Distinct Lipid Metabolism Leading to a High Concentration of Branched Chain Fatty Acids in Tissues? Schunck W-H, editor. PLoS One. 2013;8:e77313

60. Smith SB, Kawachi H, Choi CB, Choi CW, Wu G, Sawyer JE. Cellular regulation of bovine intramuscular adipose tissue development and composition. J Anim Sci. 2009;87:E72–82.

61. Noci F, French P, Monahan FJ, Moloney AP. The fatty acid composition of muscle fat and subcutaneous adipose tissue of grazing heifers supplemented with plant oil-enriched concentrates. J Anim Sci. 2007;85:1062.

62. Costa ASH, Silva MP, Alfaia CPM, Pires VMR, Fontes CMGA, Bessa RJB, et al. Genetic Background and Diet Impact Beef Fatty Acid Composition and Stearoyl-CoA Desaturase mRNA Expression. Lipids. 2013;48:369–81.

63. Noci F, Monahan FJ, French P, Moloney AP. The fatty acid composition of muscle fat and subcutaneous adipose tissue of pasture-fed beef heifers: Influence of the duration of grazing. J Anim Sci. 2005;83:1167–78.

64. Wood JD, Enser M, Fisher AV, Nute GR, Sheard PR, Richardson RI, et al. Fat deposition, fatty acid composition and meat quality: A review. Meat Sci. 2008;78:343–58.

65. Raes K, De Smet S, Demeyer D. Effect of dietary fatty acids on incorporation of long chain polyunsaturated fatty acids and conjugated linoleic acid in lamb, beef and pork meat: a review. Anim Feed Sci Technol. 2004;113:199–221.

66. Komprda T, Kuchtík J, Jarošová A, Dračková E, Zemánek L, Filipčík B. Meat quality characteristics of lambs of three organically raised breeds. Meat Sci. 2012;91:499–505.

67. Lestingi A, Facciolongo AM, Marzo DD, Nicastro F, Toteda F. The use of faba bean and sweet lupin seeds in fattening lamb feed. 2. Effects on meat quality and fatty acid composition. Small Ruminant Res. 2015;131:2–5. Elsevier B.V.

68. Liu J, Guo J, Wang F, Yue Y, Zhang W, Feng R, et al. Carcass and meat quality characteristics of Oula lambs in China. Small Ruminant Res. 2015;123: 251–9. Elsevier B.V.

69. Laudadio V, Tufarelli V. Influence of substituting dietary soybean meal for dehulled-micronized lupin (Lupinus albus cv. Multitalia) on early phase laying hens production and egg quality. Livest Sci. 2011;140:184–8. Elsevier B.V.

70. Fiorentini G, Lage JF, Carvalho IPC, Messana JD, Canesin RC, Reis RA, et al. Lipid sources with different fatty acid profile alters the fatty acid profile and quality of beef from confined Nellore steers. Asian Australas J Anim Sci. 2015;28:976–86.

71. Oliveira DM, Ladeira MM, Chizzotti ML, Machado Neto OR, Ramos EM, Goncalves TM, et al. Fatty acid profile and qualitative characteristics of meat from zebu steers fed with different oilseeds. J Anim Sci. 2011;89:2546–55.

8

The influence of in ovo injection with the prebiotic DiNovo® on the development of histomorphological parameters of the duodenum, body mass and productivity in large-scale poultry production conditions

Adrianna Sobolewska[1*], Gabriela Elminowska-Wenda[2], Joanna Bogucka[1], Agata Dankowiakowska[1], Anna Kułakowska[1], Agata Szczerba[1], Katarzyna Stadnicka[1], Michał Szpinda[2] and Marek Bednarczyk[1]

Abstract

Background: Among various feed additives currently used in poultry nutrition, an important role is played by bioactive substances, including prebiotics. The beneficial effect of these bioactive substances on the gastrointestinal tract and immune system give rise to improvements in broiler health and performance nutrition, thus increasing the productivity of these birds. An innovative method for introducing bioactive substances into chickens is the in ovo injection into eggs intended for hatching. The aim of the study was to evaluate the development of histomorphological parameters of the duodenum and productivity in chickens injected in ovo with the prebiotic DiNovo® (extract of *Laminaria* species of seaweed, BioAtlantis Ltd., Ireland) on d 12 of incubation, under large - scale, high density poultry production conditions.

Results: There was no significant impact of the injection of DiNovo® prebiotic on the production parameters of broiler chickens (body weight, FCR, EBI and mortality) obtained on d 42 of rearing. No significant impact of the DiNovo® injection on the duodenum weight and length was observed, as well as on the CSA, diameter and muscular layer thickness of the duodenum. The in ovo injection of DiNovo® significantly increased the width of the duodenal villi ($P < 0.05$) and crypt depth ($P < 0.01$) of chickens on d 21 of rearing. Other histomorphological parameters of duodenal villi at d 42 of chickens rearing such as: the height, width, and cross section area of villi were significantly greater in chickens from the control group compared to those from the DiNovo® group ($P < 0.05$ and $P < 0.01$).

Conclusions: In conclusion, this study demonstrates that injection of DiNovo® prebiotic into the air chamber of egg significantly influences the histomorphological parameters on d 21 of rearing without negatively affecting productivity in chickens at the end of rearing.

Keywords: Chicken, Duodenum, Histomorphological parameters, In ovo, Prebiotic, Productivity

* Correspondence: sobolewska@utp.edu.pl
[1]Department of Animal Biochemistry and Biotechnology, University of Science and Technology in Bydgoszcz, Mazowiecka 28 Street, 85-084 Bydgoszcz, Poland
Full list of author information is available at the end of the article

Background

Both the worldwide and domestic production of poultry meat has been increasing dynamically, and in 2014 Poland became the leader of that production in the EU [1]. Commercial breeding programs, balanced nutrition and good health status of the birds result in high effectiveness of poultry production. The parameters demonstrating an economic effect of rearing broilers are: FCR (Feed Conversion Ratio) and EBI (European Broiler Index).

A major factor affecting the efficiency of animal husbandry is proper nutrition that provides properly balanced nutrients. Feeding can affect not only the growth and development of the birds, but also - to some extent - the functioning of the immune system, primarily through the use of appropriate feed additives, such as various bioactive substances. Among various feed additives currently used in poultry nutrition, an important role belongs to prebiotics. Prebiotics have been shown to exert beneficial effects on the gastrointestinal tract of broilers [2] and to enhance feed efficiency, thus improving the productivity of these birds [3]. The use of prebiotics and probiotics in the diet of broilers and laying hens was a response to the prohibition of the use in feed antibiotic growth stimulators by the EU (Regulations (EC) No. 1831/2003 and 1334/2003).

Prebiotics are components of feed derived from sugars, including raffinose family oligosaccharides (RFOs), galactooligosaccharides (GOS) and ß-glucans that selectively stimulate both the growth and activity of the desired intestinal microflora [4]. ß-glucans are naturally occurring polysaccharides that can be synthesized by many prokaryotic and eukaryotic organisms [5]. These compounds may be a constituent of cell walls in plants, fungi and various microorganisms [6]. The prebiotic used in this study, DiNovo® (BioAtlantis Ltd., Ireland), is an extract of *Laminaria* spp. containing specific quantities of laminarin and fucoidan. Laminarins have shown promising immunomodulatory activities. Fucoidan was proven to have also antiviral and antibacterial properties which result in improved health, a lower mortality and enhanced productivity of animals [5, 7–10]. Both bioactives stimulated proliferation of beneficial microflora and improved digestibility of nutrients in monogastrics compared with non-fucoidan diet [11, 12]. The activity of prebiotics in the gastrointestinal tract is somewhat related to pH adjustment, which results in a beneficial effect on the composition of the intestinal microflora. To date, the methods of prebiotic supplementation used have been limited to administration with feed or water. An innovative method for introducing bioactive substances into chickens is the in ovo injection into eggs intended for hatching. This technique is based on the introduction - on the appropriate day of embryonic development - of bioactive substances into the air chamber of the egg or directly into the developing embryo [13]. A thorough understanding of various stages of the embryonic development in birds allows the optimal time of injection to be defined [14, 15]. The use of in ovo techniques to introduce prebiotics and probiotics to chickens provides a means of modulating the immune system at early embryonic stages. Substances administered in ovo during the embryonic development of birds reach the intestines and affect the development of the gastrointestinal tract before hatching. Villaluenga et al. [16] demonstrated that the optimal time for the injection of a prebiotic is the 12th day of embryonic development. In comparison with injections on d 1, 8 and 17, a significantly higher number of bifidobacteria was observed in the gut. Moreover, on d 12, the chorioallantoic membrane is already fully developed and vascularized, while the embryo is surrounded by the amniotic fluid that remains in contact with the embryonic gastrointestinal tract, which allows the transport of substances from the air chamber into the intestine [17].

The intestine is highly specialized in the hydrolysis and absorption of nutrients, and constitutes the paramount barrier between the host's external and internal environment. The integration of digestive, absorptive and immune functions of the gastrointestinal tract, as well as the ability to regulate these functions are of key importance for animals, including the productivity of livestock [18]. In the final phase of digestion and absorption of nutrients, a substantive role is played by intestinal villi lined with epithelium, composed of various cells [19]. The intestinal epithelium covering villi is invaginated into the lamina propria forming tubular glands called intestinal crypts. The crypts are comprised of populations of continuously proliferating stem cells. These cells are responsible for the formation of various types of intestinal epithelial cells. Among the most abundant cells are enterocytes that migrate to the top of villi and incorporate into it towards the intestinal lumen. These cells are responsible for the transport of nutrients from the intestinal lumen into blood vessels [20]. After several hours of life, enterocytes are replaced with new cells; while over time, the depleted cells peel off into the lumen.

The histomorphological parameters of all sections of the small intestine, such as the height of intestinal villi, the crypt depth and the ratio between these two values, are some of the indicators of the health and functional status of the intestine in chickens. An increase in height of intestinal villi and the appropriate ratio between the height of villi and crypt depth are a measure of the intensity of recovery processes of intestinal epithelial cells [21, 22]. Both shorter villi and deeper crypts lead to an increase in the secretion of digestive enzymes and to a decrease in the absorption of nutrients, and may result in a lower productivity of animals [21]. Simultaneous shortening of the villi and deepening of crypts may reduce the productivity of the flock because shorter villi reduce the total surface area of the intestinal absorption which results in poorer absorption of nutrients, and deeper crypt contributing to an increased secretion of digestive enzymes [23]. In contrast to mammals, the small intestine in birds is relatively shorter and the passage of the content is faster, therefore the digestive processes are more intense. Moreover, the supply of feed to the currently bred broilers, from the first d after hatching to the end of rearing, is often conducted 24 h a day. Therefore, thick muscular layers (*muscularis mucosae* and the intestinal *muscularis externa*) induce contractions of villi with longitudinal and transverse folds of the mucosa, thus permitting the appropriate motor activity of the intestine. This is accompanied by an enhanced use of nutrients through a more effective mixing of the intestinal content and a better contact with digestive enzymes resulting in the faster absorption of nutrients.

In terms of digestion, as the first loop of the small intestine, the duodenum is a very important one. The pancreas is located within this loop. The posterior end of the duodenum becomes the jejunum that subsequently passes into the ileum, but there is no clearly marked boundary between the jejunum and ileum [24]. Unlike in mammals, the avian duodenum does not include Brunner's glands since the submucosa is particularly hypoplastic.

The aim of the study was to evaluate the productivity and development of the histomorphological parameters of the duodenum on d 21 and 42 of rearing in chickens injected in ovo with the prebiotic DiNovo® on d 12 of incubation.

Methods

The experiment was conducted on Ross 308 broiler breeder eggs incubated in large - scale, high density commercial hatchery conditions (Drobex - Agro Ltd., Solec Kujawski, Poland) in Petersime incubators. On d 12 of incubation, eggs were candled, and the infertile ones or those containing dead embryos were discarded. A total of 54,000 eggs, containing living embryos were randomly divided into 2 equal groups: a control group and an experimental group, injected with the prebiotic DiNovo® (DiNovo® Group). DiNovo® is an extract of *Laminaria* spp. containing laminarin and fucoidan (BioAtlantis Ltd., Ireland). The control group was injected with 0.2 mL of sterile physiological saline, while the eggs of the experimental group were injected with DiNovo®, 0.88 mg/egg dissolved in 0.2 mL of physiological saline. The solutions were delivered into the air chamber of every egg, and the hole in the egg shell was sealed with an automatic system dedicated for the in ovo injection of prebiotics.

Animals

After hatching, the chickens were reared on the same farm in separate broiler houses (with the same environmental conditions) and fed on commercial diets (starter, grower, finisher) for 42 d. They were fed on and watered ad libitum. Either group (control group - CG, and experimental group - DiNovo® group) consisted of 25,000 chickens. The rearing experiment was conducted on the experimental farm of the Drobex - Agro company and lasted for 42 d, upon the approval of the Polish Local Ethical Commission (No 22/2012. 21.06.2012) and in accordance with the animal welfare recommendations of the European Union directive 86/609/EEC, providing adequate husbandry conditions with continuous monitoring of stocking density, litter, ventilation etc.

The cumulative feed intake for the whole period of rearing was measured and feed conversion ratio (kg feed intake/kg live mass gain) was calculated. The European Broiler Index (EBI) according to the following formulae was also calculated:

$$EBI = \frac{Viability\ (\%) \times ADG\ (g/chick/d)}{FCR\ (kg\ feed/kg\ gain) \times 10}$$

ADG = average daily gain

Viability (%) = chicks remaining at the end of period (%)

On the day of slaughter (d 42 of life), all chickens from both groups were transported to the Drobex - Agro slaughterhouse, and their mean body weight was calculated before slaughter in accordance with the methodology and technology used in that establishment.

Histomorphological samples

The material for the morphological and histological analysis of the duodenum was collected from 21- and 42-day-old chickens of each. Before slaughter, a total of 100 chickens (a representative selection) from each group were weighed, and their mean body weight was calculated. Subsequently, 15 chickens per group, with the body weight similar to the mean for the group were selected. After slaughter, the small intestine was removed out and the duodenum was dissected, measured and weighed. Samples for histomorphometrical analyses (approx. 2 cm) were taken from the midway of the duodenum.

Histomorphological examination

The sampled segment of the duodenum was carefully washed with 0.9% saline and then fixed in 4% formalin buffered with $CaCO_3$. The fixed samples were dehydrated, cleared and permeated with paraffin in a tissue processor (Thermo Shandon, Chadwick Road, Astmoor, Runcorn, Cheshire, United Kingdom), and subsequently embedded in paraffin blocks using an embedding system

(Medite, Burgdorf, Germany). Thus, formed blocks were cut on a rotary microtome (Thermo Shandon, Chadwick Road, Astmoor, Runcorn, Cheshire, United Kingdom) into slices of 10 μm thick which were successively placed on microscope slides coated with ovoalbumin with an addition of glycerol.

Staining methods
Before staining, the specimens were deparaffinized and rehydrated. The specimens were then stained using the periodic acid - Shiff reagent (PAS) method for the morphometric analysis of the duodenum.

Histomorphological measurements
Using a Carl Zeiss microscope (Jena, Germany) equipped with a ToupCam™ digital camera and the MultiScan 14.02 computer software for microscopic image analysis (Computer Scanning Systems II, Warsaw, Poland), the following measurements were done: the height and width of intestinal villi and the depth of intestinal crypts. The measurement of the height of intestinal villi was conducted on 10 duodenal villi per one individual. The height was measured from the top to the base of the villus at the entrance to the intestinal crypt. The width of the villus was measured at half of its length. Subsequently, the villus surface area was calculated using the formula proposed by Sakamoto et al. [25]: $(2\pi) \times (VW / 2) \times (VH)$, where VW = villus width, and VH = villus height. The crypt depth was defined as the depth of the invagination between adjacent villi. This parameter was measured in10 crypts [26].

In order to measure and calculate the thickness of the muscular layers and the cross - sectional area (CSA) of the duodenum, microscopic slides were imaged on a Kaiser rePro image capture system using a Canon EOS 70D digital SLR camera equipped with a Canon 100 mm f/2.8 L EF MACRO IS USM lens. For the calculation of the above listed parameters, the NIS ELEMENTS AR software (Nikon, Japan) was used. System precalibration was based on the microscopic reference line captured in the same conditions as the analyzed slides. Linear measurements of the thickness of the muscular layer of the duodenum were conducted on three consecutive slices of that segment by selecting two extreme points. The cross sectional area (CSA) of the duodenum was estimated on the base of the ellipse automatically generated from 5 different points localized on the circumference of its tunica muscularis. The diameter of the duodenum was calculated based on the cross - sectional area measured.

Statistics
The obtained results were subjected to one - way analysis of variance (body weight before slaughter, mortality, FCR, EBI) and two-way analysis of variance (histomorphological measurements of the duodenum) using the

SAS Institute Inc. 2013 computer program. SAS/ STAT(r) 9.4 User\'s Guide. Cary, NC: SAS Institute Inc. The arithmetic mean ($\bar{x}$) and standard deviation (SD) were calculated. The significant differences between groups were tested using Duncan's multiple range test.

Results
Table 1 presents the evaluation results of the production parameters of broiler chickens (body weight before slaughter, FCR, EBI and mortality) obtained on d 42 of rearing. There were no significant differences in the case of the above mentioned parameters between the control and DiNovo® group. However, considering such a large number of broiler chickens in the experience, we can talk about the tendency a more favorable impact of the DiNovo® prebiotic injected at 12 d of incubation for the production traits of studied birds. The experimental group indicated a greater body weight before slaughter and the EBI ratio and a lower rate of the FCR ratio compared to the control group.

Table 2 presents the mean body weight of chickens for histological studies and the morphological parameters of the duodenum on d 21 and 42 of rearing in chickens from both investigated groups. The mean body weight values in the control and DiNovo® groups were similar in both terms of slaughter and did not differ significantly. No significant impact of the DiNovo® injection on the duodenum weight and length was observed, as well as on the CSA, diameter and muscular layer thickness of the duodenum (Fig. 1a, b; Fig. 2a, b).

The histomorphological parameters of duodenal villi on d 21 and 42 of rearing are also presented in Table 2 and Figs. 1a, b and 2a, b. The in ovo injection of DiNovo® significantly increased the width of duodenal villi of chickens on d 21 of rearing ($P < 0.05$). This resulted in a greater surface area of villi in these birds, however this was not confirmed statistically. DiNovo® prebiotic, used in the study, significantly increased the crypt depth on d 21 ($P < 0.01$), in contrast to the last day of rearing, wherein crypts in the experimental group were significantly shorter in comparison to the control group ($P < 0.01$). Other histomorphological parameters of duodenal villi at 42 d of chickens rearing such as: the height, width, and

Table 1 Productivity parameters of chickens on d 42 of rearing

Parameters	D 42		P-value
	Control	DiNovo®	
Body weight before slaughter, g, $n =$ [a]	2140	2210	0.1998
Mortality, %	4.29	4.36	0.5970
FCR	1.79	1.72	0.3041
EBI	288	308	0.0971

[a]Control group, $n = 21{,}934$ individuals
DiNovo® group, $n = 22{,}980$ individuals

Table 2 Body weight of the chickens (g) and histomorphological measurements of the duodenum on d 21 and 42 of rearing

Parameters	D 21		D 42	
	Control	DiNovo®	Control	DiNovo®
Body weight of the chickens with histological samples taken, g, $n = 15$	811 ± 23.2	845 ± 15.1	2170 ± 30.3	2190 ± 51.6
Duodenal weight, g	7.40 ± 0.2	8.03 ± 0.4	15.30 ± 0.6	15.28 ± 0.6
Duodenal length, cm	25.79 ± 3.9	25.30 ± 3.0	33.60 ± 2.6	33.16 ± 3.2
Duodenal CSA (cross-sectional area), mm^2	23.81 ± 4.27	23.73 ± 2.82	35.34 ± 4.66	36.62 ± 6.14
Duodenal diameter (excluding tunica serosa), µm	5484 ± 493.94	5484 ± 330.71	6690 ± 444.08	6808 ± 537.65
Muscularis thickness, µm	148 ± 25.38	149 ± 22.15	194 ± 13.50	191 ± 31.31
Villus height, µm	1316 ± 43.2	1302 ± 43.2	1536[a] ± 56.8	1383[b] ± 32.3
Villus width, µm	109[b] ± 3.6	126[a] ± 5.7	115[A] ± 5.5	99[B] ± 2.3
Villus surface area, µm^2	443,997 ± 13.654	514,529 ± 34.306	558,730[A] ± 41.777	429,764[B] ± 11.648
Crypt depth, µm	108[B] ± 4.2	146[A] ± 1.8	227[A] ± 3.7	113[B] ± 4.6

Values with different letters differ significantly between treatments (a-b $P < 0.05$, A-B $P < 0.01$)

surface area of villus were significantly greater in chickens from the control group compared to those from the DiNovo® group ($P < 0.05$ and $P < 0.01$).

Figure 1a, b; Fig. 2a, b present microscopic images of the cross sections of the duodenum on d 21 and 42 of rearing in the chickens from the control group and the DiNovo® group. The muscular layer thickness, CSA (ellipse), diameter of the duodenum, height and width of the intestinal villi, and crypt depth are marked in each image.

Discussion

In the present study, the effect of in ovo injection of the prebiotic DiNovo® on the morphometric parameters of the duodenum in broiler chickens on d 21 and 42 of rearing was examined. The obtained results are problematic to compare with the literature data because most studies have been focused on the impact of prebiotics as additives in feedstuffs, and not administered in ovo, as achieved in this study [27–31]. Only Tako et al. [32] and Cheled-Shoval et al. [33] studied the effect of various substances injected during chicken embryogenesis on small intestine morphometry. The first group injected in ovo into the amniotic fluid carbohydrates and β-hydroxymethylbutyrate (HMB) on d 17.5 of egg incubation (19E – d of embryonic development, 20E, hatch and 3 d). The authors observed an increased surface area of the intestinal villi at all tested time points of the embryonic development and on d 1 and 3 after hatching in chickens provided with HMB. In turn, the smallest surface area of intestinal villi was observed in the control group with 1.5 to 3 times lower values than in the group receiving HMB. In our study we have not achieved such a positive effect on the last day of rearing chickens (42 d of rearing). There was, however, a

clear tendency of beneficial effects of used prebiotic on the surface area of duodenal villi at d 21 of life. We can therefore assume that the effect of the DiNovo® injection occurs in an earlier period of rearing chickens that was also observed in the study by Bogucka et al. [34], who analyzed the effect of various bioactive substances given in ovo on the histomorphology of chickens small intestine in the first days after hatching. Similarly, Cheled-Shoval et al. [33] used the in ovo technique 3 d before hatching to administer a preparation of mannan oligosaccharide (MOS), and examined the morphology of the intestine of chickens (Cobb 500) on d 1 after hatching. These authors observed a positive impact of the prebiotic on the height and width of intestinal villi and the depth of intestinal crypts. However, in our study a significant effect ($P < 0.05$) of in ovo injection of DiNovo® on the width of duodenal villi and crypt depth ($P < 0.01$) on d 21 of rearing compared to the control group has been found. Significantly deeper crypts in this group may indicate intense renewal of the intestinal epithelial cells, which in turn can exert a positive effect on the function of the intestinal absorption and secretion. Evaluating the impact of the MOS on the intestinal muscular layer by Cheled-Shoval et al. [33], the authors found that the thickness of the muscular layer was significantly greater ($P < 0.05$) in chickens from the experimental group (MOS) compared with the control group. However, in our study there were no significant differences in the thickness of muscular layer in the duodenum in both on d 21 and 42 of chickens life (Table 2).

A study by Houshmand et al. [35] focused on the effect of the prebiotic MOS (Bio-MOS) and a probiotic (*Bacillus subtilis* and *Clostridium butyricum*) on the morphology of the duodenum and jejunum in cockerels (Cobb) on d 21 and 42 of rearing. The study revealed significantly higher villi in the duodenum in 21-day-old

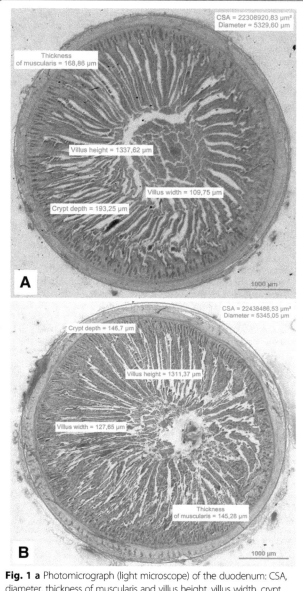

Fig. 1 a Photomicrograph (light microscope) of the duodenum: CSA, diameter, thickness of muscularis and villus height, villus width, crypt depth in the control group on d 21. **b** Photomicrograph (light microscope) of the duodenum: CSA, diameter, thickness of muscularis and villus height, villus width, crypt depth in the DiNovo® group on d 21

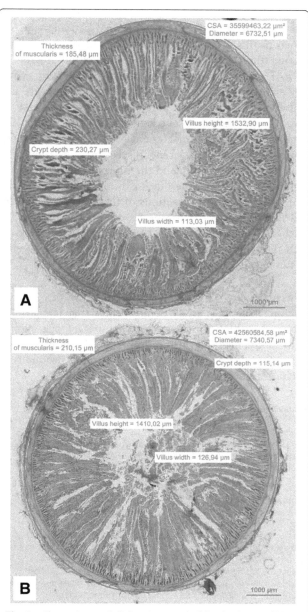

Fig. 2 a Photomicrograph (light microscope) of the duodenum: CSA, diameter, thickness of muscularis and villus height, villus width, crypt depth in the control group on d 42. **b** Photomicrograph (light microscope) of the duodenum: CSA, diameter, thickness of muscularis and villus height, villus width, crypt depth in the DiNovo® group on d 42

chickens that received the prebiotic with feed (Starter), as compared with the chickens from the control group and those receiving the probiotic. No statistically significant differences in the crypt depth were observed between the study groups. The same results were obtained in cockerels on d 42 of life. Similarly to our study, Houshmand et al. [35] did not report significant differences in either the length of the duodenum or the weight of the small intestine and duodenum (% of body weight) on both d 21 and d 42 of rearing. However, the duodenum on d 21 and 42 was approximately by 1 cm and 4 cm shorter, respectively than in this study.

The analysis of the morphometric results in the current study (Table 2) with reference to the productivity parameters (Table 1), i.e. body weight on slaughter and FCR does not produce clear results. It seems that by d 21 of rearing, the wider villi and deeper crypts in the DiNovo® group positively affected the digestive and absorptive potential of these chickens. Perhaps, in the second half of rearing, despite the significantly poorer histomorphometrical properties of the duodenum in the DiNovo® group, the better digestive and absorptive potential from the first period of life of these chickens (up to d 21) contributed to the final productivity obtained in this group. As a result, the DiNovo® group indicated their body weight before slaughter greater by 70 g, FCR improved by approximately 0.07 units, and EBI greater by 20 points. Taking into account the fact that the values of these parameters were calculated in over twenty thousand chickens from each group, the obtained results are reliable and have a high implementation value. The apparent discrepancy between the productivity parameters and the morphology of the duodenum on d 42 may be due to the lack of data on the morphology of further sections of the intestine (jejunum and ileum). Furthermore, the morphometric parameters of the small intestine are characterized by a relatively high intra-species variability and high dynamics fettered by numerous factors.

Conclusions

In conclusion, this study demonstrates that injection of DiNovo® prebiotic into the air chamber of egg significantly influences the histomorphological parameters on d 21 of rearing without negatively affecting productivity in chickens at the end of rearing.

Abbreviations

ADG: Average daily gain; CSA: The cross - sectional area; E: Day of embryonic development; EBI: European Broiler Index; FCR: Feed Conversion Ratio; GOS: Galactooligosaccharides; HMB: β-hydroxymethylbutyrate; MOS: Mannan oligosaccharide; PAS: Periodic acid–Schiff; RFO: Raffinose family of oligosaccharides; SD: Standard deviation

Acknowledgments

The research leading to these results has received funding from the European Union's Seventh Framework Programme managed by REA Research Executive Agency http://ec.europa.eu/research/rea (FP7/2007–2013) under grant agreement number: 315198. This research was undertaken as part of a project entitled "Thrive-Rite: Natural Compounds to enhance Productivity, Quality and Health in Intensive Farming Systems". Further details are provided on the consortium's website (www.thriverite.eu) and the EU Commission's webpage: http://cordis.europa.eu/project/rcn/104395_en.html).

Funding

This project was funded from the European Union's Seventh Framework Programme managed by REA Research Executive Agency http://ec.europa.eu/research/rea (FP7/2007-2013) under grant agreement number: 315198.

Authors' contributions

GEW, AS data and results interpretation, drafted the manuscript. JB histological study coordination, aided in data interpretation, performed the statistical analysis. AD performed the statistical analysis, histological analysis. AK, AS histological analysis. KS participated in study design and coordination. MS results interpretation, assistance in preparing the manuscript in English. MB designed research. All authors read and approved the final manuscript.

Competing interests

The authors declare that they have no competing interests.

Author details

[1]Department of Animal Biochemistry and Biotechnology, University of Science and Technology in Bydgoszcz, Mazowiecka 28 Street, 85-084 Bydgoszcz, Poland. [2]Department of Normal Anatomy, the Ludwik Rydygier Collegium Medicum in Bydgoszcz, the Nicolaus Copernicus University in Torun, 24 Karłowicza Street, Bydgoszcz 85-092, Poland.

References

1. Krajowa Rada Drobiarstwa – Izba Gospodarcza w Warszawie. http://www.krd-ig.com.pl/sytuacjanarynkumiesadrobiowegowpolsceiwuew2014oraz prognozyna2015rok,832,l1.html. Accessed 25 May 2015.
2. An BK, Cho BL, You SJ, Paik HD, Chang HI, Kim SW, et al. Growth performance and antibody response of broiler chicks fed yeast derived [beta]-glucan and single-strain probiotics. Asian Australas J Anim Sci. 2008; 21:1027–32.
3. Cavazzoni V, Adami A, Castrovilli C. Performance of broiler chickens supplemented with Bacillus coagulans as probiotics. Br Poult Sci. 1998;39:526–9.
4. Gibson GR, Roberfroid MB. Dietary modulation of the human colonic microbiota: introducing the concept of prebiotics. J Nutr. 1995;125:1401–12.
5. Petravié-Tominac V, Zechner-Krpan V, Grba S, Srečec S, Panjkota-Krbavčić I. Biological effects of yeast β-glucans. ACS. 2010;75(4):149–58.
6. Akramienė D, Anatolijus K, Didžiapetrienė J, Kėvelaitis E. Effects of β-glucans on the immune system. Medicina (Kaunas). 2007;43(8):597–607.
7. Maiorano G, Sobolewska A, Cianciullo D, Walasik K, Elminowska-Wenda G, Sławińska A, et al. Influence of in ovo prebiotic and synbiotic administration on meat quality of broiler chickens. Poult Sci. 2012;91:2693–969.
8. Bednarczyk M, Stadnicka K, Kozłowska I, Abiuso C, Tavaniello S, Dankowiakowska A, et al. Influence of different prebiotics and mode of their administration on broiler chicken performance. Animal. 2016. doi:10.1017/8175173111600017.
9. Heim G, Walsh AM, Sweeney T, Doyle DN, O'Shea CJ, Ryan MT, et al. Effect of seaweed-derived laminarin and fucoidan and zinc oxide on gut morphology, nutrient transporters, nutrient digestibility, growth performance and selected microbial populations in weaned pigs. Br J Nutr. 2014;111:1577–85.
10. Leonard SG, Sweeney T, Bahar B, O'Doherty JV. Effect of maternal seaweed extract supplementation on suckling piglet growth, humoral immunity, selected microflora, and immune response after an ex vivo lipopolysaccharide challenge. J Anim Sci. 2012;90:505–14.
11. Lynch MB, Sweeney T, Callan JJ, O'Sullivan JT, O'Doherty JV. The effect of dietary Laminaria hyperborea derived laminarin and fucoidan on nutrient digestibility, nitrogen utilisation, intestinal microflora and volatile fatty acid concentration in pigs. J Sci Food Agric. 2010;90(3):430–7.
12. Walsh AM, Sweeney T, O'Shea CJ, Doyle DN, O'Doherty JV. Effect of supplementing varying inclusion levels of laminarin and fucoidan on growth performance, digestibility of diet components, selected faecal microbial populations and volatile fatty acid concentrations in weaned pigs. Anim Feed Sci Technol. 2013;183:151–9.
13. Ciesiolka D, Gulewicz P, Martinez-Villaluenga C, Pilarski R, Bednarczyk M, Gulewicz K. Products and biopreparations from alkaloid-rich lupin in animal nutrition and ecological agriculture. Folia Biol (Krakow). 2005;53:59–66.
14. Williams C. In ovo vaccination for disease prevention. Int Poult Prod. 2007;15(8):7–9.
15. Williams CJ. In ovo vaccination and chick quality. Int Hatch Pract. 2011; 19(2):7–13.
16. Villaluenga CM, Wardeńska M, Pilarski R, Bednarczyk M, Gulewicz K. Utilization of the chicken embryo model for assessment of biological activity of different oligosaccharides. Folia Biol (Kraków). 2004;52(3–4):135–42.
17. Jankowski J. Hodowla i użytkowanie drobiu. Warszawa: Powszechne Wydawnictwo Rolnicze I Leśne; 2012.
18. Bengmark S. Gut microenvironment and immune function. Curr Opin Clin Nutr Metab Care. 1999;2:83–5.
19. Yamauchi K. Review on chicken intestinal villus histological alterations related with intestinal function. J Poult Sci. 2002;39:229–42.
20. Potten CS, Loeffler M. A comparative model of the crypts of the small intestine of the mouse provides insight into the mechanisms of cell migration and the proliferation hierarchy. J Theor Biol. 1987;127:381–91.
21. Fan Y, Croom J, Christensen V, Black B, Bird A, Daniel L, et al. Jejunal glucose uptake and oxygen consumption in turkey poults selected for rapid growth. Poult Sci. 1997;76:1738–45.
22. Samanya M, Yamauchi KE. Histological alterations of intestinal villi in

chickens fed dried *Bacillus subtilis* var. *natto*. Comp Biochem Physiol A Mol Integr Physiol. 2002;133:95–104.

23. Xu ZR, Hu CH, Xia MS, Zhan XA, Wang MQ. Effects of dietary fructooligosaccharides on digestive enzyme activities, intestinal microflora, and morphology of male broilers. Poult Sci. 2003;82:1030–6.

24. Kierszenbaum AL, Tres LL. Histology and cell biology: an introduction of pathology. 3rd ed. 2012.

25. Sakamoto K, Hirose H, Onizuka A, Hayashi M, Futamura N, Kawamura Y, et al. Quantitative study of changes in intestinal morphology and mucus gel on total parenteral nutrition in rats. J Surg Res. 2000;94:99–106.

26. Uni Z, Noy Y, Sklan D. Posthatch development of small intestinal function in the poult. Poult Sci. 1999;78:215–22.

27. Awad W, Bohm J, Razzazi-Fazeli E, Ghareeb K, Zentek J. Effect of addition of a probiotic microorganism to broiler diets contaminated with deoxynivalenol on performance and histological alterations of intestinal villi of broiler chickens. Poult Sci. 2006;85(6):974–9.

28. Awad WA, Ghareeb K, Abdel-Raheem S, Bohm J. Effects of dietary inclusion of probiotic and synbiotic on growth performance, organ weights, and intestinal histomorphology of broiler chickens. Poult Sci. 2009;88(1):49–56.

29. Solis de los Santos F, Farnell MB, Téllez G, Balog JM, Anthony NB, Torres-Rodriguez A, et al. Effect of prebiotic on gut development and ascites incidence of broilers reared in a hypoxic environment. Poult Sci. 2005;84: 1092–100.

30. Solis de los Santos FS, Donoghue AM, Farnell MB, Huff GR, Huff WE, Donoghue DJ. Gastrointestinal maturation is accelerated in turkey poults supplemented with a mannan-oligosaccharide yeast extract (Alphamune). Poult Sci. 2007;86(5):921–30.

31. Uni Z, Ferket PR. Enhancement of development of oviparous species by in ovo feeding. Patent number US 6. 2003. p. 592–878.

32. Tako P, Ferket PR, Uni Z. Effects of *in ovo* feeding of carbohydrates and β-hydroxy-β-methylbutyrate on the development of chicken intestine. Poult Sci. 2004;83:2023–8.

33. Cheled-Shoval SL, Amit-Romach E, Barbakov M, Uni Z. The effect of in ovo administration of mannan oligosaccharide on small intestine development during the pre-and posthatch periods in chickens. Poult Sci. 2011;90:2301–10.

34. Bogucka J, Dankowiakowska A, Elminowska-Wenda G, Sobolewska A, Szczerba A, Bednarczyk M. Effect of prebiotics and synbiotics delivered *in ovo* on the broiler small intestine histomorphology during the first days after hatching. Folia Biol (Kraków). 2016;64(3):131–43.

35. Houshmand M, Azhar K, Zulkifli I, Bejo MH, Kamyab A. Effects of non-antibiotic feed additives on performance, immunity and intestinal morphology of broilers fed different levels of protein. South African J Anim Sci. 2012; 42(No.1):22–32.

9

The community structure of *Methanomassiliicoccales* in the rumen of Chinese goats and its response to a high-grain diet

Wei Jin, Yanfen Cheng and Weiyun Zhu[*]

Abstract

Background: The newly proposed methanogenic order '*Methanomassiliicoccales*' is the second largest archaeal population in the rumen, second only to the *Methanobrevibacter* population. However, information is limited regarding the community of this new order in the rumen.

Methods: This study used real-time PCR and 454 pyrosequencing to explore the abundance and community composition of *Methanomassiliicoccales* in the rumen of Chinese goats fed a hay (0% grain, $n = 5$) or a high grain (65% grain, $n = 5$) diet.

Results: Real-time PCR analysis showed that the relative abundance of *Methanomassiliicoccales* (% of total archaea) in the goat rumen was significantly lower in the high-grain-diet group ($0.5\% \pm 0.2\%$) than that in the hay-diet group ($8.2\% \pm 1.1\%$, $P < 0.05$). The pyrosequencing results showed that a total of 208 operational taxonomic units (OTUs) were formed from ten samples at 99% sequence identity. All the sequences were identified as Methanomassiliicoccaceae at the family level, and most of the sequences ($96.82\% \pm 1.64\%$) were further classified as Group 8, 9, and 10 at the *Methanomassiliicoccales* genus level in each sample based on the RIM-DB database. No significant differences were observed in the number of OTUs or Chao1's, Shannon's or Pielou's evenness indexes between the hay- and high-grain-diet groups ($P \geq 0.05$). PCoA analysis showed that diet altered the community of *Methanomassiliicoccales*. At the genus level, the relative abundances of Group 10 (67.25 ± 12.76 vs. 38.13 ± 15.66, $P = 0.012$) and Group 4 (2.07 ± 1.30 vs. 0.27 ± 0.30, $P = 0.035$) were significantly higher in the high-grain-diet group, while the relative abundance of Group 9 was significantly higher in the hay-diet group (18.82 ± 6.20 vs. 47.14 ± 17.72, $P = 0.020$). At the species level, the relative abundance of Group 10 sp. (67.25 ± 12.76 vs. 38.13 ± 15.66, $P = 0.012$) and Group 4 sp. MpT1 (2.07 ± 1.30 vs. 0.27 ± 0.30, $P = 0.035$) were significantly higher in the high-grain-diet group, while the relative abundance of Group 9 sp. ISO4-G1 was significantly higher in the hay-diet group (12.83 ± 3.87 vs. 42.44 ± 18.47, $P = 0.022$).

Conclusions: Only a few highly abundant phylogenetic groups dominated within the *Methanomassiliicoccales* community in the rumens of Chinese goats, and these were easily depressed by high-grain-diet feeding. The relatively low abundance suggests a small contribution on the part of *Methanomassiliicoccales* to the rumen methanogenesis of Chinese goats.

Keywords: Community structure, Diet, Goat, *Methanomassiliicoccales*, Pyrosequencing

* Correspondence: zhuweiyun@njau.edu.cn
Jiangsu Province Key Laboratory of Gastrointestinal Nutrition and Animal Health; Laboratory of Gastrointestinal Microbiology, College of Animal Science and Technology, Nanjing Agricultural University, Nanjing 210095, China

Background

Recently, a large group of archaea closely related to the *Thermoplasmatales* has been proposed as a new methanogenic order 'Methanomassiliicoccales' [1, 2]. Two synonymous names, 'Rumen Cluster C' (RCC) and 'Methanoplasmatales', were previously used for this new order before the formal name was proposed [2–4].

St-Pierre and Wright [5] reviewed the representation of gastrointestinal methanogens in a variety of host species based on 16S r RNA gene clone library analysis and found wide variations in the abundance of *Methanomassiliicoccales* in the rumen. In most cases, *Methanomassiliicoccales* accounted for less than 5% of the archaea in the rumens of sheep, water buffalo, yaks, dairy cattle, and other ruminant species [5]. In several cases, including reindeer (Svalbard), beef cattle (Hereford-cross, potato diet), and sheep (Merino, Queensland, Australia), *Methanomassiliicoccales* accounted for more than 50% of archaea in the rumen [5]. Recently, Seedorf et al. [6] used 454 pyrosequencing to study the methanogenic communities in the rumens of New Zealand sheep and cattle fed with pasture (ryegrass and white clover). They reported a mean relative abundance of *Methanomassiliicoccales* of about 10.4%, but this value exceeded 40% in some treatment groups [6]. This high degree of abundance suggests the significant involvement of the *Methanomassiliicocales* in rumen methanogenesis, but their exact role remains to be established. Interestingly, increasing evidence shows that the *Methanomassiliicoccales* do not possess the genes required for the first six steps of hydrogenotrophic methanogenesis [7]. Instead, methane is produced by via the hydrogen-mediated reduction of methyl-group-containing compounds (e.g., methylamines and methanol) [8–10].

Methanomassiliicoccales-affiliated 16S rRNA genes have been found in a wide range of anoxic environments [11], including marine habitats, rice paddy fields, anaerobic digestors, and the intestinal tracts of termites, mammals, and humans. Most recently, Söllinger et al. [10] confirmed a clade-specific habitat preference on the part of *Methanomassiliicoccales* and divided this order into two clades, the 'gastrointestinal tract' (GIT) clade and the 'environmental' clade, based on phylogenetic and genomic analyses of methanogenic sequences from wetlands and animal intestinal tracts. The authors also reported larger genomes and a larger number of genes encoding anti-oxidative enzymes in the 'environmental' clade than in the 'GIT' clade representatives.

The taxonomy of *Methanomassiliicoccales* remains incomplete due to the lack of type strains, which has limited the study of the community of this new archaeal order [12]. Seedorf et al. [12] constructed a 16S rRNA gene database (RIM-DB) for the phylogenetic analysis of archaea from the gastrointestinal tract. This database clustered the sequences of *Methanomassiliicoccales* into twelve groups at the genus level, thereby enabling the analysis of the community structure of this order. Using this database, Seedorf et al. [6] reported that the members of *Methanomassiliicoccales* in New Zealand sheep and cattle were mainly located in Groups 10, 11, and 12.

China has more than 140 million goats (China Statistical Yearbook 2015), which is an important contributor to the production of the greenhouse gas methane. For the development of methane-mitigation tools, it is important to understand the methanogen community and identify the dominant methanogens in the rumen of Chinese goats [6]. St-Pierre and Wright reviewed the diversity of gut methanogens in herbivorous animals and concluded that the population structure of methanogens was strongly influenced by diet and host species [5]. Thus, we hypothesized that the community of the new established methanogenic order 'Methanomassiliicoccales' in the rumens of Chinese goats was distinct from that in the other ruminants examined in the previous studies and also affected by the two main dietary types in China. In the present study, we used real-time PCR and 454 pyrosequencing to determine the abundance and the composition of *Methanomassiliicoccales* in the rumens of Chinese goats based on the RIM-DB database. Our findings differ from those previously reported for New Zealand sheep and cattle because the *Methanomassiliicoccales* community in the rumens of Chinese goats consists mainly of members of Groups 8, 9, and 10.

Methods

Animal, diet and experimental design

This animal experiment was carried out at the experimental station of Nanjing Agricultural University in Jiangsu Province, China. The experimental design was previously described in detail [13]. Briefly, ten male goats with rumen fistula (Boer × Yangtze River Delta White; 2–3 years old) were used and placed in individual pens (1.2 m× 1.2 m) with free access to water. These animals were randomly assigned to one of two groups and fed either a hay diet (0% grain, $n = 5$) or high-grain diet (65% grain, $n = 5$). The body weights of the goats were 29.8 ± 0.9 kg for the hay-diet group and 30.0 ± 1.1 kg for the high-grain-diet group at the beginning of the trial. The diets (750 g dry matter per animal per day) were offered in equal amounts at 08:30 and 16:30 daily during the trial (7 wks). The nutrient compositions of the diets are presented in Table 1. The metabolic energy intake was slightly above that required for the maintenance of goats in the hay group and permitted a growth rate of 200 g/d in the high-grain group (30 kg body wt).

Sample collection

On d 50, the goats were slaughtered at a local slaughterhouse 4–5 h after the morning feeding. Immediately

Table 1 Ingredient and nutrient composition of the diets

Item	Hay diet	High grain diet
Ingredient composition, % DM		
Chinese wildrye	81.00	30.00
Alfalfa	15.00	0
Corn meal	0	45.00
Wheat meal	0	20.00
Soybean	0	1.10
$CaCO_3$	0.50	0.95
NaCl, Salt	0.80	0.65
$CaHPO_4$	1.70	1.20
Mineral and vitamin supplement	1.00	1.00
$NaHCO_3$	0	0.10
Nutrient composition		
Metabolic energy, MJ/kg DM	8.31	11.31
Crude protein, % DM	10.06	10.06
Crude fat, % DM	3.55	3.59
Crude fiber, % DM	30.17	11.18
Neutral detergent fiber, % DM	57.01	25.23
Acid detergent fiber, % DM	35.72	13.55
Crude ash, % DM	10.62	6.52
Starch, % DM	ND[a]	58.23

[a]Not determined, but considered equal to 0

after slaughter, a representative sample (solid and liquid) from five sections of whole-rumen digesta was collected and homogenised in a soybean milk blender. Then samples were frozen and stored at −80 °C.

DNA extraction

DNA was extracted from a representative 1 mL of rumen content using a FastPrep®-24 Instrument (MP Biomedicals, South Florida, USA) and processed (bead beat) at a setting of 5 for 2 min, combined with cetyltrimethylammonium bromide and phenol-chloroform-isopentanol extraction [14]. The DNA in the solution was precipitated with ethanol. Then, the pellets were suspended in 50 μL Tris-EDTA buffer. The sample DNA concentrations were quantified using a NanoDrop ND-1000 Spectrophotometer (Nyxor Biotech, Paris, France) and adjusted to yield similar DNA concentrations. The total DNA sample was divided into two portions for 454 pyrosequencing and real-time PCR.

Real-time PCR

Real-time PCR was performed to quantify the archaea and *Methanomassiliicoccales* according to the method described by Jeyanathan et al. [15] using an Applied Biosystems 7300 Real-Time PCR system (Applied Biosystems, California, USA). The primers for total archaea were 915f 5'-AAG AAT TGG CGG GGG AGC AC, 1386r 5'-GCG

GTG TGT GCA AGG AGC. The primers for *Methanomassiliicoccales* were 762f 5'- GAC GAA GCC CTG GGT C, 1099r 5'- GAG GGT CTC GTT CGT TAT. The reaction mixture (20 μL) consisted of 10 μL of SYBR® Premix Ex Tag TM (TaKaRa, Dalian, China), 0.2 μmol/L of each primer, and 2 μL of the template DNA. Real-time PCR amplification for archaea was initiated at 95 °C for 30 s, followed by 40 cycles at 95 °C for 5 s, 59 °C for 30 s, and 72 °C for 30 s. Real-time PCR amplification for *Methanomassiliicoccales* was initiated at 95 °C for 30 s, followed by 40 cycles at 95 °C for 5 s, 56 °C for 30 s, and 72 °C for 30 s. Melting curve analysis was also conducted over a range of 60–95 °C to assess the specificity of the amplification products.

The genomic DNA of the rumen contents was amplified with the two pairs of primers for archaea and *Methanomassiliicoccales*. The reaction mixture (50 μL) consisted of 0.25 μL of TaKaRa Ex Taq® (TaKaRa, Dalian, China), 5 μL 10 × Ex Taq Buffer (Mg^{2+} Plus), 4 μL dNTP Mixture, 0.2 μmol/L of each primer, and 1 μL (10 ~ 20 ng) of the template DNA. PCR amplification for archaea (for *Methanomassiliicoccales*) was initiated at 95 °C for 5 min, followed by 35 cycles at 95 °C for 30 s, 59 °C (56 °C) for 30s, and 72 °C for 30 s, as well as a final extension at 72 °C for 7 min. The amplicons were cloned into *Escherichia coli* JM109 using the pGEM-T Easy vector (Promega). The plasmid containing the amplicon was used to construct a standard for the estimation of the 16S rRNA gene copy number for the total archaea or *Methanomassiliicoccales*. The concentration of the plasmids was quantified using a Qubit dsDNA HS Assay Kit (Invitrogen, Eugene, Oregon, USA) on a Qubit 2.0 Fluorometer (Invitrogen, Carlsbad, CA, USA). The plasmid DNA standard was prepared according to Koike et al. [16]. The copy number of each standard plasmid was calculated based on the DNA concentration and molecular weight of the cloned plasmid.

A 10-fold dilution series of the standard plasmid for the related target was also run with the samples; all samples and standards were run in triplicate. The total numbers of *Methanomassiliicoccales* and archaea per mL were determined using ABI SDS software (Applied Biosystems, Foster City, CA, USA), based on the standard curve, the dilution factor, and the volumes of the DNA extracts.

Roche 454 pyrosequencing for 16S rRNA gene amplicon of *Methanomassiliicoccales*

The *Methanomassiliicoccales* were low in abundance in the goat rumen samples. Therefore, nested PCR was used to amplify the 16S rRNA gene to obtain a sufficient concentration of PCR product for 454 pyrosequencing. For total methanogens, the 86f/1340r primer pair was used as the outer primers [17]. The PCR reactions were performed in a 20 µL mixture containing 4 µL of 5× FastPfu Buffer, 2 µL of 2.5 mmol/L dNTPs, 0.8 µL of each primer (5 µmol/L), 0.4 µL of FastPfu Polymerase (TransGen Biotech, Beijing, China), and 10 ng of template DNA. The cycling parameters were as follows: 95 °C for 2 min, followed by 20 cycles at 95 °C for 30 s, 55 °C for 30 s, and 72 °C for 30 s, as well as a final extension at 72 °C for 5 min. The primers for *Methanomassiliicoccales* (762f/1099r), described above, were used as the inner primers. The PCR reactions were performed in a 20 µL mixture containing 4 µL of 5 × FastPfu Buffer, 2 µL of 2.5 mmol/L dNTPs, 0.8 µL of each primer (5 µmol/L), 0.4 µL of FastPfu Polymerase (TransGen Biotech, Beijing, China), and 2 µL of the first round PCR products (diluted 20-fold before use). The cycling parameters were as follows: 95 °C for 2 min, followed by 28 cycles at 95 °C for 30 s, 55 °C for 30 s, and 72 °C for 30 s, as well as a final extension at 72 °C for 5 min. Negative controls were used for both PCR runs (outer primer run and inner primer run) to avoid contamination. After purification using the AxyPrep DNA Gel Extraction Kit (Axygen Biosciences, Union City, CA, U.S.) and quantification using QuantiFluor™ -ST (Promega, U.S.), a mixture of amplicons was used for pyrosequencing on a Roche 454 GS FLX+ Titanium platform (Roche 454 Life Sciences, Branford, CT, U.S.) according to standard protocols [18] at the Majorbio Bio-Pharm Technology Co., Ltd., Shanghai, China.

The raw data were submitted to the Sequence Read Archive, under accession number SRP053014.

Amplicon sequence data analysis

The analysis was performed following the procedure described by Seedorf et al. [6]. Briefly, data were processed in QIIME [19]. The reads were quality filtered and assigned to the corresponding samples via barcodes, using the *split library.py* script with the "-r -z truncate_remove −s 20" options. Acacia with the default settings was used to correct the error of the sequences. Chimeras were checked using the *parallel_identify_chimeric_seqs.py* script with the parameters "− d 4 and −n 2" and RIM-DB as a reference database [12] and removed using the *filter_fasta.py* script. Sequences were clustered into operational taxonomic units (OTUs) by the *pick_otus.py* script with the default clustering method UCLUST at a distance of 0.01. Representative sequences of OTUs were obtained and classified using the *parallel_assign_taxonomy_blast.py* script with RIM-DB as a reference database for taxonomic assignment. OTU tables were generated using the *make_otu_table.py* script. OTUs that contained less than ten sequences and were not classified as *Methanomassiliicoccales* were removed from the OTU tables. The Chao1 richness index was calculated according to Chao and Bunge [20]. Shannon diversity indices were

calculated according to Shannon [21]. Pielou's evenness index was calculated according to Pielou [22]. The principal coordinates analysis (PcoA) was performed via the unweighted UniFrac distance method [23]. Before the following analysis was performed, the OTU tables were filtered. OTUs that occurred in fewer than three samples were removed from the OTU tables. The predominant OTUs (≥5% of the relative abundance in one of the 10 samples) were used to form the heat map. Hierarchical clustering based on the predominant OTUs was performed with HemI 1.0 software as described by Deng et al. [24].

Phylogenetic tree construction

The representative sequences of highly abundant OTUs (relative abundance > 5% in one of the 10 samples) were aligned by MEGA 7 [25] (http://www.megasoftware.net/). Sixty-three sequences from Groups 8, 9, and 10 and from isolates and enrichment cultures were selected from the RIM-DB. The phylogenetic analysis was performed on the 762 bp-to-1099 bp region of the 16S rRNA genes via the Maximum Likelihood method based on the Kimura 2-parameter model. The tree was resampled 500 times. The dendrogram was rooted with four *Methanopyrus* sequences.

Statistical analyses

The statistical calculations were carried out with appropriate tests using the SPSS software package (SPSS v. 20.0, SPSS Inc., Chicago, IL, USA). The data are shown as means ± standard deviations. The goat was the experimental unit for all comparisons, and diet was regarded as the fixed effect. The normality of the distribution of variables was tested via the Shapiro–Wilk test. The independent samples t-test procedure was used to analyze the variables found to have a normal distribution. The variables found to have a non-normal distributions were analyzed using the Kruskal–Wallis test procedure. Significance was defined as $P < 0.05$.

Results

The pH values of the rumen contents for the hay-diet animals (6.12 ± 0.09) were significantly higher than those for the high-grain-diet animals (5.33 ± 0.09, $P < 0.05$).

The relative abundance of *Methanomassiliicoccales* in the goat rumen as measured via real-time PCR

The real-time PCR results showed that the relative abundance of *Methanomassiliicoccales* (vs. total archaea) in the rumen of goats fed the hay diet was 8.20% ± 2.43%, which was higher than that (0.50% ± 0.38%) in the rumen of goats fed the high-grain diet ($P = 0.002$).

Diversity of *Methanomassiliicoccales* in goat rumen

Across all ten samples, 55,914 trimmed sequences were obtained. The average length of the trimmed sequences was 321 bp. Of these sequences, 51,345 were classified as *Methanomassiliicoccales*, and 180 were classified as Methanomicrobia or Methanobacteria or remained unclassified. Across all reads from the ten samples, a total of 208 OTUs were formed after the removal the OTUs containing less than ten sequences. All the sequences were classified as Methanomassiliicoccaceae at family level, and most of the sequences (96.82% ± 1.64%) were further classified as Group 8, 9, and 10 at the genus level of *Methanomassiliicoccales* based on the RIM-DB (Fig. 1).

Differences in diversity and evenness between the hay- and high-grain-diet groups were determined via an OTU-level-based calculation of the number of OTUs, as well as the Chao1, Shannon, and Pielou's evenness indexes. No significant differences were noted between the two diet groups regarding for the number of OTUs (108.0 ± 19.9 vs. 86.4 ± 6.5, $P = 0.050$), Chao1 index (133.5 ± 22.9 vs. 115.9 ± 27.4, $P = 0.302$), Shannon index (3.98 ± 0.72 vs. 3.46 ± 0.19, $P = 0.181$), or Pielou's evenness index (0.85 ± 0.12 vs. 0.77 ± 0.04, $P = 0.261$).

The influences of diets on the *Methanomassiliicoccales* community

At the genus level (Fig. 1), the relative abundances of Group 10 (67.25 ± 12.76 vs. 38.13 ± 15.66, $P = 0.012$) and Group 4 (2.07 ± 1.30 vs. 0.27 ± 0.30, $P = 0.035$) were significantly higher in the high-grain-diet group, while the relative abundance of Group 9 was significantly higher in the hay-diet group (18.82 ± 6.20 vs. 47.14 ± 17.72, $P = 0.020$). There were no significant differences observed for Group 8 (10.00 ± 6.91 vs. 12.30 ± 3.65, $P = 0.529$) or unclassified Methanomassiliicoccaceae (1.51 ± 1.24 vs. 1.57 ± 0.86, $P = 0.0.934$) between the two diet groups.

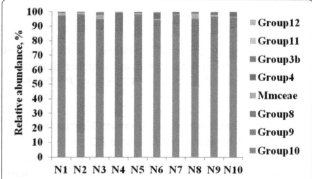

Fig. 1 The relative abundance of *Methanomassiliicoccales* at the genus level based on the RIM-DB. N1 ~ 10, Name of each animal; Mmceae, Methanomassiliicoccaceae

At the species level (Fig. 2), the relative abundance of Group 10 sp. (67.25 ± 12.76 vs. 38.13 ± 15.66, $P = 0.012$) and Group 4 sp. MpT1 (2.07 ± 1.30 vs. 0.27 ± 0.30, $P = 0.035$) were significantly higher in the high-grain diet group, while the relative abundance of Group 9 sp. ISO4-G1 was significantly higher in the hay-diet group (12.83 ± 3.87 vs. 42.44 ± 18.47, $P = 0.022$). There was no significant difference observed for Group 8 sp. WGK1 (10.00 ± 6.91 vs. 12.30 ± 3.65, $P = 0.529$), Group 9 sp. (5.99 ± 2.34 vs. 4.70 ± 2.10, $P = 0.387$) and unclassified Methanomassiliicoccaceae (1.51 ± 1.24 vs. 1.57 ± 0.86, $P = 0.0934$) between the two diet groups.

At the OTU level (Fig. 3), among the dominant OTUs (≥5% of the relative abundance in at least one of the samples), only the relative abundances of OTU1468 (belonging to Group 10, 30.28% ± 7.93% vs. 6.61% ± 8.29%, $P = 0.002$) was significantly higher in the high-grain-diet group.

Principal coordinate analysis (PCoA) plots based on unweighted UniFrac distance metrics showed that the community of *Methanomassiliicoccales* was altered by the hay- and high-grain-diet using PC1 and PC2 (69.58% and 16.64%, respectively, of the explained variance, Fig. 4). The hay-diet group displayed relatively larger individual differences, as compared with the high-grain diet group.

Identification and phylogenetic analysis of dominant OTUs in *Methanomassiliicoccales*

Thirteen predominant OTUs contributed 71.41% (±8.15%) of the total relative abundance of *Methanomassiliicoccales* in the hay-diet group and 79.79% (±3.84%) in the high-grain-diet group. A phylogenetic analysis of the representative sequences from the 13 predominant OTUs revealed that all 13 OTUs were situated in Groups 8, 9, and 10 (Fig. 5). The representative sequences of the two most highly abundant OTUs in each group were blasted in GenBank. The most closely related sequences of these OTUs were identified, and all were uncultivated representatives. OTU1331 and OTU1431 had the highest mean relative abundance (17.84% ± 27.11% and 15.43% ± 17.87%) in the hay-diet group and were situated in *Methanomassiliicoccales* Group 9. OTU1331 and OTU1431 were 98% and 97% similar to sheep rumen mixed culture ISO4-G1 (CP013703), respectively. OTU1331 was widely distributed among ruminants; it was 100% similar to water buffalo rumen clone BBC-21 (JF9517736), *Budorcas* rumen clone f142b (KM650116), sika deer rumen clone SDmet98 (KC454162), cattle rumen clone QTPC108 (JF807170), yak rumen clone QTPYAK64 (JF807240), and other clones. OTU1431 showed 99% similarity to the above clone sequences. OTU511 and OTU1468 had the highest relative abundance (25.07% ± 15.63% and 30.28% ± 7.93%) in the high-grain-diet groups and were situated in Group10.

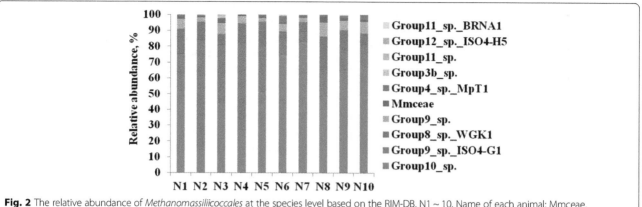

Fig. 2 The relative abundance of *Methanomassiliicoccales* at the species level based on the RIM-DB. N1 ~ 10, Name of each animal; Mmceae, Methanomassiliicoccaceae

OTU511 was 96% similar to *Candidatus Methanoplasma termitum* strain MpT1 and 99% similar to the well-known sheep rumen clone NZCRCC002 (HM624057), feedlot cattle clone ON-CAN.16 (DQ123886), and other clones. No cultured representatives are currently available for Group 10. OTU1468 was 94% similar to *Ca. M. termitum* strain MpT1 and 99% similar to water buffalo rumen clone G-56 (AB906274), sika deer rumen clone SDcsmet90 (KC454244), and other clones.

Discussion

A comprehensive understanding of the archaeal communities in the rumen is important for rumen methane mitigation and the improvement of animal performance. For the *Methanomassiliicoccales,* as a new archaeal order, information on community and taxonomy remains limited. St-Pierre and Wright [5] reported substantial variability in the abundance of *Methanomassiliicoccales* in the rumens of different animals. The present study used real-time PCR, as well as a pyrosequencing approach and the RIM-DB database, to describe the abundance and community of *Methanomassiliicoccales* in the rumens of Chinese goats. These rumens had a relatively low abundance of *Methanomassiliicoccales* in the archaeal community, which complicated the sequencing of the 16S rRNA. For this reason, nested PCR was employed in the present study for the amplification of the 16S rRNA gene sequences of *Methanomassiliicoccales* in the goat rumen contents. However, it should be noted that performing nested PCR prior to in-depth

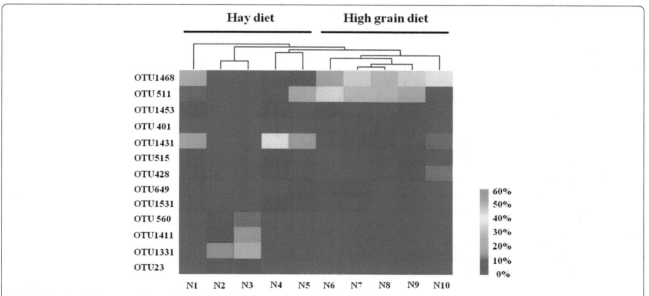

Fig. 3 Hierarchical clustering of samples based on the relative abundance of each OTU in each sample. Only OTUs with a relative abundance of at least 5% in one of the ten samples were included. The names of animals are shown below the heat map (N1 ~ 10). Heat map colors represent the relative abundance of OTUs (%)

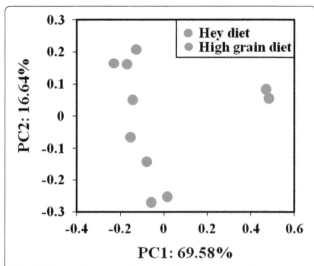

Fig. 4 Effects of diet on *Methanomassiliicoccales* community structure. A principal coordinate analysis (PcoA) plot of unweighted unifrac dissimilarities is shown

sequencing may introduce biases regarding estimated microbial diversity and community structure, especially for communities with relatively higher levels of diversity [26]. Compared with standard PCR, nested PCR may yeild fewer OTUs, lower Shannon and Chao1 indeces, as well as increases or decreases in the relative abundances of some taxa [26].

Pyrosequencing targeted at the 16S rRNA genes of this new methanogenic order revealed a diverse community of *Methanomassiliicoccales* in the goat rumen, which was consistent with previous findings for sheep, cattle, and red deer in New Zealand obtained using the denaturing gradient gel electrophresis (DGGE) method [14]. To date, only one pure strain and a few enriched cultures have been described in this order, and no type strains are available for most of the phylogenetic groups of *Methanomassiliicoccales*. This limited further taxonomic analysis of the sequences in the present study. Most sequences were assigned to one of three groups (Groups 8, 9, and 10) based on the RIM-DB [12], which was inconsistent with the findings of Seedorf et al. [6, 12], in which most of the ruminal sequences from New Zealand sheep and cattle were assigned to Groups 10, 11, and 12. This difference should be taken into account when the methane mitigation tools, such as vaccines or small-molecule inhibitors, are developed for various ruminants.

The most highly abundant OTUs in the goat rumen were uncultured. The two most highly abundant OTUs in the hay-diet group were different from the two most highly abundant OTUs in the high-grain-diet group. A high-grain diet is a well-known general cause of pH decreases in the rumen, which could suppress methanogens that are sensitive to low pH values [27]. Thus, the

four species may have specific properties that allowed their survival in their corresponding habitats. The genome of *Ca. M. alvus* encodes a choloylglycine hydrolase that imparts resistance to bile salts encountered in the GIT [28]. Another adaptation to GIT could be inferred by the presence of a conserved amino acid domain corresponding to COG0790. *Ca. M. alvus* encoded 28 genes related to the proteins within this domain, *Ca. M. intestinalis* encoded six such genes, and *M. luminyensis* encoded one, which suggests differences in the adaptation of these three species to the digestive tract [28]. Söllinger et al. [10] reported larger genomes and a larger number of genes encoding anti-oxidative enzymes in the members of *Methanomassiliicoccales* from the 'environmental' clade than in those from the GIT clade, indicating that the GIT members were restricted and specialized to the GIT habitat. Further isolation and genomic analysis may help to reveal the physiological properties of these novel methanogens, as well as their genomic adaptation to their habitats.

The high-grain diet decreased the relative abundance of *Methanomassiliicoccales* in the rumen, possibly due to its pH-lowering effect in the rumen, suggesting that the growth of *Methanomassiliicoccales* was depressed. Methylamines and methanol are the substrates of *Methanomassiliicoccales* [1, 8, 9]. The suppression of the growth of *Methanomassiliicoccales* in the high-grain-diet group might cause an accumulation of these methyl compounds in the rumen. Ametaj et al. [29] observed an accumulation of methylamines and methanol in the rumen liquid with the increasing proportions of cereal grain in the diet. It is worthwhile to note the effect of the accumulation of these methyl compounds on the physiology of the rumen and the health of the host animals.

The relative abundances of *Methanomassiliicoccales* in the rumen contents of Chinese goats were $0.5 \pm 0.2\%$ with the high-grain diet and $8.2 \pm 1.1\%$ with the hay diet. St-Pierre and Wright [5] described a wide variation (~80.8%) in the relative abundance of ruminal *Methanomassiliicoccales* based on host species, diet, and geographical locations. Seedorf et al. [6] reported that *Methanosphaera* and *Methanomassiliicoccales* OTUs were highly abundant in lucerne-fed sheep as compared to pasture-fed sheep. This high abundance may be related to the levels of the precursors of the substrates in the animal's diets, but this requires further confirmation. The exact factors contributing to the high proportion of *Methanomassiliicoccales* in some animals remains unclear.

The communities of *Methanomassiliicoccales* were altered at OTU level by diet using PcoA analysis and hierarchical clustering, respectively. Apart from the low pH, the observed differences may be related to changes in the community structure of the other prominent

Fig. 5 Phylogeny of *Methanomassiliicoccales* sequences and highly abundant OTUs. Representative sequences of highly abundant OTUs (bold, ≥5% of the relative abundance in one of the ten samples) were aligned by MEGA 7. The GenBank accession numbers of these highly abundant OTUs are KX525649–61. Sixty-three sequences from Groups 8, 9, and 10 and isolates and enrichment cultures were selected in the RIM-DB. Phylogenetic analysis was performed on the 762 bp-to-1099 bp region of the 16S rRNA genes via the Maximum Likelihood method based on the Kimura 2-parameter model. The tree was resampled 500 times, and only bootstrap values of ≥70% are shown. The dendrogram was rooted with four *Methanopyrus* sequences. The scale bar indicates 0.02 inferred nucleotide substitutions per position. *Mmc., Methanomassiliicoccus; Mmp., Methanomethylophilus*

microbial organism groups (bacteria, protozoa, and fungi). Previous studies have reported a close relationship between *Methanomassiliicoccales* species and protozoa [30] and fungi [31]. Most of the cultured *Methanomassiliicoccales* species were enriched with bacteria that were difficult to remove, which suggested a close interaction between them. Information is limited regarding the relationships between *Methanomassiliicoccales* species and other rumen microorganisms, but this information is important in the effect to reveal the factors affecting the *Methanomassiliicoccales* community. A co-culture approach may be a feasible way of studing these relationships [31].

Conclusion

The data presented here have shown that at the genus level, only a few groups were dominant in the Chinese goat rumen. Overall, *Methanomassiliicoccales* were present at a low relative abundance among the methanogen community; therefore, they are probably less important contributors to the rumen methane emission of the Chinese goats. Furthermore, the abundance of *Methanomassiliicoccales* in the rumen was easily disturbed by dietary changes. A better understanding of the community of *Methanomassiliicoccales* in the rumen could help to identify the role of this new order in the rumen.

Acknowledgements
Not applicable.

Funding
This work was supported by the Natural Science Foundation of China (31301999, 31101735), and the Fundamental Research Funds for the Central Universities (KYZ201412).

Authors' contributions
WJ conceived and performed this study and drafted this manuscript. YC analyzed the pyrosequencing data and revised this manuscript. WZ conceived this study and revised the manuscript. All authors read and approved the final manuscript.

Competing interests
The authors declare that they have no competing interests.

References
1. Dridi B, Fardeau ML, Ollivier B, Raoult D, Drancourt M. *Methanomassiliicoccus luminyensisgen* nov., sp. nov., a methanogenic archaeon isolated from human faeces. Int J Syst Evol Microbiol. 2012;62(8):1902–7. doi:10.1099/ijs.0.033712-0.
2. Iino T, Tamaki H, Tamazawa S, Ueno Y, Ohkuma M, Suzuki K, et al. *Candidatus Methanogranum caenicola*: a novel methanogen from the anaerobic digested sludge, and proposal of *Methanomassiliicoccaceae* fam. nov. and *Methanomassiliicoccales* ord. nov., for a methanogenic lineage of the class Thermoplasmata. Microbes Environ. 2013;28(2):244–50. doi:10.1264/jsme2.ME12189.
3. Janssen PH, Kirs M. Structure of the archaeal community of the rumen. Appl Environ Microbiol. 2008;74(12):3619–25. doi:10.1128/AEM.02812-07.
4. Paul K, Nonoh JO, Mikulski L, Brune A. Methanoplasmatales, Thermoplasmatales-related archaea in termite guts and other environments, are the seventh order of methanogens. Appl Environ Microbiol. 2012;78(23):8245–53. doi:10.1128/AEM.02193-12.
5. St-Pierre B, Wright ADG. Diversity of gut methanogens in herbivorous animals. Animal. 2013;7 Suppl 1:49–56. doi:10.1017/S1751731112000912.
6. Seedorf H, Kittelmann S, Janssen PH. Few highly abundant operational taxonomic units dominate within rumen methanogenic archaeal species in New Zealand sheep and cattle. Appl Environ Microbiol. 2015;81(3):986–95. doi:10.1128/AEM.03018-14.
7. Li Y, Leahy SC, Jeyanathan J, Henderson G, Cox F, Altermann E, et al. The complete genome sequence of the methanogenic archaeon ISO4-H5 provides insights into the methylotrophic lifestyle of a ruminal representative of the Methanomassiliicoccales. Stand Genomic Sci. 2016;11(1):59. doi:10.1186/s40793-016-0183-5.
8. Brugère JF, Borrel G, Gaci N, Tottey W, O'Toole PW, Malpuech-Brugère C. Archaebiotics: proposed therapeutic use of archaea to prevent trimethylaminuria and cardiovascular disease. Gut Microbes. 2014;5(1):5–10. doi:10.4161/gmic.26749.
9. Lang K, Schuldes J, Klingl A, Poehlein A, Daniel R, Brunea A. New Mode of Energy Metabolism in the Seventh Order of Methanogens as Revealed by Comparative Genome Analysis of"Candidatus Methanoplasma termitum". Appl Environ Microbiol. 2015;81(4):1338–52. doi:10.1128/AEM.03389-14.
10. Söllinger A, Schwab C, Weinmaier T, Loy A, Tveit AT, Schleper C, et al. Phylogenetic and genomic analysis of Methanomassiliicoccales in wetlands and animal intestinal tracts reveals clade-specific habitat preferences. FEMS Microbiol. Ecol. 2016;92(1):fiv149. doi:10.1093/femsec/fiv149.
11. Borrel G, O'Toole PW, Harris HM, Peyret P, Brugère JF, Gribaldo S. Phylogenomic data support a seventh order of Methylotrophic methanogens and provide insights into the evolution of Methanogenesis. Genome Biol Evol. 2013;5(10):1769–80. doi:10.1093/gbe/evt128.
12. Seedorf H, Kittelmann S, Henderson G, Janssen PH. RIM-DB: a taxonomic framework for community structure analysis of methanogenic archaea from the rumen and other intestinal environments. Peer J. 2014;2:e494. doi:10.7717/peerj.494.
13. Liu JH, Xu TT, Liu YJ, Zhu WY, Mao SY. A high grain diet causes massive disruption of ruminal epithelial tight junctions in goats. Am J Physiol Regul Integr Comp Physiol. 2013;305:R232–41. doi:10.1152/ajpregu.00068.2013.
14. Wright ADG, Dehority BA, Lynn DH. Phylogeny of the rumen ciliates Entodinium, Epidinium and Polyplastron (Litostomatea: Entodiniomorphida) inferred from small subunit ribosomal RNA sequences. J Euk Microbiol. 1997;44(1):61–7. doi:10.1111/j.1550-7408.1997.tb05693.x.
15. Jeyanathan J, Kirs M, Ronimus RS, Hoskin SO, Janssen PH. Methanogen community structure in the rumens of farmed sheep, cattle and red deer fed different diets. FEMS Microbiol Ecol. 2011;76(2):311–26. doi:10.1111/j.1574-6941.2011.01056.
16. Koike S, Handa Y, Goto H, Sakai K, Miyagawa E, Matsui H, et al. Molecular monitoring and isolation of previously uncultured bacterial strains from the sheep rumen. Appl Environ Microbiol. 2010;76(6):1887–94. doi:10.1128/AEM.02606-09.
17. Wright ADG, Williams AJ, Winder B, Christophersen C, Rodgers S, Smith K. Molecular diversity of rumen methanogens from sheep in Western Australia. Appl Environ Microbiol. 2004;70(3):1263–70. doi:10.1128/AEM.70.3.1263-1270.2004.
18. Tamaki H, Wright CL, Li X, Lin Q, Hwang C, Wang S, et al. Analysis of 16S rRNA amplicon sequencing options on the Roche/454 next-generation titanium sequencing platform. PLoS One. 2011;6(9):e25263. doi:10.1371/journal.pone.0025263.
19. Caporaso JG, Kuczynski J, Stombaugh J, Bittinger K, Bushman FD, Costello EK, et al. QIIME allows analysis of high-throughput community sequencing data. Nat Methods. 2010;7(5):335–6. doi:10.1038/nmeth.f.303.
20. Chao A, Bunge J. Estimating the number of species in a stochastic abundance model. Biometrics. 2002;58(3):531–9. doi:10.1111/j.0006-341X.2002.00531.x.
21. Shannon CE. A mathematical theory of communication. Bell Syst Tech J. 1948;27:379–423. doi:10.1002/j.1538-7305.1948.tb01338.x.
22. Pielou E. The measurement of diversity in different types of biological collections. J Theor Biol. 1966;13:131–44. doi:10.1016/0022-5193(66)90013-0.
23. Lozupone C, Knight R. UniFrac:anewphylogenetic methodforcomparing microbial communities. Appl Environ Microbiol. 2005;71:8228–35. doi:10.1128/AEM.71.12.8228-8235.
24. Deng W, Wang Y, Liu Z, Cheng H, Xue Y. HemI: A Toolkit for Illustrating Heatmaps. PLoS One. 2014;9(11):e111988. doi:10.1371/journal.pone.0111988.
25. Kumar S, Stecher G, Tamura K. MEGA7: Molecular Evolutionary Genetics Analysis version 7.0 for bigger datasets. Mol Biol Evol. 2016;33:1870–4. doi:10.1093/molbev/msw054.
26. Yu G, Fadrosh D, Goedert JJ, Ravel J, Goldstein AM. Nested PCR biases in interpreting microbial community structure in 16S rRNA gene sequence datasets. PLoS One. 2015;10(7):e0132253. doi:10.1371/journal.pone.0132253.
27. Lana RP, Russell JB, Van Amburgh ME. The role of pH in regulating ruminal methane and ammonia production. J Anim Sci. 1998;76(8):2190–6. doi:10.2527/1998.7682190x.
28. Borrel G, Parisot N, Harris HMB, Peyretaillade E, Gaci N, Tottey W, et al. Comparative genomics highlights the unique biology of Methanomassiliicoccales, a Thermoplasmatales-related seventh order of methanogenic archaea that encodes pyrrolysine. BMC Genomics. 2014;15:679–702. doi:10.1186/1471-2164-15-679.
29. Ametaj BN, Zebeli Q, Saleem F, Psychogios N, Lewis MJ, Dunn SM, et al. Metabolomics reveals unhealthy alterations in rumen metabolism with increased proportion of cereal grain in the diet of dairy cows. Metabolomics. 2010;6:583–94. doi:10.1007/s11306-010-0227-6.
30. Irbis C, Ushida K. Detection of methanogens and proteobacteria from a single cell of rumen ciliate protozoa. J Gen Appl Microbiol. 2004;50:203–12. doi:10.2323/jgam.50.203.

10

Nutritional value of a partially defatted and a highly defatted black soldier fly larvae (*Hermetia illucens* L.) meal for broiler chickens: apparent nutrient digestibility, apparent metabolizable energy and apparent ileal amino acid digestibility

Achille Schiavone[1,4]*⬛, Michele De Marco[1], Silvia Martínez[2], Sihem Dabbou[1,3], Manuela Renna[3], Josefa Madrid[2], Fuensanta Hernandez[2], Luca Rotolo[3], Pierluca Costa[1], Francesco Gai[4] and Laura Gasco[3,4]

Abstract

Background: The study aimed to determine the apparent total tract digestibility coefficients (ATTDC) of nutrients, the apparent metabolizable energy (AME and AMEn) and the amino acid (AA) apparent ileal digestibility coefficients (AIDC) of a partially defatted (BSFp) and a highly defatted (BSFh) black soldier fly larvae meal. The experimental diets were: a basal diet and two diets prepared by substituting 250 g/kg (w/w) of the basal diet with BSFp or BSFh, respectively.

Results: Significant differences were found between BSFp and BSFh meals for ATTDC of the nutrients: BSFp resulted more digestible than BSFh, except for ATTDC of CP which did not differed between meals, while a statistical trend was observed for ATTDC of DM and EE. The AME and AMEn values were significantly ($P < 0.05$) different between the two BSF meals, with higher levels for BSFp (16.25 and 14.87 MJ/kg DM, respectively). The AIDC of the AA in BSFp ranged from 0.44 to 0.92, while in BSFh they ranged from 0.45 to 0.99. No significant differences were observed for the AA digestibility (0.77 and 0.80 for BSFp and BSFh, respectively), except for glutamic acid, proline and serine that were more digestible in the BSFh meal ($P < 0.05$).

Conclusions: Defatted BSF meals can be considered as an excellent source of AME and digestible AA for broilers with a better efficient nutrient digestion. These considerations suggested the effective utilization of defatted BSF larvae meal in poultry feed formulation.

Keywords: Amino acid, Apparent digestibility, Black soldier fly meal, Broiler chicken, Metabolizable energy

Background

World population is expected to grow by over a third, reaching over 9 billion people in 2050 having as main consequence that the world will have to produce 70% more food [1]. Consequently, livestock production (in particular that of poultry and swine) will grow exponentially reaching up to double of the current production. Therefore, the foremost gamble will be to guarantee the global capacity to provide enough animal feed (in particular protein ingredients) trying to avoid as much as possible competition with human food demand. For this purpose, insects have been already proposed as a high quality, efficient and sustainable alternative protein source for poultry [2], fish [3–5] or swine [6]. One of the most promising insects species identified for industrial production in the Western world is the black soldier fly

* Correspondence: achille.schiavone@unito.it
[1]Department of Veterinary Sciences, University of Turin, Largo Paolo Braccini 2, 10095 Grugliasco, TO, Italy
[4]Institute of Science of Food Production, National Research Council, Largo Paolo Braccini 2, 10095 Grugliasco, TO, Italy
Full list of author information is available at the end of the article

(BSF, *Hermetia illucens* L.). This insect is normally reared on materials that are unsuitable for human nutrition (e.g., by-products from food processing, organic waste) [7, 8], reaching high growth rates with a good feed conversion [9]. Using different rearing substrates derived from food manufacturing (beet molasses, potato steam peelings, spent grains and beer yeast, bread remains, and cookie remains) and differing in lipid and fat contents, BSF showed a feed conversion rate that ranged between 1.4 and 2.6 [9]. BSF is able to convert these substrates in high quality protein, the content of which ranges from 38% to 46% of DM [9]. The amino acids (AA) composition of BSF is rich in methionine and lysine (9.05 and 22.3 g/kg DM, respectively) [9], and is reported to be similar or even superior than that of soybean [10]. However, nowadays knowledge about the suitability of the use of BSF as poultry feed ingredient is scarce and little-to-date. At this regard, acceptability by animals, feed conversion rates, animal health issues and quality of the obtainable animal-derived food products are of particular interest and not yet investigated sufficiently [11]. In general, little information on the digestibility of insects in livestock species is currently available. Published data on the apparent metabolizable energy (AME) of BSF for broilers are limited as also those on the ileal amino acid digestibility [12]. That limitation consisted on this information is often obtained from a whole insect meal without separation of the fat fraction from the protein one. As it is known, for an efficient and optimal broiler feed formulation the best situation is to manage fat and protein sources separately. For this purpose, and because insect farming is going to be a growing market, nowadays BSF manufacturers have started to produce defatted BSF meals. Defatting of insects can be obtained mechanically by cutting the frozen insect larvae and then pressing them to enable the leakage of intracellular fat [13], or chemically using petroleum ether extraction of the insect meal [14]. The defatting process results in meals with larger protein values (476 to 583 g/kg DM for defatted BSF meals) [13, 15], exceeding those usually found in soybean meals [14, 16, 17]. Henry et al. [5] suggested the importance of defatting insect meals, then using insect protein concentrate as animal feed ingredient and the lipids both for animal nutrition and the production of biodiesel [18, 19].

In order to provide new information useful to facilitate poultry feed formulation, the aim of this study was to determine the total tract digestibility coefficients (ATTDC), the AME, the nitrogen-corrected AME (AMEn) and the AA apparent ileal digestibility coefficients (AIDC) of a partially and a highly defatted BSF meal for broiler chickens.

Methods
The study was performed at the poultry facility of the Department of Agricultural, Forest and Food Sciences of the University of Turin (Italy). The experimental protocol was designed according to the guidelines of the current European and Italian laws on the care and use of experimental animals (European Directive 86/609/EEC, put into law in Italy with D.L. 116/92) and was approved by the Ethical Committee of the Department of Veterinary Sciences of the University of Turin (Italy) (protocol number 01/28/06/2016).

Insect meals
Two BSF meals, differing for their fat content, were used in this trial. Both insect meals were obtained from Hermetia Deutschland GmbH & Co. KG, Baruth/Mark (Germany). The two BSF meals derived from larvae which were fed with cereal by-products. At collection, the larvae weighed around 150–220 mg. The collected larvae were dried for 20 h in an oven at low temperature (60 °C) and ground to a meal. BSF meals were partially and highly defatted. The process was performed using high pressure and without solvents.

Pre-experimental period
A total of one hundred one-day-old male broiler chickens (Ross 308) were raised in a floor pen till d 21 and fed a commercial broiler starter diet (227 g/kg of crude protein (CP); 13.4 MJ/kg of AME). All the birds were vaccinated at hatching against Newcastle disease, Marek disease, infectious bronchitis and coccidiosis. At d 21, sixty birds of uniform body weight (822.0 ± 47.7 g) were chosen and randomly distributed over fifteen cages (4 birds per cage). The cages (120 cm × 60 cm) were placed in an insulated room provided with devices to control temperature, light and humidity. Each cage had a linear feeder at the front and a nipple drinker at the back. Health status and mortality of the birds were monitored daily throughout the whole experimental period. From d 21, the birds were fed a commercial broiler diet (190 g/kg of CP; 13.6 MJ/kg AME) until the assay diets were introduced on d 26. During the pre-experimental period, feeds and water were provided *ad libitum*.

Digestibility trial
On d 26, five replicate cages were randomly assigned to one out of three experimental diets. A basal diet, based on corn and soybean meal, was formulated (Table 1), and two experimental diets were subsequently formulated by substituting 250 g/kg (w/w) of the basal diet with BSFp or BSFh, respectively. All diets contained titanium oxide (TiO, 5 g/kg) as an indigestible marker to calculate the ileal digestibility of amino acids. The diet adaptation period lasted 6 d. Starting at the age of 32 d, the excreta were collected per cage over a period of four days in order to evaluate the total tract digestibility. Fresh feeds and water were available ad libitum during the whole experimental period. Feed intake per cage was

Table 1 Composition (g/kg as fed) of the basal diet

Ingredients	
Maize meal	582.9
Soybean meal	350.0
Soybean oil	36.0
Dicalcium phosphate	12.4
Calcium carbonate	11.2
Sodium chloride	2.0
Sodium bicarbonate	1.5
Trace mineral-vitamin premix[a]	4.0
Calculated analysis	
AME, MJ/kg	12.3
Crude protein	214
Methionine	1.0
Lysine	10.1
Threonine	7.1
Calcium	8.8
Phosphorous	5.8

AME apparent metabolizable energy

[a]Mineral-vitamin premix (Final B Prisma, IZA SRL), given values are supplied per kg diet: 2,500,000 IU of vitamin A; 1,000,000 IU of vitamin D_3; 7,000 IU of vitamin E; 700 mg of vitamin K; 400 mg of vitamin B_1; 800 mg of vitamin B_2; 400 mg of vitamin B_6; 4 mg of vitamin B_{12}; 30 mg of biotin; 3,111 mg of Ca pantothenate acid; 100 mg of folic acid; 15,000 mg of vitamin C; 5,600 mg of vitamin B3; 10,500 mg of Zn, 10,920 mg of Fe; 9,960 mg of Mn; 3850 mg of Cu; 137 mg of I; 70 mg of Se

measured throughout the experiment and the excreta were sampled daily during the test period. The total fresh excreta per cage was weighed daily, frozen at –20 °C and lyophilized. Four days excreta per cage were pooled for further analysis.

On d 35, all the birds were euthanized by intravenous injection of sodium pentobarbital and the content of the lower half of the ileum was collected, according to the procedures described by Ravindran et al. [20]. The ileum was defined as the portion of small intestine extending from Meckel's diverticulum to a point 40 mm proximal to the ileo-cecal junction. The ileal content for each cage was pooled, lyophilized, ground to pass through a 0.5-mm sieve, and stored at –20 °C in airtight containers until laboratory analyses were conducted.

Chemical analysis

Insect meals, diet samples and dried excreta were ground to pass through a 0.5-mm sieve and stored in airtight plastic containers. They were analyzed for DM (# 930.15), ash (# 924.05), CP (# 984.13) and ether extract (EE, # 920.39) according to AOAC procedures [21]. Gross energy (GE) was measured using an adiabatic calorimetric bomb (C7000, IKA, Staufen, Germany). Chitin was analyzed as D-Glucosamine [22] using a modification of method described by Madrid et al. [23] for AA.

The uric acid (UA) content in the excreta samples was determined spectrophotometrically (UNICAN UV–Vis Spectrometry, Helios Gamma, United Kingdom) according to Marquardt method [24]. The CP amount of excreta was calculated as follows: CP = (total nitrogen – UA-nitrogen) × 6.25. All analyses were carried out on three replicates for each sample.

The apparent digestibility trial was performed, using the total excreta collection method, to determine ATTDC for DM, organic matter (OM), CP, EE, GE and the AME.

Ileal content samples from each cage were analyzed for DM and AA. In order to perform the AA determination, samples of the diets, ileal digesta and insect larvae meals were prepared using a 22-h hydrolysis step in 6 NHCl at 112 °C under a nitrogen atmosphere. The AA in hydrolysate was determined by HPLC (Waters Alliance System with a Waters 1525 Binary HPLC pump, Waters 2707 autosampler and Waters 2475 multi λ Fluorescence Detector, Milford, USA) after derivatization, according to the procedure described by Madrid et al. [23]. Tryptophan was not determined. Diets samples and ileal content samples were analyzed for TiO content. TiO content was measured on a UV spectrophotometer (UNICAN UV–vis Spectrometry, Helios Gamma, United Kingdom) following the method of Short et al. [25].

Calculations

The ATTDC of the dietary nutrients, the AME and the AIDC of the AA were calculated using two different methods [20, 26].

The ATTDC of the dietary nutrients of the insect larvae meals were calculated as follows:

$$\text{ATTDC X}_{\text{diet}} = [(\text{total X ingested} - \text{total X excreted}) / \text{total X ingested}]$$

$$\text{ATTDC X}_{\text{insect larvae meal}} = [\text{ATTDC X of insect larvae meal diet} - (\text{ATTDC X of basal diet} \times 0.75)] / 0.25.$$

where X represents DM, OM, CP, EE or GE.

The AME values of the insect larvae meals were calculated using the following formula with appropriate corrections made according to the differences in the DM content:

$$\text{AME}_{\text{diet}}(\text{MJ/kg}) = [(\text{feed intake} \times \text{GE diet}) - (\text{excreta output} \times \text{GE excreta})] / \text{feed intake}$$

$$\text{AME}_{\text{insect larvae meal}}(\text{MJ/kg}) = [\text{AME of insect larvae meal diet} - (\text{AME basal diet} \times 0.75)] / 0.25$$

Correction for zero nitrogen (N) retention was made using a factor of 36.54 KJ per gram N retained in the body in order to estimate the N-corrected apparent metabolizable energy (AMEn) [27]. N-retention was calculated using the following formula:

$$N_{retention} = [(\text{feed intake} \times \text{N diet})$$
$$- (\text{excreta output} \times \text{N excreta})] / \text{feed intake (kg)}$$

The AIDC of the AA of the insect larvae meals was calculated, using TiO as the indigestible marker, as follows:

$$\text{AIDC of AAX}_{diet} = (\text{AAX} / \text{TiO})d - (\text{AAX} / \text{TiO})i / (\text{AAX} / \text{TiO})d$$

$$\text{AIDC of AAX}_{\text{insect larvae meal}}$$
$$= [(\text{AIDC AAX of the insect larvae meal diet}$$
$$\times \text{AAX of the insect larvae meal diet})$$
$$- (\text{AIDC AAX of the basal diet}$$
$$\times \text{AAX of the basal diet} \times 0.75)]$$
$$/ (\text{AAX of the insect larvae meal diet} \times 0.25).$$

where:

$(\text{AA} / \text{TiO})d$ = ratio of the AA and TiO concentrations in the diet;

$(\text{AA} / \text{AIA})i$
    = ratio of the AA and TiO concentrations in the ileal digesta;

AAX : represents each AA evaluated.

Statistical analyses

The statistical analysis of the ATTDC, AME and AIDC of BSFp and BSFh meals was performed with SPSS 17 for Windows (SPSS, Inc., Chicago, IL, USA 2008) [28]. The experimental unit was the cage. Data were analyzed using Mann–Whitney U test. Before testing for group differences, non-normality of the data distribution and homogeneity of variances were assessed using the Shapiro-Wilk and Levene test, respectively. Differences were considered to be significant at $P \le 0.05$. A statistical trend was considered at $P \le 0.10$.

Results

Chemical and amino acid compositions of diets and insect meals

The proximate composition and GE of the two BSF meals and the three experimental diets are summarized in Table 2. The chemical composition of the two BSF meals differed mainly in terms of CP and EE contents. As expected, the BSFp meal showed a lower CP content than the BSFh meal (553 and 655 g/kg DM, respectively). On the contrary, the BSFh meal, being highly defatted, showed a lower EE content than the BSFp meal

(46 and 180 g/kg DM, respectively). The GE contents of the BSFp and BSFh meals were 24.4 and 21.2 MJ/kg DM, respectively. The two defatted BSF meals revealed a relevant chitin content (50 and 69 g/kg DM for BSFp and BSFh, respectively).

The AA compositions of the two BSF larvae meals and of the three experimental diets are presented in Table 3. As expected, the AA content was different in the two BSF meals, and BSFh showed higher content than BSFp for all AA (both indispensable and dispensable). In both meals, leucine was the most abundant indispensable AA (28.6 and 36.7 g/kg DM in BSFp and BSFh, respectively), whereas glutamic acid was the most abundant dispensable one (48.7 and 63.7 g/kg DM in BSFp and BSFh, respectively). Both meals were deficient of cysteine (0.1 and 0.2 g/kg DM in BSFp and BSFh, respectively). BSFh, having higher CP content than BSFp, was the highest source of lysine, threonine and methionine (25.2, 21.8 and 8.56 g/kg DM, respectively).

Apparent nutrient digestibility

The ATTDC of the nutrients, as well as the AME and AMEn of the BSFp and BSFh meals are reported in Table 4. The BSFp meal resulted more digestible than the BSFh one. Significant differences were found between the two BSF meals for the ATTDC of EE and GE ($P < 0.05$), a statistical trend was observed for the ATTDC of DM and OM ($P < 0.10$), while no significant difference was found for ATTDC of CP ($P = 0.834$).

Apparent metabolizable energy

The two BSF meals also differed ($P < 0.05$) for AME and AMEn (Table 4). In particular, BSFp showed mean AME and AMEn values of 16.25 and 14.87 MJ/kg DM, respectively, while for BSFh, the values of AME and AMEn were 11.55 and 9.87 MJ/kg DM, respectively.

Apparent ileal amino acid digestibility

The determined values for the AIDC of the AA are shown in Table 5. The AIDC of the AA in the BSFp meal ranged from 0.44 to 0.92, while in the BSFh meal values ranged from 0.45 to 0.99. Overall, no significant differences were found between the two BSF meals for the AIDC of indispensable AA. The AIDC of lysine was

Table 2 Analyzed chemical composition of the two BSF meals and of the three experimental diets

Items	BSFp meal	BSFh meal	Basal diet	BSFp diet	BSFh diet
Dry matter, g/kg diet	942	985	889	897	906
Organic matter, g/kg DM	901	907	814	817	824
Crude protein, g/kg DM	553	655	246	343	374
Ether extract, g/kg DM	180	46	71	103	72
Gross energy, MJ/kg DM	24.4	21.2	18.7	20.0	19.4
Chitin, g/kg DM	50	69	-	17	21

Table 3 Amino acid concentration (g/kg DM) of the two BSF meals and of the three experimental diets

Items	BSFp meal	BSFh meal	Basal diet	BSFp diet	BSFh diet
Indispensable amino acids					
Arginine	21.5	27.0	14.5	16.0	17.7
Histidine	12.3	16.3	5.73	7.64	8.66
Isoleucine	18.5	24.0	9.27	12.0	13.1
Leucine	28.6	36.7	17.8	20.6	22.9
Lysine	21.0	25.2	12.1	14.5	15.8
Methionine	6.46	8.56	2.43	3.52	4.46
Phenylalanine	16.6	21.8	10.5	12.0	13.4
Threonine	17.2	21.8	8.06	11.0	12.1
Valine	27.2	34.5	10.2	15.4	16.7
Dispensable amino acids					
Alanine	34.5	43.7	10.2	17.8	20.1
Aspartic acid	37.2	48.8	22.5	26.6	30.0
Cysteine	0.1	0.2	2.49	2.35	2.74
Glycine	23.5	30.3	9.03	13.6	15.3
Glutamic acid	48.7	63.7	40.1	41.4	46.2
Proline	30.6	32.7	12.6	16.1	19.0
Serine	20.3	26.8	11.8	14.6	16.5
Tyrosine	26.4	34.1	5.84	11.5	13.5

0.80 for both BSF meals while the AIDC of methionine were 0.83 and 0.78 for BSFp and BSFh, respectively. Among the dispensable AA, no significant differences were found except for glutamic acid (0.81 vs. 0.85), proline (0.65 vs. 0.82) and serine (0.71 vs. 0.77) that were more digestible in the BSFh meal ($P < 0.05$).

Discussion

To the best of our knowledge, this study is the first one testing the ATTDC of nutrients, the AME and AMEn, and the AA AIDC of a partially defatted and a highly defatted black soldier fly larvae meal for broiler chickens.

Table 4 Apparent digestibility coefficients of the total tract (ATTDC) of the nutrients, AME and AMEn of the two BSF meals for broilers ($n = 5$)

Items	BSFp meal	BSFh meal	SEM	P-value
Dry matter	0.63	0.59	0.010	0.092
Organic matter	0.69	0.64	0.012	0.057
Crude protein	0.62	0.62	0.019	0.834
Ether extract	0.98	0.93	0.951	0.008
Gross energy	0.61	0.50	0.021	0.012
AME, MJ/kg DM	16.25	11.55	0.811	0.008
AMEn, MJ/kg DM	14.87	9.87	0.860	0.008

AME apparent metabolizable energy, *AMEn* N-corrected apparent metabolizable energy

Chemical and amino acid compositions of diets and insect meals

The compositional data of our study showed that the two BSF meals obtained after defatting are good sources of dietary protein. Both meals showed a higher CP content than soybean meal. In particular, the BSFh meal with its 655 g/kg of DM, was close to the CP content of meat and fish meals [29, 30]. The BSFp meal had similar CP content of a defatted BSF larvae meal reported by Cullere et al. [31]. The EE content found in this experiment for BSFp was higher than those reported by Cullere et al. [31] who used defatted BSF larvae meals (156 g/kg DM) in broiler quails. These differences could be mainly due to the insect rearing substrate [6, 32] and the defatting process [33, 34], which can both influence the variability in the amounts of EE. Using three rearing substrates differing in their nutrient composition [mixture of middlings (EE: 5.9% DM), dried distillers' grains with solubles (EE: 8.4% DM), and dried sugar beet pulp (EE: 1.1% DM)], Tschirner and Simon [35] reported that the EE content of BSF larvae reached 30.8%, 38.6% and 3.4% of DM, respectively. Fasakin et al. [14] also showed that defatting maggot meal, either oven dried or sun dried, definitely influenced its nutrient composition (EE contents varied from 7.00% to 7.40% DM, respectively). Kroeckel et al. [13] used a defatted BSF meal obtained by cutting the frozen pupae to enable the leakage of intracellular fat from the larvae using tincture press

Table 5 Apparent ileal digestibility coefficients (AIDC) of amino acids of the two BSF meals for broilers ($n = 5$)

Items	BSFp meal	BSFh meal	SEM	P-value
Indispensable amino acids				
Arginine	0.79	0.80	0.015	0.841
Histidine	0.64	0.63	0.022	0.834
Isoleucine	0.83	0.87	0.018	0.293
Leucine	0.84	0.89	0.019	0.141
Lysine	0.80	0.80	0.013	0.753
Methionine	0.83	0.78	0.023	0.293
Phenylalanine	0.82	0.86	0.020	0.249
Threonine	0.73	0.77	0.022	0.675
Valine	0.90	0.91	0.015	0.833
Mean	0.80	0.81	0.016	0.917
Dispensable amino acids				
Alanine	0.92	0.99	0.026	0.249
Aspartic acid	0.82	0.80	0.012	0.234
Cysteine	0.44	0.45	0.017	0.834
Glycine	0.66	0.65	0.020	0.548
Glutamic acid	0.81	0.85	0.013	0.046
Proline	0.65	0.82	0.041	0.008
Serine	0.71	0.77	0.019	0.035
Tyrosine	0.92	0.95	0.019	0.833
Mean	0.74	0.79	0.014	0.114
Overall mean[a]	0.77	0.80	0.015	0.673

[a]Average digestibility of 17 amino acids

(pression at 450 bar and 60 °C for 30 min). These authors reported that the EE and CP contents for such a BSF meal were 118 and 475 g/kg DM, respectively, with high methionine and lysine contents (21.8 and 71.2 g/kg CP). Tschirner and Simon [35] showed that the best fat reduction of BSF larvae (from 30.8 to 16.6% of DM) was achieved by pressing them at 250 bar and 50 °C for 30 min. In addition, Purshke et al. [33] demonstrated that pressure, time and their interaction had the most significant effects on defatting mealworm (*Tenebrio molitor* L.) larvae. In the same context, Sheppard et al. [36] stated that the CP content of larvae meal could be increased by over 60% when protein and fat are separated. According to Castell [37], the defatting process increases the CP and decreases the lipid value of the insect meal, which may lead to total or partial destruction of AA such as cysteine, tryptophan and methionine. Our study showed that the only difference between the two meals was the fat content and that the AA concentration proportionally grew in relation to the CP content.

Based on the current knowledge, very few studies are available in the literature on the chitin content of defatted BSF larvae meals. Kroeckel et al. [13] reported a chitin amount of 96 g/kg DM in defatted BSF larvae. However, Diener et al. [38] indicated that the exoskeleton of the BSF pre-pupae contains approximately 87 g/kg DM of the polysaccharide chitin. Our results on chitin contents of both BSFp and BSFh were lower than those reported by the above mentioned authors.

Our results also confirmed that the AA of highly or partially defatted BSF meals were higher than those reported for full-fat BSF ones [12, 39]. As expected, in the defatted meal the protein was more concentrated and therefore the amount of AA was higher. In fact, the AA level in BSFp was about 20% lower than in BSFh meal. The relative percentage of each AA remained unchanged in both BSFp and BSFh meals compared with full-fat BSF meal [6, 39]. Performic acid oxidation was not performed prior to acid hydrolysis, so the values of cysteine and methionine could be undervalued.

In their study about full-fat BSF meal, De Marco et al. [12] concluded that AA level in BSF meal was lower than in other insect meals. The results obtained in the current trial showed that the defatting process managed to match, and even surpass, the AA profile of other insect meals, for example obtainable from *T. molitor* [12] and other species of insects rich in AA [5, 6, 32, 40].

AA in both defatted BSF meals used in this trial were at higher levels than in plant protein sources (pea protein, full-fat soya bean, sunflower meal and soya bean meal) [20, 30, 41, 42]. The BSFh meal showed similar levels of some AA (valine, methionine, lysine, aspartic acid, serine and alanine) than those found in meat meal (26.9–29.0; 8.1–10.2; 24.6–30.0; 41.5–45.7; 25.6–26.4; 38.8–43.5 g/kg DM, respectively) [20, 30]. In BSFh meal, the amounts of some AA (histidine, alanine, tyrosine and valine) were higher than those found for fishmeal (13.9–16.2; 40.5–41.4; 19.9–21.5; 30.4–33.0 g/kg DM, respectively), while the rest of AA remained at levels below those usually found in fishmeal [20, 30]. Therefore, the insect meal defatting process provided a new ingredient characterized by high quality AA profile. Such profile is comparable to that of other animal protein sources, namely meat and fish meal, usually used in poultry diets.

Apparent nutrient digestibility

In the present study, significant differences were found between BSFp and BSFh meals in the ATTDC for EE and GE and a statistical trend was observed for the ATTDC of DM and OM. Nevertheless, no differences were found for ATTDC of CP. Overall, the ATTDC of the nutrients were moderate for the two defatted BSF larvae meals, except for EE. Furthermore, the ATTDC of EE in our study was comparable to that reported by De Marco et al. [12] for full-fat BSF meal. Our results are also in line with those reported by Cullere et al. [31] who found EE apparent digestibility of BSF diets close to

90% in broiler quails. Till now few researches have dealt with the use of defatted BSF meals in animal nutrition. Regarding the use of defatted insect meals in fish feed, Kroeckel et al. [13] reported that at inclusion levels higher than 33%, a defatted BSF pre-pupae meal decreased diet palatability, protein digestibility and growth performance of juvenile turbot if compared to a control diet.

Fasakin et al. [14] indicated that defatting maggot meal enabled an increase of its dietary inclusion level without affecting the growth performance of African catfish. However, a recent study on Atlantic salmon showed that highly defatted BSF meal (170 g/kg DM), dried at a conventional temperature, surprisingly reduced fish growth compared to a slightly defatted BSF meal (255 g/kg DM) dried at a low temperature [15]. The effectiveness of the defatted BSF meal seems related to a better efficient nutrient digestion, even if the metabolic mechanisms related to this effect require a more in-depth investigation. Regarding poultry, Cullere et al. [31] showed that a defatted BSF meal could be introduced in the diet for growing broiler quails at 10–15%, partially replacing conventional soybean meal and soybean oil, with no negative effects on productive performance, mortality and carcass traits. Elwert et al. [43] demonstrated that increasing levels of defatted BSF meal resulted in decreasing feed consumption and body weight of broiler chickens, but no significant effect on feed conversion ratio was observed. Recently Schiavone et al. [44] reported that the partial or total replacement of soybean oil by BSF fat in broiler diets did not affect growth performance, feed choice and carcass characteristics, while the fatty acid profile of breast meat was greatly affected by the BSF fat inclusion level. All these researchers reported interesting results about the suitability of defatted BSF meals as diet ingredients for poultry and different fish species.

Although chickens have been shown to produce chitinase in the proventriculus and hepatocytes [45], the digestibility of chitin seems to be limited [46]. The moderate ATTDC of the nutrients in the present study can be attributed to the level of chitin supplied by the BSF meals, which was reported to inhibit nutrient absorption from the intestinal tract and thereby to reduce fat and protein absorption in broiler chickens [2, 47–49]. In this context, De Marco et al. [12] speculated that the chitin contained in the exoskeleton of the BSF meal may negatively influence the apparent digestibility coefficient of the total tract of nutrients in broiler chickens. Indeed, Marono et al. [50] indicated that chitin is the main factor affecting the in vitro protein digestibility of BSF meal and showed that CP digestibility was negatively correlated to the chitin content. Moreover, it has been shown that high concentrations (up to 45%) [51] of the chitin present in the cuticular exoskeleton of insects negatively affect the feed intake and reduce protein digestibility [52].

Apparent metabolizable energy
Regarding AME, the BSFp meal showed significantly higher levels of AME and AMEn than the BSFh meal, due to its higher content of EE. De Marco et al. [12] reported a higher level of AME and AMEn in full-fat BSF larvae meal. Elwert et al. [43] calculated the content of AMEn in three different defatted BSF meals at three different levels of fat. The data obtained showed AMEn of 12.3, 12.4 and 12.7 MJ / kg, respectively. In view of these results, the AME and AMEn values of defatted BSF are relevant, and this can result in higher economic efficiency in broiler meat production.

Apparent ileal amino acid digestibility
Broilers were able to utilize the AA of both partially and highly defatted BSF meals with high AIDC (about 80%) and, as expected, AIDC were not affected by the defatting process, except for proline. AIDC for AA in the defatted BSF meals were similar to those reported for soybean by Valencia et al. [42], Ravindran et al. [20] and Huang et al. [53] for 21, 42 and 49 d old broiler chickens, respectively. Also, the results of AIDC were within the range of values reported by Valencia et al. [42] for protein concentrates (pea protein and soya protein concentrates). Furthermore, in the present study the AIDC of AA were higher than those reported for other animal protein sources (feather meal, meat meal and meat and bone meals) and were similar to those obtained for fish meal [20], also for the most limiting AA such as lysine, methionine and threonine. Histidine was the AA which showed the lowest AIDC for the defatted BSF meals; however, the obtained value was similar to that reported for other animal protein sources [20]. The high content of AA with good AIDC indicates that the tested defatted BSF meals could find a place as invaluable protein-rich sources in broiler feeding.

Conclusions
The findings of this study suggest that partially and highly defatted BSF larvae meals can be suitable ingredients for broiler chickens diets. The two tested BSF meals, being rich in AA with high AIDC, could find a place as a valuable AA rich resource in the feed industry for broiler chickens. Furthermore, this study provides data on the apparent metabolizable energy of BSF meals that is helpful in the formulation of broiler diets. However moderate ATTDC of nutrients was observed, consequently digestibility issues require further studies, as the chitin content may not be the only factor able to affect nutrient digestibility. Regarding the chemical composition, BSFh resulted more appealing than BSFp due

to its higher protein content. Furthermore BSFh, containing a lower EE amount, would be less prone to lipid oxidation than BSFp, which may improve the storage time of the meal. Further research efforts are necessary necessary to deeply investigate the impact of defatted larvae meal on broiler chickens growth performance, meat quality traits and consumer acceptance.

Abbreviations
AA: Amino acid; AIDC: Apparent ileal digestibility coefficient; AME: Apparent metabolizable energy; AMEn: N-corrected apparent metabolizable energy; ATTDC: Apparent total tract digestibility coefficients; BSFh: Highly defatted black soldier fly larvae meal; BSFp: Partially defatted black soldier fly larvae meal; CP: Crude protein; d: Day; DM: Dry matter; EE: Ether extract; GE: Gross energy; OM: Organic matter; TiO: Titanium oxide; UA: Uric acid

Acknowledgements
The authors would like to thank Mr. Heinrich Katz, the owner of Hermetia Baruth GmbH, Baruth/Mark (Germany), for providing the two black soldiers fly meals. The authors are also grateful to Mrs. Chiara Bianchi, Mrs. Vanda Malfatto, Dr. Paolo Montersino, Mr. Dario Sola and Mr. Mario Colombano for bird care and technical support.

Funding
The research was supported by the University of Turin (Ex 60% 2014–2015).

Authors' contributions
AS, MDM and LG conceived and designed the experiment. AS, MDM, SD, MR, LR,PC and LG collected the experimental data. FG carried out the chemical analysis. SM, JM and FH carried out the AA analysis of feed and excreta. AS, MDM, SM, SD, MR, JM and LG analyzed and interpreted the data. AS, MDM, SM, SD, LG and JM wrote the first draft of the manuscript. All authors read and approved the final manuscript.

Competing interests
The authors declare that they have no competing interests.

Author details
[1]Department of Veterinary Sciences, University of Turin, Largo Paolo Braccini 2, 10095 Grugliasco, TO, Italy. [2]Department of Animal Production, University of Murcia, Campus de Espinardo, 30071 Murcia, Spain. [3]Department of Agricultural, Forest and Food Sciences, University of Turin, Largo Paolo Braccini 2, 10095 Grugliasco, TO, Italy. [4]Institute of Science of Food Production, National Research Council, Largo Paolo Braccini 2, 10095 Grugliasco, TO, Italy.

References
1. FAO (Food and Agriculture Organization of the United Nations). How to feed the world 2050: Global agriculture towards 2050. Rome: Food and Agriculture Organization of the United Nations (FAO); 2009. p. 12–3.
2. Bovera F, Piccolo G, Gasco L, Marono S, Loponte R, Vassalotti G, et al. Yellow mealworm larvae (Tenebrio molitor L.) as a possible alternative to soybean meal in broiler diets. Br Poult Sci. 2015;56(5):569–75.
3. Gasco L, Henry M, Piccolo G, Marono S, Gai F, Renna M, et al. Tenebrio molitor meal in diets for European sea bass (Dicentrarchus labrax L.) juveniles: Growth performance, whole body composition and in vivo apparent digestibility. Anim Feed Sci Technol. 2016;220:34–45.
4. Belforti M, Gai F, Lussiana C, Renna M, Malfatto V, Rotolo L, et al. Tenebrio molitor meal in rainbow trout (Oncorhynchus mykiss) diets: effects on animal performance, nutrient digestibility and chemical composition of fillets. Ital J Anim Sci. 2015;14:670–75.
5. Henry M, Gasco L, Piccolo G, Fountoulaki E. Review on the use of insects in the diet of farmed fish: past and future. Anim Feed Sci Technol. 2015;203:1–22.
6. Makkar HP, Tran G, Heuzé V, Ankers P. State of the art on use of insects as animal feed. Anim Feed Sci Technol. 2014;197:1–33.
7. Nguyen TTX, Tomberlin JK, Vanlaerhoven S. Ability of Black Soldier Fly (Diptera: Stratiomyidae) Larvae to recycle food waste. Environ Entomol. 2015;44(2):407–10.
8. Spranghers T, Ottoboni M, Klootwijk C, Ovyn A, Deboosere S, De Meulenaer B, et al. Nutritional composition of black soldier fly (Hermetia illucens) prepupae reared on different organic waste substrates. J Sci Food Agr. 2017;97:2594–600.
9. Oonincx DGAB, van Broekhoven S, van Huis A, van Loon JJA. Feed conversion, survival and development, and composition of four insect species on diets composed of food by-products. PLoS One. 2015;10(12):e0144601.
10. Veldkamp T, Van Duinkerken G, Van Huis A, Iakemond CMM, Ottevanger E, Bosch G, et al. Insects as a sustainable feed ingredient in pig and poultry diets - a feasibility study. Wageningen UR Livest. Res. 2012. Report 638.
11. Maurer V, Holinger M, Amsler Z, Früh B, Wohlfahrt J, Stamer A, et al. Replacement of soybean cake by Hermetia illucens meal in diets for layers. J Insects Food Feed. 2016;2(2):83–90.
12. De Marco M, Martínez S, Hernandez F, Madrid J, Gai F, Rotolo L, et al. Nutritional value of two insect meals (Tenebrio molitor and Hermetia illucens) for broiler chickens: apparent nutrient digestibility, apparent ileal amino acid digestibility and apparent metabolizable energy. Anim Feed Sci Technol. 2015;209:211–8.
13. Kroeckel S, Harjes AGE, Roth I, Katz H, Wuertz S, Susenbeth A, et al. When a turbot catches a fly: evaluation of a pre-pupae meal of the Black Soldier Fly (Hermetia illucens) as fish meal substitute – growth performance and chitin degradation in juvenile turbot (Psetta maxima). Aquaculture. 2012;364-365:345–52.
14. Fasakin EA, Balogun AM, Ajayi OO. Evaluation of full-fat and defatted maggot meals in the feeding of clariid catfish Clarias gariepinus fingerlings. Aquac Res. 2003;34:733–8.
15. Lock EJ, Arsiwalla T, Waagbø R. Insect larvae meal as an alternative source of nutrients the diet of Atlantic salmon (Salmo salar) postsmolt. Aquacult Nutr. 2016;22:1202–13.
16. Veldkamp T, Bosch G. Insects: a protein-rich feed ingredient in pig and poultry diets. Anim Front. 2015;5(2):45–50.
17. Hossain MA, Nahar N, Kamal M. Nutrient digestibility coefficients of some plant and animal proteins for rohu (Labeo rohita). Aquaculture. 1997;151:37–45.
18. Belforti M, Lussiana C, Malfatto V, Rotolo L, Zoccarato I, Gasco L. Two rearing substrates on Tenebrio molitor meal composition: issues on aquaculture and biodiesel production. In: Vantomme P, Munke C, van Huis A, editors. 1st International Conference "Insects to Feed the World". The Netherlands: Wageningen University, Ede-Wageningen; 2014. p. 59.
19. Surendra KC, Olivier R, Tomberlin JK, Jha R, Khanal SK. Bioconversion of organic wastes into biodiesel and animal feed via insect farming. Renew Energy. 2016;98:197–202.
20. Ravindran V, Hew LI, Ravindran G, Bryden WL. Apparent ileal digestibility of amino acids in dietary ingredients for broiler chickens. Anim Sci. 2005;8:85–97.
21. AOAC. Official methods of analysis. 18th ed. Washington: AOAC International; 2005.
22. Wang X, Chen X, Chen L, Wang B, Peng C, He C, et al. Optimizing high-performance liquid chromatography method for quantification of glucosamine using 6-aminoquinolyl-N-hydroxysuccinimidyl carbamate derivation in rat plasma: application to a pharmacokinetic study. Biomed Chromatogr. 2008;22:1265–71.
23. Madrid J, Martínez S, López C, Orengo J, López MJ, Hernández F. Effects of low protein diets on growth performance, carcass traits and ammonia emission of barrows and gilts. Anim Prod Sci. 2013;53:146–53.
24. Marquardt RR. A simple spectrophotometric method for the direct determination of uric acid in avian excreta. Poult Sci. 1983;62:2106–08.
25. Short FJ, Gorton P, Wiseman J, Boorman KN. Determination of titanium dioxide added as an inert marker in chicken digestibility studies. Anim Feed Sci Technol. 1996;59:215–21.

26. Nalle CL, Ravindran V, Ravindran G. Nutritional value of white lupins (*Lupinus albus*) for broilers: apparent metabolizable energy, apparent ileal amino acid digestibility and production performance. Animal. 2012;6:579–85.

27. Hill FW, Anderson DL. Comparison of metabolizable energy and productive energy determinations with growing chicks. J Nutr. 1958;64:587–603.

28. SPSS. Statistical Package for the Social Sciences, version 17.0. New York: Mc Graw-Hill; 2008.

29. Tran G, Sauvant D. Tables of composition and nutritional value of feed materials: pigs, poultry, cattle, sheep, goats, rabbits, horses, fish. In: Sauvant D, Pérez JM and Tran G (editors). Chemical data and nutritional value. Paris: Institut National de la Recherche Agronomique, Association Française de Zootechnie; 2004. p. 17–24.

30. Ravindran V, Hew LI, Ravindran G, Bryden WL. A comparison of ileal digesta and excreta analysis for the determination of amino acid digestibility in food ingredients for poultry. Br Poult Sci. 1999;40:266–74.

31. Cullere M, Tasoniero G, Giaccone V, Miotti-Scapin R, Claeys E, De Smet S, et al. Black soldier fly as dietary protein source for broiler quails: apparent digestibility, excreta microbial load, feed choice, performance, carcass and meat traits. Animal. 2016;10(12):1923–30.

32. Sánchez-Muros MJ, Barroso FG, Manzano-Agugliaro F. Insect meal as renewable source of food for animal feeding: a review. J Clean Prod. 2014;65:16–27.

33. Purschke B, Stegmann T, Schreiner M, Jäger H. Pilot-scale supercritical CO_2 extraction of edible insect oil from *Tenebrio molitor* L. larvae – Influence of extraction conditions on kinetics, defatting performance and compositional properties. Eur J Lipid Sci Technol. 2017;119:1–12.

34. Zhao X, Vázquez-Gutiérrez JL, Johansson DP, Landberg R, Langton M. Yellow mealworm protein for food purposes - Extraction and functional properties. PLoS One. 2016;11(2):e0147791.

35. Tschirner M, Simon A. Influence of different growing substrates and processing on the nutrient composition of black soldier fly larvae destined for animal feed. J Insects Food Feed. 2015;1(4):249–59.

36. Sheppard C, Newton GL, Burtle G. Black soldier fly prepupae a compelling alternative to fish meal and fish oil. A public comment prepared in response to a request by the National Marine Fisheries Service. Tifton: University of Georgia; 2007.

37. Castell JD. Aquaculture nutrition. In: Bilio M, Rosenthal H, Sinderman CJ, editors. Realism in Aquaculture; Advances, Constraints, Perspectives, Bredene, Belgium: European Aquaculture Society. 1986. p. 251–308.

38. Diener S, Zubrügg C, Trockner K. Conversion of organic material by black soldier fly larvae: establishing optimal feeding rates. Waste Manag Res. 2009;27:603–10.

39. Finke MD. Complete nutrient content of four species of feeder insects. Zoo Biol. 2013;32:27–36.

40. Barroso FG, de Haro C, Sánchez-Muros MJ, Venegas E, Martínez-Sánchez A, Pérez-Bañón C. The potential of various insect species for use as food for fish. Aquaculture. 2014;422/423:193–201.

41. Huang KH, Li X, Ravidran V, Bryden WL. Comparison of apparent ileal amino acid digestibility of feed ingredients measured with broilers, layers and roosters. Poult Sci. 2006;85:625–34.

42. Valencia DG, Serrano MP, Lázaro R, Jiménez-Moreno E, Mateos GG. Influence of micronization (fine grinding) of soya bean meal and full-fat soya bean on the ileal digestibility of amino acids for broilers. Anim Feed Sci Technol. 2009;150:238–48.

43. Elwert C, Knips I, Katz P. A novel protein source: maggot meal of the black soldier fly (*Hermetia illucens*) in broiler feed. In: Gierus M, Kluth H, Bulang M, Kluge H, editors. Tagung Schweine- und Geflügelernährung, November 23–25, 2010, Institut für Agrar- und Ernährungswissenschaften, Universität HalleWittenberg, Lutherstadt Wittenberg, Germany. 2010. p. 140–42.

44. Schiavone A, Cullere M, De Marco M, Meneguz M, Biasato I, Bergagna S, et al. Partial or total replacement of soybean oil by black soldier larvae (*Hermetia illucens* L.) fat in broiler diets: effect on growth performances, feed-choice, blood traits, carcass characteristics and meat quality. Ital J Anim Sci. 2017;16:93–100. doi:10.1080/1828051X.2016.1249968.

45. Suzuki M, Fujimoto W, Goto M, Morimatsu M, Syuto B, Toshihiko I. Cellular expression of gut chitinase mRNA in the gastrointestinal tract of mice and chickens. J Histochem Cytochem. 2002;50:1081–9.

46. Hossain S, Blair R. Chitin utilization by broilers and its effect on body composition and blood metabolites. Br Poult Sci. 2007;48(1):33–8.

47. Razdan A, Pettersson D. Effect of chitin and chitosan on nutrient digestibility and plasma lipid concentrations in broiler chickens. Br J Nutr. 1994;72(2):277–88.

48. Khempaka S, Chitsatchapong C, Molee W. Effect of chitin and protein constituents in shrimp head meal on growth performance, nutrient digestibility, intestinal microbial populations, volatile fatty acids, and ammonia production in broilers. J Appl Poult Res. 2011;20:1–11.

49. Vidanarachchi JK, Kurukulasuriya MS, Kim SK. Chitin, chitosan and their oligosaccharides in food industry. In: Kim SK, editor. Chitin, Chitosan, Oligosaccharides and Their Derivatives: Biological Activities and Applications. New York: CRC Press; 2010. p. 543–60.

50. Marono S, Piccolo G, Loponte R, Di Meo C, Attia YA, Nizza A, et al. In vitro crude protein digestibility of *Tenebrio molitor* and *Hermetia illucens* insect meals and its correlation with chemical composition traits. Ital J Anim Sci. 2015;14:338–43.

51. Muzzarelli RAA. Chitin. Oxford: Pergamon Press; 1977.

52. Longvah T, Mangthya K, Ramulu P. Nutrient composition and protein quality evaluation of eri silkworm (*Samia ricinii*) prepupae and pupae. Food Chem. 2011;128:400–3.

53. Huang KH, Ravindran V, Li X, Bryden WL. Influence of age on the apparent ileal amino acid digestibility of feed ingredients for broiler chickens. Br Poult Sci. 2005;46:236–45.

11

Evidence of endometrial amino acid metabolism and transport modulation by peri-ovulatory endocrine profiles driving uterine receptivity

Moana Rodrigues França[1], Maressa Izabel Santos da Silva[1], Guilherme Pugliesi[1], Veerle Van Hoeck[2] and Mario Binelli[1*]

Abstract

Background: In beef cattle, changes in the periovulatory endocrine milieu are associated with fertility and conceptus growth. A large preovulatory follicle (POF) and the resulting elevated concentrations of progesterone (P4) during diestrus positively affect pregnancy rates. Amino acids (AA) are important components of maternally derived secretions that are crucial for embryonic survival before implantation. The hypothesis is that the size of the POF and the concentration of P4 in early diestrus modulate the endometrial abundance of SLC transcripts related to AA transport and metabolism and subsequently impact luminal concentrations of AA. The follicle growth of Nelore cows was manipulated to produce two experimental groups: large POF and CL (LF-LCL group) and small POF and CL (SF-SCL group). On Day 4 (D4; Experiment 1) and Day 7 (D7; Experiment 2) after GnRH-induced ovulation (GnRH treatment = D0), the animals were slaughtered and uterine tissues and uterine washings were collected. qRT-PCR was used to evaluate the expression levels of AA transporters in D4 and D7 endometrial tissues. The concentrations of AA were quantified in D4 and D7 uterine washings by HPLC.

Results: Transcript results show that, on D4, *SLC6A6, SLC7A4, SLC17A5, SLC38A1, SLC38A7* and *SCLY* and on D7 *SLC1A4, SLC6A1, SLC6A14, SLC7A4, SLC7A7, SLC7A8, SLC17A5, SLC38A1, SLC38A7, SLC43A2* and *DDO* were more abundant in the endometria of cows from the LF-LCL group ($P < 0.05$). In addition, concentrations of AA in the uterine lumen were influenced by the endocrine profiles of the mother. In this context, D4 uterine washings revealed that greater concentrations of taurine, alanine and α-aminobutyric acid were present in SF-SCL ($P < 0.05$). In contrast, lower concentrations of valine and cystathionine were quantified on D7 uterine washings from SF-SCL cows ($P < 0.05$).

Conclusion: The present study revealed an association between the abundance of transcripts related to AA transport and metabolism in the endometrium and specific periovulatory endocrine profiles related to the receptive status of the mother. Such insights suggest that AAs are involved in uterine function to support embryo development.

Keywords: Amino acids, Beef cattle, Sex steroids, Uterus

* Correspondence: binelli@usp.br
[1]Department of Animal Reproduction, School of Veterinary Medicine and Animal Science, University of São Paulo, 225, Duque de Caxias Norte Ave. Jd. Elite, 13635-900 Pirassununga, SP, Brazil
Full list of author information is available at the end of the article

Background

A profitable beef cattle production system requires high reproductive efficiency. Acceptable female fertility rates depend on an in-depth understanding of endocrine, cellular and molecular mechanisms regulating pregnancy. Hormonal variations during each bovine estrous cycle induce uterine changes that are crucial for uterine receptivity for conceptus development and implantation. Elevated levels of plasmatic progesterone (P4) immediately after conception are related to advanced conceptus elongation [1–3]. Moreover, recent studies have shown that low plasmatic concentrations of P4 are associated with a suboptimal uterine environment for conceptus (or blastocyst) development [3–5]. Conversely, a positive association exists between the probability of a successful pregnancy and plasma concentrations of P4 at D7 after estrus in both dairy and beef cows [4, 6]. However, the identities of molecules and mechanisms responsible for triggering the latter phenomena need further investigation.

In this context, information on the involvement of amino acid (AA) transport and metabolic pathways and availability to the embryo during early diestrus remains limited. During the first two weeks of pregnancy, before implantation, the conceptus depends exclusively on the intrauterine milieu created by the endometrial secretions or molecules transported into the lumen by the uterine endometrium prior to implantation and placentation [7]. Amino acids are important components of these maternally derived secretions that are crucial for embryonic survival mainly during early pregnancy [8–12]. Optimal amounts of essential and non-essential AAs are important for embryonic development, which is altered in case of suboptimal amounts of these molecules [13–15]. The regulation of AA transport and concentration in cells and tissues, including the placenta and endometrium, depends on specific transport proteins [16]. There is no consistent information regarding the regulation of protein synthesis and the activity of AA transporters; however, changes in transcriptional profiles of AA transporters reveal changes in AA availability in the uterine lumen [12]. Interestingly, AA transporter gene expression increases simultaneously in maternal endometrium and conceptus cells. This indicates transport of AAs from the maternal circulation to the endometrial lumen exclusively for embryonic development.

Amino acids seem to be indispensable for embryonic survival and development. When non-essential AAs were completely removed from the medium, blastocyst growth drastically decreased [14]; the same was observed in medium with any combination of glucose and phosphate without AAs [15]. More interestingly, P4 infusion in animals with no CL and no follicle increased AA availability in the uterus and maternal plasma of cows [17] and gene expression of AA transporters in ovine and bovine endometria [8, 12, 18].

During early embryonic development in vitro, AA requirements seem to change [19]. Until the blastocyst stage, an excess of essential AAs seems to be detrimental for embryonic development, at least in vitro [13]; however, the absence of non-essential AAs in the culture medium is unfavorable for embryonic development [14]. During the blastocyst expansion stage, few changes were observed in AA turnover during embryo culture [20], when patterns of AA depletion and release were similar to those observed in vitro for pre-elongation embryos [21]. More interestingly, AA turnover has been identified as an indicator of embryonic viability in humans and cattle [22, 23]. Another relevant finding is that enzymes present in endometrial tissue can regulate AA synthesis or degradation. These enzymes are not characterized in the bovine endometrium. In other tissues, DDO (D-aspartate oxidase) catalyzes the deamination of alanine and aspartate [24]. SCLY (selenocysteine lyase) is involved in the production of alanine and elemental selenium from selenocysteine [25].

In this context, AA uterine transport and luminal availability emerge as important variables that might play a role in maternal and embryonic well-being. However, there is limited information available on how changes in the pre- and post-ovulatory endocrine milieu affect the transport of AAs from the maternal circulation to endometrial cells and finally to the uterine lumen. We hypothesize that the size of the POF and subsequent physiological circulating concentrations of P4 in early diestrus modulate endometrial expression of AA transporter protein pathways.

We recently described a model to manipulate preovulatory follicle growth to produce groups of cyclic beef cows with distinctly different circulating preovulatory concentrations of E2 and early diestrus concentrations of P4 [26]. Based on the contrasting ovarian and endocrine characteristics of these two groups of animals, we studied the following variables on both D4 and D7 of the estrous cycle: (1) transcript abundance of AA transporters in endometrial tissues; (2) transcript abundance of enzymes related to AA metabolism in endometrial tissues; and (3) uterine luminal concentrations of AA. Previous studies indicated that the phenotype of smaller follicles and short proestrus was associated with lower receptivity and capacity to support conceptus development in comparison with a group manipulated to have a longer proestrus and ovulate a larger follicle [27]. We focused on D4 and D7 for our investigation because it is around D4-5 that the embryo moves from the oviduct to the uterus; thus, from this moment until implantation (i.e., D20 in cattle) the embryo depends exclusively on endometrial secretions for its development [28–30]. Moreover, most embryonic losses occur during the two first weeks of pregnancy [31]. Therefore, mechanistic insights retrieved from this animal model could serve as a potential basis for future strategies to fine-tune maternal receptivity towards the embryo.

Methods
Animals and reproductive management
Eighty-six Nelore (*Bos indicus*) cows started in the experiment. The females selected were cycling, pluriparous, and non-lactating and did not present any detectable reproductive disorder. To form two distinct groups of females with different POF sizes and subsequent CL volumes and plasmatic concentrations of P4 (Fig. 1), a hormonal protocol was used as described previously [26, 27, 32].

Briefly, the cows received two injections of prostaglandin F2α (PGF; 0.5 mg; Cloprostenol; Sincrocio®; Ourofino, Cravinhos, SP, Brazil) 14 d apart. Next, the ovaries were examined using transrectal ultrasonography (US) to confirm the presence of a PGF-induced CL 10 d after the second PGF administration, (D – 10; D0 = day of induction of ovulation by GnRH injection). On D – 10, each female was treated with 2 mg of estradiol benzoate (Sincrodiol®, Ourofino, Cravinhos, SP, Brazil) to stimulate the emergence of a follicular wave and received an intravaginal P4-releasing device (1 mg; Sincrogest®; Ourofino, Cravinhos, SP, Brazil). Additionally, females assigned to the large POF followed by large CL (LF-LCL) group also received a PGF injection (0.5 mg;

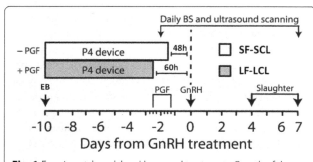

Fig. 1 Experimental model and hormonal treatments. Growth of the pre-ovulatory follicle (POF) of beef cows was programmed to generate two groups of cows, the large follicle-large CL group (LF-LCL; associated with greater receptivity to the embryo and greater fertility) and small follicle-small CL group (SF-SCL). To decrease exposure to P4 and thereby stimulate growth of the POF, animals from LF-LCL group received an injection of PGF at the moment of intravaginal P4-releasing device insertion vs. no injections in the animals from SF-SCL group. Also, removal of the P4-releasing device was 12 h earlier in the LF-LCL group. Follicle size, ovulation and CL size were accessed by ultrasound scanning of the ovaries. Blood samples were collected for P4 assay. Ovulation was induced by GnRH on D0. On D4 (Experiment 1) and D7 (Experiment 2) animal were slaughtered for samples collection. BS, blood sampling; GnRH, 1 μg of buserarine acetate im; P4, P4 progesterone-releasing device containing 1 g of P4; +PGF, cows received 0.5 mg of sodium cloprostenol on D-10; -PGF, cows did not receive Cloprostenol on D-10; EB, 2 mg of estradiol benzoate; Slaughter, endpoint for endometrial tissue and uterine washings collection (Adapted from Reference 26)

Cloprostenol; Sincrocio®) on D-10 to induce CL regression during follicle development, whereas cows assigned to the small POF followed by small CL (SF-SCL) group did not. Sixty hours prior to the induction of ovulation, the P4 devices were removed and an injection of PGF was administered to the LF-LCL females, whereas cows in the SF-SCL received PGF administration 12 h later (D – 2.5 and D – 2, respectively). Ovulation was induced on D0 by administration of a gonadotrophin-releasing hormone agonist (GnRH; 1 μg buserelin; Sincroforte®; Ourofino, Cravinhos, SP, Brazil).

Follicular growth, ovulation and CL diameter were monitored by transrectal ultrasound. Only cows that ovulated in response to the GnRH (i.e., between 24 and 36 h after GnRH injection; N = 57) were included in the data analysis. The cows were slaughtered and their reproductive tracts collected for further analysis on D4 (Experiment 1; LF-LCL n = 16; SF-SCL n = 8) or D7 (Experiment 2; LF-LCL n = 18 SF-SCL n = 18).

Ultrasound Exams and CL Measurements
Transrectal ultrasound exams were performed using an Aloka SSD-500 device attached to a 5 MHz linear probe. The presence of an active CL on D-10, the time of ovulation, the size of the CL and the properties of the dominant follicle and POF were noted. Postmortem CLs were dissected and weighed separately, and the length, width and height were recorded. A calculated diameter for each CL was estimated by the mean of these three values. The CL volume was estimated using the formula for the volume of a sphere ($V = 4/3\pi R^3$), where R was equal to half the mean diameter.

Plasma Progesterone Measurement
Blood was collected from the jugular vein on D4 and D7. The blood samples were centrifuged at 1500 × g for 30 min at 4 °C. Plasma aliquots were stored at –20 °C until analysis. Plasmatic concentrations of P4 were measured by radioimmunoassay analysis (Coat-A-Count®; Siemens Medical Solutions Diagnostics, Munich, Germany) as described previously [33].

Endometrial tissue and uterine fluid collection
Postmortem uteri were collected, and the uterine horns ipsilateral to the CL were washed with 20 mL PBS. Subsequently, the washings were centrifuged (1000 × g) for 30 min at 4 °C, and the supernatant was collected, snap frozen in liquid nitrogen and stored at -80 °C. Subsequently, endometrial tissue from the washed uteri was dissected from the intercaruncular region of the ipsilateral uterine horn, frozen in liquid nitrogen and stored at -80 °C.

Quantification of amino acids

The concentrations of free AA in uterine flushings were determined by high-performance liquid chromatography (HPLC) using an HPLC Luna Column 3u C18(2) 100A 250 × 4.6 mm (00G-4251-E0 - Phenomenex, Torrance, CA, USA). The protocol for AA quantification was adapted from [34]. Briefly, 100 μL of internal standards of free AAs and methanol were added to 200 μg of uterine flushing samples (n = 12 LF-LCL, n = 7 SF-SCL for Experiment 1; n = 9 LF-LCL, n = 10 SF-SCL for Experiment 2). The dissolved samples were deproteinized using Vivaspin 500 (MWCO 3000 Da, Sartorius, Goettingen, Germany) followed by centrifugation (15,000×g for 45 min at 4 °C) and derivatization and were subsequently analyzed by programmed chromatography. The samples were analyzed in simplicate. The results were obtained by comparing each AA peak to its corresponding peak on a multilevel (3 levels) standard curve, based on certified standards.

Analysis of gene expression levels in endometrial tissue

Total RNA was extracted from endometrial samples using the RNeasy Mini Kit (Qiagen Laboratories, Germantown, MD, USA) following manufacturer instructions. The concentration and purity of mRNA were estimated using NanoVue (GE Healthcare Life Sciences, Buckinghamshire, England). Total RNA extracts were stored at −80 °C until cDNA synthesis. One microgram of total RNA was used for reverse transcription using the High-Capacity cDNA Reverse Transcription kit (Life Technologies, Carlsbad, CA, USA), according to manufacturer's instructions.

Quantification of endometrial gene expression was obtained by qPCR analysis using a StepOne Plus® apparatus from Applied Biosystems. Transcript abundance was determined for several AA-transport-related genes. The genes tested were *SLC1A1*, *SLC1A4*, *SLC1A5*, *SLC6A1*, *SLC6A6*, *SLC6A14*, *SLC7A2*, *SLC7A4*, *SLC7A5*, *SLC7A7*, *SLC7A8*, *SLC7A11*, SLC17A5, SLC17A9, *SLC36A2*, *SLC38A1*, *SLC38A4*, *SLC38A6*, *SLC38A7*, *SLC43A2*, *SCLY*, and *DDO*. Each primer pair was analyzed, considering the probabilities of hairpin, homodimer and heterodimer formation, using Oligo Analyzer 3.1 software (IDT®; http://www.idtdna.com/analyzer/Applications/OligoAnalyzer/). Sequence specificities were tested using the software *Basic Local Alignment Search Tool* (Blast) (http://blast.ncbi.nlm.nih.gov). PCR product identity was confirmed by sequencing (Table 1). Transcript abundance was compared between tissues from animals in the LF-LCL and SF-SCL groups from Experiment 1 (*n* = 8 per group) and Experiment 2 (*n* = 8 for LF-LCL and *n* = 9 for SF-SCL group). The selection of reference genes was performed using geNorm software (www.qbaseplus.com) [35]. After selection performed by the software, RPS18 was selected as

an endogenous control for endometrium on D4, and cyclophilin, β-actin and GAPDH were chosen for endometrium on D7. Transcript abundance was calculated as the relative abundance between the target gene and the geometric average of selected housekeeping genes and given as an arbitrary value.

Statistical analysis

Cows from each group were ranked according to the plasmatic concentration of P4 at D7, P4 at D7/P4 at D2 ratio, CL size at D7, CL weight, follicle size at D2, D1 and D0 and preovulatory follicle size as previously described [36]. The samples selected for analysis were according to this ranking. Further analyses were conducted in 8 animals per group (D4) and 8 and 9 animals (D7) from the LF-LCL and SF-SCL groups, respectively. The data were tested for normality of residuals using the Shapiro-Wilk test and for homogeneity of variance using the F-max text (SAS; Version 9.2; SAS Institute). The data were analyzed independently for Experiments 1 and 2, as they were not conducted contemporaneously. Discrete dependent variables (diameter of the dominant follicle on D2 and D0, POF diameter, CL volume and weight measured postmortem and plasmatic concentrations of P4 on D4 or D7) were analyzed by one-way ANOVA for the effect of group using the PROC GLM procedure (SAS; Version 9.2; SAS Institute). The amino acid quantification data were tested for the presence of outliers by Dixon's test, and outliers were removed before the normality of residuals test. Uterine flushing concentrations of amino acid values that did not follow the normality of residuals were transformed for analysis. On D4, uterine flushing concentrations of threonine, alanine, methionine sulfone internal standard, tyrosine, leucine and lysine were transformed to natural log and isoleucine, ornithine and lysine were transformed to rank. On D7, taurine, proline, α-aminobutyric acid, valine, cysteine, isoleucine and lysine were transformed to natural log and β-alanine was transformed to rank. Concentrations of amino acid in uterine flushings and relative gene expression were analyzed by Student's *t*-test. Means were considered significantly different when they presented a P value of 0.05 or less. Means with P values between 0.06 and 0.1 were considered as approaching significance.

Results

Animal Model

The hormonal strategy that we employed successfully produced groups of cows presenting distinctly different periovulatory ovarian and endocrine characteristics, as expected and described earlier [26, 32, 36–38]. Specifically, for Experiment 1, cows assigned to the LF-LCL group had larger follicle diameters on D2, D1 and D0 compared to animals from the SF-SCL group ($P < 0.01$; Table 2). Pre-ovulatory follicles were also larger for

Table 1 Target genes, primer sequence and amplicon information

Gene	Gene Symbol	Representative ID	Sequence	Amplicon size, bp
Solute Carrier Family 1 member 1	SLC1A1[a]	NM_174599.2	F: AAGGAGTTGGAGCAAATGGA	152
			R: AACGAGATGGTATCGGACTTG	
Solute Carrier Family 1 member 4	SLC1A4[a]	NM_001081577.1	F: ATCTTGATAGGCGTGGTTTC	132
			R: GCAACACTGGTTCTCTCTATAA	
Solute Carrier Family 1 member 5	SLC1A5[a]	NM_174601.2	F: TCCAAATCTGCCCAGTCCTCAACT	141
			R: TTCCCATGATTCCCTATGCCCTGA	
Solute Carrier Family 6 member 1	SLC6A1[a]	NM_001077836.1	F: GTGTCTCCATTTCCTGGTTT	150
			R: GCACAGCACTGAAGATGAA	
Solute Carrier Family 6 member 14	SLC6A14[a]	NM_001098461.1	F: GATGCTGCCACACAGATATT	154
			R: TAGCAAATCCAGCGAACAC	
Solute Carrier Family 6 member 6	SLC6A6[a]	NM_174610.2	F: GAAAGCCGTGACGATGAT	158
			R: GGTGAAAGCCCTTCCTTAG	
Solute Carrier Family 7 member 2	SLC7A2[a]	XM_010820288.1	F: GATGCTGGAGGGACTAGAT	146
			R: AGATACCCAGGCACAGAA	
Solute Carrier Family 7 member 4	SLC7A4[a]	NM_001192042.1	F: GACCCACGGACTCTAGTTTA	156
			R: GGTTGAGCGATACCTATTGTG	
Solute Carrier Family 7 member 5	SLC7A5[a]	NM_174613.2	F: TGTGGTCCGATAGGCATAGA	151
			R: ACAACGGTGGATGCTGTT	
Solute Carrier Family 7 member 7	SLC7A7[a]	NM_001075151.1	F: CCTCCAGGTCCTATGTATGT	144
			R: CAGCCAGCAGGAGATAGA	
Solute Carrier Family 7 member 8	SLC7A8[a]	NM_001192889.1	F: GAGATTGGATTGGTCAGTGG	155
			R: GCTCCCACAACTGTGATAAG	
Solute Carrier Family 7 member 11	SLC7A11[a]	XM_010826337.1	F: GTACAGGGATTGGCTTCATC	144
			R: CTGGCACAACTTCCAGTATT	
Solute Carrier Family 17 member 5	SLC17A5[a]	NM_001205974.1	F: CTGCAGTCCCTTATTTAGGC	144
			R: GGAATATCGCAGGTCCAATC	
Solute Carrier Family 17 member 9	SLC17A9[a]	NM_001100378.2	F: GTCTAGACACACCAAGG	141
			R: GGGAAGGTCTCCTTAAA	
Solute Carrier Family 36 member 2	SLC36A2[a]	NM_001206180.1	F: GACGACCAAGGGATAAC	157
			R: AGAGTGGAGGAGATGAA	
Solute Carrier Family 38 member 1	SLC38A1[a]	XM_002687321.3	F: TCACGGTTCGATCTTCTTTATT	131
			R: ATCCTTCATGGAGGGTATGA	
Solute Carrier Family 38 member 4	SLC38A4[a]	NM_001205943.1	F: CTGTGCCCATAGTGCTATTC	139
			R: GGCACAAGGATGACCAAA	
Solute Carrier Family 38 member 6	SLC38A6[a]	XM_010823227.1	F: CACCTCAATACTGCCCATATAC	150
			R: GACGCCACACTGTCATAAA	
Solute Carrier Family 38 member 7	SLC38A7[a]	NM_001100355.1	F: TATAGCTGTGATGGCAAAGG	153
			R: CTCAGGAAGCTGGATATTT	
Solute Carrier Family 43 member 2	SLC43A2[a]	NM_001075546.1	F: AAGGCCCAGGATGAGAT	149
			R: AAGCAGGAGACAGCAAAG	
Cyclophilin	PPIA[b]	NM_178320.2	F: GCCATGGAGCGCTTTGG	65
			R: CCACAGTCAGCAATGGTGATCT	
β-actin	ACTB[#]	NM_173979.3	F: GGATGAGGCTCAGAGCAAGAGA	78
			R: TCGTCCCAGTTGGTGACGAT	

Table 1 Target genes, primer sequence and amplicon information *(Continued)*

D-aspartate oxidase	*DDO*[a]	NM_173908.2	F: CTGGCGTATCCAATGTAACC	158
			R: CCTCAAACCCACTTTCTCTC	
Glyceraldehyde 3-phosphate dehydrogenase	*GAPDH*[b]	NM_001034034.2	F: GCCATCAATGACCCCTTCAT	70
			R: TGCCGTGGGTGGAATCA	
Ribosomal protein S18 Selenocysteine lyase	*RPS18*[b]	NM_001033614.2	F: GCCTGAGAAACGGCTACCAC	171
	SCLY[a]	NM_001083804.1	R: CACCAGACTTGCCCTCCAAT	151
			F: AGGAAGGCCAAGGAGATTAT	
			R: CGTGGACTTTGTGGAAGTG	

(F) Primer forward sequence; (R) Primer reverse sequence. [a]indicates primer sequences obtained by PrimerQuestQM software (IDT technologies).[b][64]

animals from the LF-LCL group ($P < 0.01$; Table 2), but plasmatic concentrations of P4 on D4 were similar for animals from both groups ($P > 0.05$; Table 2). For Experiment 2, cows assigned to the LF-LCL group had larger POF diameters than animals from the SF-SCL group ($P < 0.01$; Table 2). Furthermore, the larger POFs resulted in larger and heavier CLs on D7 ($P < 0.05$; Table 2). Moreover, plasmatic concentrations of P4 on D7 were also greater in cows from the LF-LCL group. More details on ovarian and endocrine responses from the LF-LCL and SF-SCL groups on D4 and D7 were published elsewhere [26, 39].

Gene Expression
On D4, the abundances of *SLC6A6*, *SLC7A4*, *SLC17A5*, *SLC38A1*, *SLC38A7* and *SCLY* were on average 64.34, 70.64, 42.36, 56.24, 35,54 and 41.45% greater, respectively,

in the endometrium of LF-LCL cows than in the SF-SCL endometrial tissue ($P \leq 0.05$; Table 3; Fig. 7). These solute carriers are responsible for alanine, serine, proline, taurine, β-alanine, aspartate, glutamate, histidine, ornithine, lysine and glutamine transport. The enzyme SCLY is responsible for alanine synthesis from selenocysteine.

On D7, the transcript abundances of *SLC1A4*, *SLC6A1*, *SLC6A14*, *SLC7A4*, *SLC7A7*, *SLC7A8*, *SLC17A5*, *SLC38A1*, *SLC38A7*, *SLC43A2* and *DDO* were 49.62, 73.26, 68.98, 126.14, 33.32, 186.4, 73.88, 67.93, 35.91, 46.28, and 69.64% greater in the endometrium of the LF-LCL group ($P \leq 0.05$; Table 3; Fig. 7).

Amino Acid Quantification in the Uterine Washings
Glutamate, aminoadipic acid, asparagine, serine, glutamine, glycine, histidine, β-alanine, taurine, β-aminoisobutyric acid,

Table 2 Follicle, CL and P4 measurements

End-points	LF-LCL	SF-SCL	$P < F$
Experiment 1 (D4) Follicle diameter, mm			
D2	12.66 ± 0.4	8.84 ± 0.6	0.00006
D1	13.93 ± 0.6	9.58 ± 0.3	0.00003
D0	15.64 ± 0.6	11.52 ± 0.4	0.00005
Pre-ovulatory follicle, mm	15.99 ± 0.3	11,32 ± 0.2	0.00000002
CL volume on D4, cm³	2.66 ± 1.2	2.3 ± 1	0.03
CL weight on D4, g	1.06 ± 0.06	0.68 ± 0.08	0.002
Plasma P4 concentrations on D4, ng/mL	1.17 ± 0.3	0.8 ± 0.1	0.21
Experiment 2 (D7) Follicle diameter, mm			
D2	10.73 ± 0.7	7.09 ± 0.3	0.0002
D1	11.79 ± 0.6	7.73 ± 0.4	0.000001
D0	13.04 ± 0.5	9.68 ± 0.4	0.00002
Pre-ovulatory follicle, mm	13.63 ± 0.5	10.09 ± 0.3	0.000005
CL volume on D7, cm³	2.29 ± 0.2	1.53 ± 0.2	0.03
CL weight on D7, g	2.83 ± 0.2	1.53 ± 0.2	0.0001
Plasma P4 concentrations,ng/mL			
on D7	3.68 ± 0.2	2.1 ± 0.2	0.008

Cows were synchronized to have larger (LF-LCL; Exp 1 *n* = 8; Exp 2 *n* = 8) or smaller (SF-SCL; Exp 1 *n* = 8; Exp 2 *n* = 9) pre-ovulatory follicles and corpora lutea. Mean ± SEM

Table 3 Relative quantification of transcript abundance by qPCR

Gene	LF-LCL	SF-SCL	$P < F$
SLC1A1			
D4	0.1215 ± 0.0126	0.1662 ± 0.0339	0.2371
D7	0.0051 ± 0.0004	0.0044 ± 0.0005	0.2830
SLC1A4			
D4	0.0107 ± 0.0014	0.0073	0.08
D7	0.0391 ± 0.004	0.0262 ± 0.004	0.05
SLC1A5			
D4	1.25 ± 0.08	1.35 ± 0.1	0.45
D7	0.7 ± 0.07	0.9 ± 0.01	0.51
SLC6A1			
D4	0.0006 ± 0.0001	0.0005 ± 0.0001	0.5506
D7	0.4561 ± 0.0350	0.2633 ± 0.0282	0.0006
SLC6A6			
D4	0.0181 ± 0.0017	0.0110 ± 0.0007	0.0020
D7	0.0682 ± 0.0068	0.0711 ± 0.0090	0.8064
SLC6A14			
D4	0.021 ± 0.004	0.025 ± 0.006	0.58
D7	0.0286 ± 0.0028	0.0169 ± 0.0011	0.001
SLC7A2			
D4	0.0067 ± 0.0014	0.0068 ± 0.0012	0.9437
D7	0.0047 ± 0.0007	0.0045 ± 0.0009	0.8666
SLC7A4			
D4	0.0012 ± 0.0002	0.0007 ± 0.0002	0.052
D7	0.0077 ± 0.0011	0.0034 ± 0.0006	0.0036
SLC7A5			
D4	0.0522 ± 0.008	0.0549 ± 0.005	0.78
D7	0.0598 ± 0.01	0.0599 ± 0.007	0.99
SLC7A7			
D4	0.0167 ± 0.0012	0.0159 ± 0.0015	0.6886
D7	0.0579 ± 0.0031	0.0434 ± 0.0025	0.0021
SLC7A8			
D4	0.0153 ± 0.0021	0.0180 0.0051	0.6238
D7	0.1068 ± 0.0157	0.0373 ± 0.0071	0.0012
SLC7A11			
D4	0.0004 ± 0.0001	0.0004 ± 0.0001	0.7969
D7	0.0004 ± 0.0001	0.0005 ± 0.0001	0.8120
SLC17A5			
D4	0.0193 ± 0.0012	0.0135 ± 0.0015	0.0089
D7	0.0629 ± 0.0074	0.0362 ± 0.0024	0.0039
SLC17A9			
D4	0.0033 ± 0.0005	0.0028 ± 0.0007	0.6243
D7	0.0054 ± 0.0008	0.0048 ± 0.0009	0.6354

Table 3 Relative quantification of transcript abundance by qPCR *(Continued)*

Gene	LF-LCL	SF-SCL	$P < F$
SLC36A2			
D4	0.0005 ± 0.0001	0.0006 ± 0.0002	0.55
D7	0.0035 ± 0.0013	0.0025 ± 0.001	0.56
SLC38A1			
D4	0.066 ± 0.008	0.042 ± 0.004	0.019
D7	0.63 ± 0.08	0.38 ± 0.04	0.01
SLC38A4			
D4	0.0072 ± 0.0009	0.0086 ± 0.004	0.19
D7	0.0158 ± 0.009	0.01978 ± 0.002	0.12
SLC38A6			
D4	0.0044 ± 0.0001	0.0047 ± 0.0003	0.3745
D7	0.0077 ± 0.0005	0.0066 ± 0.0004	0.1304
SLC38A7			
D4	0.0404 ± 0.004	0.0302 ± 0.0014	0.02
D7	0.9601 ± 0.009	0.7069 ± 0.008	0.05
SLC43A2			
D4	0.6539 ± 0.0217	0.6357 ± 0.1383	0.8983
D7	0.0065 ± 0.0006	0.0044 ± 0.0006	0.0326
DDO			
D4	0.0113 ± 0.0021	0.0119 ± 0.0023	0.8501
D7	0.0208 ± 0.0028	0.0123 ± 0.0011	0.0123
SCLY			
D4	0.0090 ± 0.0006	0.0064 ± 0.0003	0.0009
D7	0.0122 ± 0.0011	0.0110 ± 0.0005	0.3279

Mean ± standard error of the mean of the relative abundance of target genes and endogenous controls for large follicle-large CL group (LF-LCL; Exp 1 $n = 8$; Exp 2 $n = 8$) and small follicle-small CL group (SF-SCL; Exp 1 $n = 8$; Exp 2 $n = 9$)

threonine, alanine, proline, α-aminobutyric acid, tyrosine, valine, methionine, cystathionine, cysteine, isoleucine, leucine, phenylalanine, tryptophan, ornithine and lysine were the AAs detected in uterine flushings on D4 and D7 (Table 4).

When the levels in the uterine washings were compared between groups, D4 data showed that concentrations of taurine, alanine and α-aminobutyric acid in uterine flushings were higher (68.71, 70 and 44.81%, respectively) in SF-SCL washings compared to their LF-LCL counterparts ($P ≤ 0.05$; Table 4; Figs. 2, 3, 4 and 7).

Conversely, on D7, the concentrations of valine and cystathionine in uterine flushings were greater (95 and 74.18%, respectively) in the LF-LCL uterine washings (Table 4; Figs. 5, 6 and 7).

Discussion

Disappointing fertility success in the beef cow industry is mainly caused by excessive rates of embryonic mortality during early pregnancy, which has been linked to

Table 4 Concentrations of amino acids in uterine washings

Amino acid	LF-LCL, μmol/g	SF-SCL, μmol/g	P < F
GLU			
D4	115.30 ± 9.23	157.40 ± 23.49	0.066
D7	127.93 ± 13.97	153.13 ± 21.92	0.38
AAD			
D4	91.52 ± 0.60	91.63 ± 1.19	0.93
D7	91.54 ± 0.82	92.71 ± 0.93	0.22
ASN			
D4	115.63 ± 2.80	126.66 ± 7.56	0.12
D7	139.83 ± 6.55	141.68 ± 6.43	0.84
SER			
D4	6.76 ± 1.03	11.04 ± 2.80	0.19
D7	17.34 ± 1.96	15.15 ± 3.12	0.63
GLN			
D4	95.74 ± 2.51	108.58 ± 7.08	0.06
D7	120.05 ± 6.16	122.52 ± 7.91	0.81
GLY			
D4	119.36 ± 12.39	164.37 ± 31.04	0.13
D7	339.70 ± 41.51	352.53 ± 43.62	0.84
HYS			
D4	11.21 ± 0.51	12.81 ± 0.61	0.08
D7	12.42 ± 0.74	12.75 ± 0.82	0.80
B-ALA			
D4	70.44 ± 0.60	71.18 ± 0.82	0.58
D7	80.95 ± 3.88	73.78 ± 1.06	0.08
TAU			
D4	117.20 ± 9.57	197.90 ± 32.56	0.01
D7	460.47 ± 106.32	334.93 ± 41.22	0.69
THR			
D4	6.88 ± 2.00	8.77 ± 2.79	0.66
D7	10.10 ± 1.65	9.83 ± 1.73	0.92
ALA			
D4	55.59 ± 5.42	95.70 ± 21.07	0.04
D7	76.94 ± 8.97	91.15 ± 11.40	0.36
PRO			
D4	6.27 ± 0.98	9.72 ± 2.21	0.13
D7	29.23 ± 7.78	23.63 ± 3.93	0.81
AAAB			
D4	19.82 ± 1.38	28.71 ± 2.85	0.03
D7	32.48 ± 7.75	23.29 ± 2.38	0.82
TYR			
D4	6.22 ± 0.99	8.37 ± 1.99	0.16
D7	6.20 ± 0.74	6.77 ± 0.78	0.62
VAL			
D4	19.11 ± 1.91	23.90 ± 3.56	0.23

Table 4 Concentrations of amino acids in uterine washings (Continued)

D7	33.11 ± 7.81	16.73 ± 2.27	0.03
MET			
D4	22.16 ± 0.26	22.36 ± 0.29	0.64
D7	22.73 ± 0.56	22.70 ± 0.26	0.96
CYS			
D4	127.83 ± 6.44	142.07 ± 10.47	0.27
D7	309.37 ± 49.54	177.92 ± 17.55	0.02
ILE			
D4	18.09 ± 2.25	21.58 ± 3.25	0.21
D7	15.40 ± 0.95	15.80 ± 1.13	0.93
PHE			
D4	120.92 ± 16.94	120.33 ± 25.84	0.98
D7	117.49 ± 23.60	142.63 ± 16.80	0.39
TRP			
D4	388.73 ± 35.18	433.83 ± 76.04	0.54
D7	403.82 ± 32.67	398.52 ± 33.38	0.91
ORN			
D4	98.07 ± 1.86	96.62 ± 2.29	0.51
D7	97.36 ± 2.33	96.71 ± 1.83	0.82
LYS			
D4	4.61 ± 1.04	4.90 ± 1.58	0.65
D7	13.44 ± 3.69	6.73 ± 0.42	0.70

Mean ± standard error of the mean of the concentration of amino acid in μmol/g of uterine washings for large follicle-large CL group (LF-LCL; Exp 1 n = 12; Exp 2 n = 9) and small follicle-small CL group (SF-SCL; Exp 1 n = 7; Exp 2 n = 10)

inadequate endometrial receptivity [40, 41]. During early pregnancy, the mother needs to provide the optimal uterine microenvironment for the embryo in order to facilitate initial embryonic-maternal interactions leading to subsequent implantation [42]. However, the exact definition of a microenvironment optimal for the proper development of embryos during early pregnancy remains to be elucidated. As previous studies pointed towards the importance of greater concentrations of AA in the uterus to increase fertility outcome in dairy cows [9], this study attempted to investigate the link between AA metabolic pathways and uterine function to support embryo growth during early diestrus in beef cows. Therefore, we characterized AA transport and metabolic pathways in the endometrium and AA levels in the uterine lumen on D4 and D7 after induction of ovulation in response to different endocrine profiles. Furthermore, the study was conducted in cyclic, non-inseminated cows based on the previous report by Forde et al. [43]. In that report, the authors indicated that from the first week of diestrus until after D13, the pregnant cow endometrium undergoes molecular changes similar to those

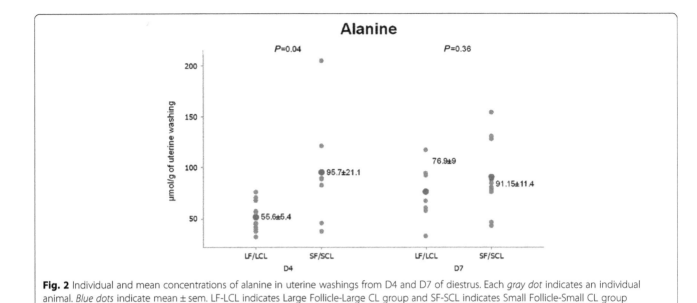

Fig. 2 Individual and mean concentrations of alanine in uterine washings from D4 and D7 of diestrus. Each *gray dot* indicates an individual animal. *Blue dots* indicate mean ± sem. LF-LCL indicates Large Follicle-Large CL group and SF-SCL indicates Small Follicle-Small CL group

of the cyclic cow endometrium. Thus, because sample collection was conducted on D4 and D7, it is reasonable to assume that the treatment effects observed in the present report would be similar were the animals pregnant.

In this context, a previously described in vivo receptivity model was used here [26, 32, 36, 38], aiming to define the AA signatures of the receptive endometrial tissues and histotroph. Using the same model, we were able to manipulate two fundamental aspects of receptivity: compared to the LF-LCL group, fertility was lower in the SF-SCL group (i.e., the low receptivity group) [27], and the concentration of PGR in the endometrium was greater in the same group on D7 of the estrous cycle [37]. More

specifically, data on the bovine endometrial tissue transcriptome associated with AA transport have been integrated with information on uterine flushing AA profiles from highly receptive (LF-LCL) versus low-receptive (SF-SCL) groups of cows. In beef cattle, altered endocrine patterns during follicle growth are known to influence the follicular size and steroidogenic capacity before ovulation [44]. Thereby, corpora lutea originating from larger follicles are characterized by larger sizes, resulting in greater circulating concentrations of P4 compared to those originating from smaller follicles [45, 46]. In our study, POF influenced CL size and plasmatic levels of P4 on D7 but not on D4. It is important to mention that CL is a transient endocrine organ that starts to develop

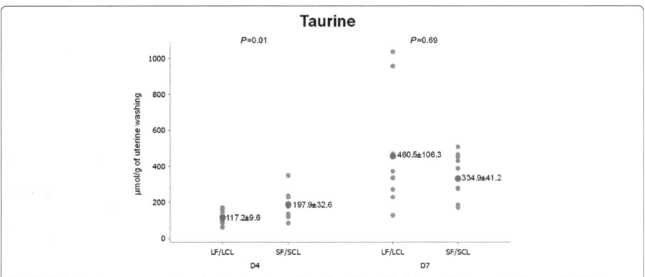

Fig. 3 Individual and mean concentrations of taurine in uterine washings from D4 and D7 of diestrus. Each *gray dot* indicates an individual animal. *Blue dots* indicate mean ± sem. LF-LCL indicates Large Follicle-Large CL group and SF-SCL indicates Small Follicle-Small CL group

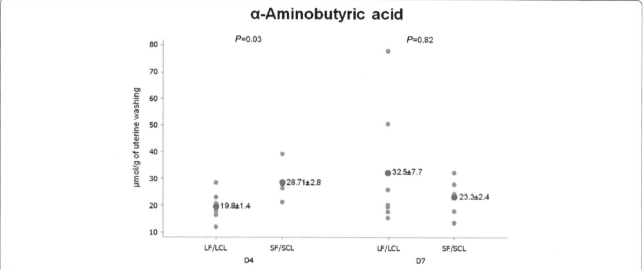

Fig. 4 Individual and mean concentrations of α-aminobutyric acid in uterine washings from D4 and D7 of diestrus. Each *gray dot* indicates an individual animal. *Blue dots* indicate mean ± sem. LF-LCL indicates Large Follicle-Large CL group and SF-SCL indicates Small Follicle-Small CL group

after ovulation, when theca and granulosa cells differentiate into luteal cells [47]. The D4 and D7 CL are considered to be in different stages of development [48, 49]. Specifically, at D4, the CL is under early development and did not achieve a large enough area to show differences in P4 production as it did on D7. Interestingly, gene expression related to P4 production indicated a greater capacity for P4 production at D7 than at D4 [49]. In cattle, POF size is directly related to E2 production [50, 51], suggesting that POF size and probably circulating concentrations of E2 influence the uterine environment and endometrial gene expression on D4 and D7.

The uterine flushing AA signatures show that on D4, concentrations of taurine, alanine and α-aminobutyric acid in uterine luminal flushings were greater in the SF-SCL group, which in this animal model represents the low-receptive uterus. The mRNA of *SCLY*, the enzyme related to alanine production, was up-regulated in the LF-LCL group. Moreover, *SLC6A6* mRNA, which transports β-alanine and mainly taurine, was more abundant in the endometrium of cows from the LF-LCL group. A collective interpretation of these data suggests that (1) alanine biosynthesis was stimulated in the LF-LCL group and that (2) alanine transport was stimulated in the LF-LCL group, mainly in a lumen-to-endometrium

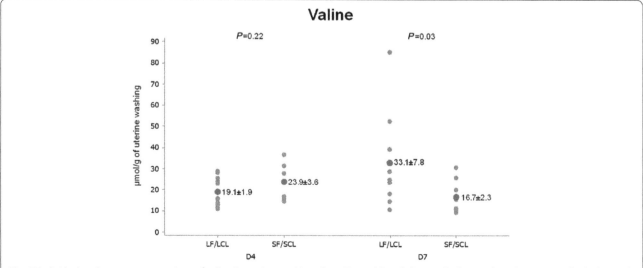

Fig. 5 Individual and mean concentrations of valine in uterine washings from D4 and D7 of diestrus. Each *gray dot* indicates an individual animal. *Blue dots* indicate mean ± sem. LF-LCL indicates Large Follicle-Large CL group and SF-SCL indicates Small Follicle-Small CL group

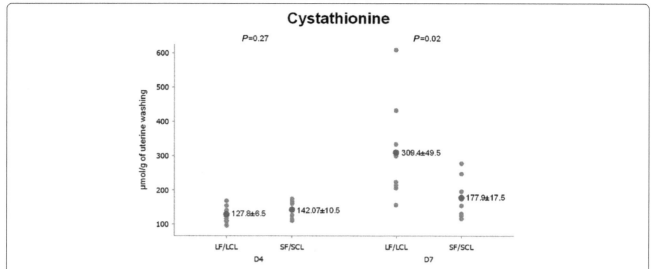

Fig. 6 Individual and mean concentrations of cystathionine in uterine washings from D4 and D7 of diestrus. Each *gray dot* indicates an individual animal. *Blue dots* indicate mean ± sem. LF-LCL indicates Large Follicle-Large CL group and SF-SCL indicates Small Follicle-Small CL group

direction. Stimulating accumulation of alanine in the endometrium may be important for endometrial cellular functions, such as proliferation, to guarantee endometrial receptivity. Alternatively, in accordance with data on cardiomyocytes, excess supplementation of AA may downregulate their transporters [52]. This relationship could explain the findings in the SF-SCL group.

Although early bovine embryos are purported to use all available AAs, alanine is unique in that it is not used and is even secreted by the embryo [53]. This suggests

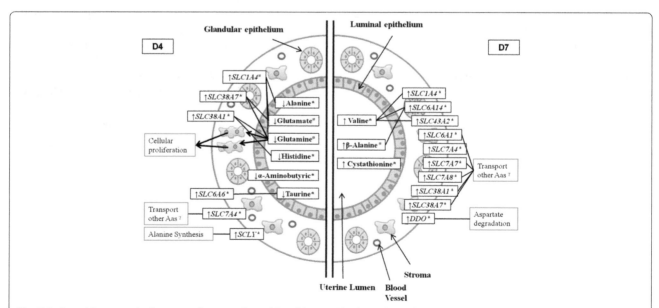

Fig. 7 Amino acid transport in the uterus of cows on D4 and D7 of diestrus. This figure shows the comparative abundance of transcripts related to amino acid (AA) transport and metabolism and the luminal concentration of AA between more receptive endometrium (Large Follicle-Large Corpus Luteum group) and less receptive endometrium (Small Follicle-Small Corpus Luteum group) on D4 and D7 after estrus. On D4, the transport of AA seems to occur preferentially from the uterine lumen towards endometrial cells, because despite elevated expression of genes related to AA transporters in endometrium there is lower availability of AA in uterine washings. Such direction of transport benefit events such as cell proliferation, which requires AA. On D7, AA availability in uterine lumen and abundance of genes related to AA transport are both stimulated in the more receptive endometrium. This phenotype is consistent with a greater provision of substrates to support embryonic needs for growth. ↑, up-regulated in LF-LCL group in comparison to SF-SCL group; ↓ down-regulated in LF-LCL group in comparison to SF-SCL group; γ, carriers related to transport of AAs similarly abundant in the lumen of both groups. solid lines connect a transporter with its cognate substrate(s); *, $P \leq 0.05$; #, $P < 0.1$; *SLC*, Solute carrier protein; *SCLY*, Selenocysteine Lyase; *DDO*, D-aspartate Oxidase

that regulation of this AA, which is available in greater concentrations in low-receptive than in high-receptive uteri, plays no major role in early embryonic development. Taurine, which was more abundant in the SF-SCL group, seems to be beneficial for embryonic development [54]; however, the addition of an elevated dose of cysteamine, a taurine precursor, to embryo culture media was toxic to the bovine embryo [55]. Supplementation of α-aminobutyric acid in the embryo culture medium did not affect porcine embryo cell viability [56].

Similar to the alanine transporter, *SLC38A1* is a transporter with high affinity for glutamine and is more abundant in high-receptive endometrial tissue than in low-receptive tissue, whereas glutamine tended to be more abundant in uterine washings of the low-receptive group. These data can be associated with the major proliferative activity found in the LF-LCL endometrium on D4, which was related to adequate uterus conditioning for receiving the embryo [39]. This proliferative activity may also be responsible for the observation of an increase in transporter abundance without an associated increase in the concentration of the target substrate in the uterine lumen. Indeed, glutamine is essential for cell proliferation in other tissues [57].

On D7, regulation of AA transporter transcript abundance and cognate substrates is complex, and several scenarios were observed. For example, *SLC1A4* was more abundant in the endometrium of LF-LCL cows, although its direct targets, serine, tyrosine, threonine, were present in similar concentrations in the uterine lumen of cows of both groups. Likewise, *SLC38A1* and *SLC38A7* transcripts were up-regulated in the endometrium of the LF-LCL group on D7. These transporters have high affinity for glutamine transport, but this AA was present in similar concentrations in the uterine lumen of both groups. In contrast, regardless of a similar abundance of the transporter *SLC7A11* between groups on D4 and D7, cystathionine, a product of serine and homocysteine condensation by cystathionine β-synthase [58], and valine were present in greater concentrations in the uterine washings of cows from the LF-LCL group compared to cows from the SF-SLC group. In a third scenario, SLC6A14 and one of its transporting targets, valine, were stimulated in the LF-LCL group. Correspondingly, in the human cervix, an increase in SLC6A14 mRNA was associated with a parallel increase in protein [59]. Interestingly, after the 4-cell stage, valine, along with leucine, isoleucine, and methionine, was favorable for the cleavage rates of the diploids after compaction and increased the total number of cells in the blastocyst and inner cell mass [60]. SLC7A8 transcript was up-regulated in the LF-LCL endometrium on D7, and this carrier transports neutral AAs such as valine, which is also up-regulated in the LF-LCL group. *SLC7A8* is also responsible for alanine transport. Despite the fact that the concentration of alanine in uterine washings is similar for both groups, the transcript abundance of *DDO*, an enzyme related to alanine degradation, is up-regulated in LF-LCL group. Perhaps there is a requirement to strictly control luminal concentrations of alanine on D7.

In this study, AA concentrations were quantified using uterine washings, composed of histotroph diluted in 20 mL of PBS. Such a method of quantification is appropriate for estimating AA availability in the uterus.

Collectively, the present data suggest that in the more receptive uterine phenotype, on D4, AA transport and metabolic pathways are directed to supply endometrial requirements for growth and function. Coincidentally, around D4, embryos have a limited demand for AAs, and they are in transit from the oviduct to the uterus. Thus, the AA composition of the histotroph reflects the endometrial demands for proper function more than a milieu for embryonic development. More interestingly, on D7, when the embryo is expected to be in the uterus, AAs are more abundant in the uterine secretions of the LF-LCL group, probably to serve as a supply to meet embryonic requirements. Correspondingly, transcript abundance data show a greater number of AA transporters that are up-regulated in the more receptive endometrium on D7. This phenotype is probably associated with the greater AA requirement for the subsequent stages of embryonic development (i.e., hatching and elongation). In terms of sex steroid regulation of such processes, the phenotype on D4 reflects classical actions of estradiol that include stimulation of endometrial cell proliferation, while the phenotype on D7 is associated with actions of progesterone, such as glandular secretory activity and histotroph formation to support embryonic growth [4, 39, 61–63]. A summarized view of the AA transport systems in the endometrium in association with receptivity during early diestrus is presented in Fig. 7.

To the best of our knowledge, this is the first investigation that relates early diestrus temporal changes in AA transport and availability in the uterine lumen to different periovulatory endocrine patterns that characterize fertility phenotypes. This study points to an important link between AA metabolic pathways in the endometrium and uterine receptivity. In addition, these data can serve as a basis for novel studies in bovine reproduction biotechnology with the aim of developing new tools to improve beef cattle fertility and profitability.

Conclusions

This study showed that the transcript abundance of AA transporters in the endometrium is linked with the receptive state of the endometrial tissue at both D4 and D7. Additionally, the AA patterns in uterine flushings differ between contrasting receptivity status of cows, both at D4 and at D7. The latter data indicate that AA metabolism and transport may be a potential key that needs to be regulated in order to fine-tune the maternal receptive state during early diestrus.

Abbreviations
AA: Amino acid; AAAB: α-amino butyric acid; AAD: α-amino adipic acid; ALA: Alanine; ANOVA: Analysis of variance; ASN: Asparagine; ASNS: Asparagine synthase; B-ALA: β-alanine; cDNA: Complementar deoxyribonucleic acid; CL: Corpus luteum; CYS: Cysteine; DDO: D-aspartate oxidade; E2: Estradiol; GLN: Glutamine; GLU: Glutamate; GLY: Glycine; GnRH: Gonadotrophin releasing hormone; GPT: Glutamic-pyruvate transaminase; HPLC: High capacity liquid chromatography; HYS: Histidine; ILE: Isoleucine; LF-LCL: Large follicle-large CL group; LYS: Lysine; MET: Methionine; ORN: Ornithine; P4: Progesterone; PBS: Phosphate buffer solution; PHE: Phenylalanine; POF: Pre-ovulatory follicle; PRO: Proline; PROC GLM: General linear models procedure; qRT-PCR: Real-time reverse transcriptase-polymerase chain reaction; RNA: Ribonucleic acid; SCLY: Selenocysteine lyase; SER: Serine; SF-SCL: Small follicle-small CL group; SLC: Solute carrier protein; TAU: Taurine; THR: Threonine; TRP: Tryptophan; TYR: Tyrosine; VAL: Valine

Acknowledgements
Authors wish to thank FAPESP, CNPq and CAPES for financial support and scholarships, Ourofino, Cravinhos, SP, Brazil for providing drugs and hormones, Dr. Messias Alves da Trindade Neto, faculty, post-doctoral fellows, students and staff of the School of Veterinary Medicine and Animal Science of the University of São Paulo for technical support.

Funding
FAPESP 2014/01727-4 to MISS
CNPq- 481199/2012-8 and FAPESP- 2011/03226-4 to MB
CNPq 140527/2013-3 to MRF
The funding bodies had no participation on the study, collection, analysis, interpretation of data nor in writing the manuscript.

Authors' contributions
MFR contributed to animal management, qPCR analysis, statistical analysis of transcripts and amino acids data and was a major contributor in writing the manuscript. MISS contributed to animal management, performed samples preparation for qPCR analysis and performed qPCR analysis. GP performed animals and reproductive management, statistical analysis of follicle, corpus luteum and progesterone data and contributed in writing the manuscript. VVH contributed in writing the manuscript. MB was the PI and contributed to experiment design and writing the manuscript. All authors read and approved the final manuscript.

Authors' information
Not applicable.

Competing interests
The authors declare that they have no competing interests.

Author details
[1]Department of Animal Reproduction, School of Veterinary Medicine and Animal Science, University of São Paulo, 225, Duque de Caxias Norte Ave. Jd. Elite, 13635-900 Pirassununga, SP, Brazil. [2]Gamete Research Centre, University of Antwerp, Antwerp, Belgium.

References
1. Garrett JE, Geisert RD, Zavy MT, Morgan GL. Evidence for maternal regulation of early conceptus growth and development in beef cattle. J Reprod Fertil. 1988;84:437–46.
2. Satterfield MC, Song G, Kochan KJ, Riggs PK, Simmons RM, Elsik CG, et al. Discovery of candidate genes and pathways in the endometrium regulating ovine blastocyst growth and conceptus elongation. Physiol Genomics. 2009;39:85–99.
3. Carter F, Forde N, Duffy P, Wade M, Fair T, Crowe MA, et al. Effect of increasing progesterone concentration from Day 3 of pregnancy on subsequent embryo survival and development in beef heifers. Reprod Fertil Dev. 2008;20:368–75.
4. Forde N, Carter F, Fair T, Crowe MA, Evans AC, Spencer TE, et al. Progesterone-regulated changes in endometrial gene expression contribute to advanced conceptus development in cattle. Biol Reprod. 2009;81:784–94.
5. Clemente M, de La Fuente J, Fair T, Al Naib A, Gutierrez-Adan A, Roche JF, et al. Progesterone and conceptus elongation in cattle: a direct effect on the embryo or an indirect effect via the endometrium? Reproduction. 2009;138:507–17.
6. Demetrio DG, Santos RM, Demetrio CG, Vasconcelos JL. Factors affecting conception rates following artificial insemination or embryo transfer in lactating Holstein cows. J Dairy Sci. 2007;90:5073–82.
7. Spencer TE, Bazer FW. Uterine and placental factors regulating conceptus growth in domestic animals. J Anim Sci. 2004;82(E-Suppl):E4–13.
8. Gao H, Wu G, Spencer TE, Johnson GA, Li X, Bazer FW. Select nutrients in the ovine uterine lumen. I. Amino acids, glucose, and ions in uterine lumenal flushings of cyclic and pregnant ewes. Biol Reprod. 2009;80:86–93.
9. Meier S, Mitchell MD, Walker CG, Roche JR, Verkerk GA. Amino acid concentrations in uterine fluid during early pregnancy differ in fertile and subfertile dairy cow strains. J Dairy Sci. 2014;97:1364–76.
10. Hugentobler SA, Diskin MG, Leese HJ, Humpherson PG, Watson T, Sreenan JM, et al. Amino acids in oviduct and uterine fluid and blood plasma during the estrous cycle in the bovine. Mol Reprod Dev. 2007;74:445–54.
11. Hugentobler SA, Sreenan JM, Humpherson PG, Leese HJ, Diskin MG, Morris DG. Effects of changes in the concentration of systemic progesterone on ions, amino acids and energy substrates in cattle oviduct and uterine fluid and blood. Reprod Fertil Dev. 2010;22:684–94.
12. Forde N, Simintiras CA, Sturmey R, Mamo S, Kelly AK, Spencer TE, et al. Amino Acids in the Uterine Luminal Fluid Reflects the Temporal Changes in Transporter Expression in the Endometrium and Conceptus during Early Pregnancy in Cattle. PLoS One. 2014;9:e100010.
13. Steeves TE, Gardner DK. Temporal and differential effects of amino acids on bovine embryo development in culture. Biol Reprod. 1999;61:731–40.
14. Daniel JC, Krishnan RS. Amino acid requirements for growth of the rabbit blastocyst in vitro. J Cell Physiol. 1967;70:155–60.
15. Kim JH, Niwa K, Lim JM, Okuda K. Effects of phosphate, energy substrates, and amino acids on development of in vitro-matured, in vitro-fertilized bovine oocytes in a chemically defined, protein-free culture medium. Biol Reprod. 1993;48:1320–5.
16. Smith RJ, Dean W, Konfortova G, Kelsey G. Identification of novel imprinted genes in a genome-wide screen for maternal methylation. Genome Res. 2003;13:558–69.
17. Groebner AE, Rubio-Aliaga I, Schulke K, Reichenbach HD, Daniel H, Wolf E, et al. Increase of essential amino acids in the bovine uterine lumen during preimplantation development. Reproduction. 2011;141:685–95.
18. Gao H, Wu G, Spencer TE, Johnson GA, Bazer FW. Select nutrients in the ovine uterine lumen. ii. glucose transporters in the uterus and peri-implantation conceptuses. Biol Reprod. 2009;80:94–104.
19. Partridge RJ, Leese HJ. Consumption of amino acids by bovine preimplantation embryos. Reprod Fertil Dev. 1996;8:945–50.
20. Donnay I, Leese HJ. Embryo metabolism during the expansion of the bovine blastocyst. Mol Reprod Dev. 1999;53:171–8.
21. Morris DG, Humpherson PG, Leese HJ, Sreenan JM. Amino acid turnover by elongating cattle blastocysts recovered on days 14-16 after insemination. Reproduction. 2002;124:667–73.
22. Houghton FD, Hawkhead JA, Humpherson PG, Hogg JE, Balen AH, Rutherford AJ, et al. Non-invasive amino acid turnover predicts human embryo developmental capacity. Hum Reprod. 2002;17:999–1005.
23. Sturmey RG, Bermejo-Alvarez P, Gutierrez-Adan A, Rizos D, Leese HJ, Lonergan P. Amino acid metabolism of bovine blastocysts: a biomarker of sex and viability. Mol Reprod Dev. 2010;77:285–96.

24. Setoyama C, Miura R. Structural and functional characterization of the human brain D-aspartate oxidase. J Biochem. 1997;121:798–803.

25. Mihara H, Kurihara T, Watanabe T, Yoshimura T, Esaki N. cDNA cloning, purification, and characterization of mouse liver selenocysteine lyase. Candidate for selenium delivery protein in selenoprotein synthesis. J Biol Chem. 2000;275:6195–200.

26. Mesquita FS, Pugliesi G, Scolari SC, França MR, Ramos RdS, Oliveira ML, et al. Manipulation of the periovulatory sex-steroidal milieu affects endometrial but not luteal gene expression on early diestrus Nelore cows., vol. in press. Theriogenology: Theriogenology. 2014.

27. Pugliesi G, Santos FB, Lopes E, Nogueira É, Maio JR, Binelli M. Improved fertility in suckled beef cows ovulating large follicles or supplemented with long-acting progesterone after timed-AI. Theriogenology. 2016;85:1239–48.

28. Bazer FW. Uterine protein secretions: Relationship to development of the conceptus. J Anim Sci. 1975;41:1376–82.

29. Roberts RM, Bazer FW. The functions of uterine secretions. J Reprod Fertil. 1988;82:875–92.

30. Kane MT, Morgan PM, Coonan C. Peptide growth factors and preimplantation development. Hum Reprod Update. 1997;3:137–57.

31. Dunne LD, Diskin MG, Sreenan JM. Embryo and foetal loss in beef heifers between day 14 of gestation and full term. Anim Reprod Sci. 2000;58:39–44.

32. França MR, Mesquita FS, Lopes E, Pugliesi G, Van Hoeck V, Chiaratti MR, et al. Modulation of periovulatory endocrine profiles in beef cows: consequences for endometrial glucose transporters and uterine fluid glucose levels. Domest Anim Endocrinol. 2015;50:83–90.

33. Garbarino EJ, Hernandez JA, Shearer JK, Risco CA, Thatcher WW. Effect of lameness on ovarian activity in postpartum holstein cows. J Dairy Sci. 2004;87:4123–31.

34. White JA, Hart RJ, Fry JC. An evaluation of the Waters Pico-Tag system for the amino-acid analysis of food materials. J Automat Chem. 1986;8:170–7.

35. Vandesompele J, De Preter K, Pattyn F, Poppe B, Van Roy N, De Paepe A, et al. Accurate normalization of real-time quantitative RT-PCR data by geometric averaging of multiple internal control genes. Genome Biol. 2002;3:RESEARCH0034.

36. Ramos RS, Oliveira ML, Izaguirry AP, Vargas LM, Soares MB, Mesquita FS, et al. The periovulatory endocrine milieu affects the uterine redox environment in beef cows. Reprod Biol Endocrinol. 2015;13:39.

37. Araújo ER, Sponchiado M, Pugliesi G, Van Hoeck V, Mesquita FS, Membrive CM, et al. Spatio-specific regulation of endocrine-responsive gene transcription by periovulatory endocrine profiles in the bovine reproductive tract. Reprod Fertil Dev. 2016;28:1533–44.

38. Ramos RS, Mesquita FS, D'Alexandri FL, Gonella-Diaza AM, Papa PC, Binelli M. Regulation of the polyamine metabolic pathway in the endometrium of cows during early diestrus. Mol Reprod Dev. 2014;81:584–94.

39. Mesquita FS, Ramos RS, Pugliesi G, Andrade SC, Van Hoeck V, Langbeen A, et al. The Receptive Endometrial Transcriptomic Signature Indicates an Earlier Shift from Proliferation to Metabolism at Early Diestrus in the Cow. Biol Reprod. 2015;93:52.

40. Minten MA, Bilby TR, Bruno RG, Allen CC, Madsen CA, Wang Z, et al. Effects of fertility on gene expression and function of the bovine endometrium. PLoS One. 2013;8:e69444.

41. Diskin MG, Murphy JJ, Sreenan JM. Embryo survival in dairy cows managed under pastoral conditions. Anim Reprod Sci. 2006;96:297–311.

42. Cavagna M, Mantese JC. Biomarkers of endometrial receptivity–a review. Placenta. 2003;24(Suppl B):S39–47.

43. Forde N, Carter F, Spencer TE, Bazer FW, Sandra O, Mansouri-Attia N, et al. Conceptus-induced changes in the endometrial transcriptome: how soon does the cow know she is pregnant? Biol Reprod. 2011;85:144–56.

44. Stegner JE, Kojima FN, Bader JF, Lucy MC, Ellersieck MR, Smith MF, et al. Follicular dynamics and steroid profiles in cows during and after treatment with progestin-based protocols for synchronization of estrus. J Anim Sci. 2004;82:1022–8.

45. Vasconcelos JL, Sartori R, Oliveira HN, Guenther JG, Wiltbank MC. Reduction in size of the ovulatory follicle reduces subsequent luteal size and pregnancy rate. Theriogenology. 2001;56:307–14.

46. Pfeifer LF, Mapletoft RJ, Kastelic JP, Small JA, Adams GP, Dionello NJ, et al. Effects of low versus physiologic plasma progesterone concentrations on ovarian follicular development and fertility in beef cattle. Theriogenology. 2009;72:1237–50.

47. Richards JS, Russell DL, Robker RL, Dajee M, Alliston TN. Molecular mechanisms of ovulation and luteinization. Mol Cell Endocrinol.

1998;145:47–54.

48. Redmer DA, Grazul AT, Kirsch JD, Reynolds LP. Angiogenic activity of bovine corpora lutea at several stages of luteal development. J Reprod Fertil. 1988;82:627–34.

49. Park HJ, Park SJ, Koo DB, Kong IK, Kim MK, Kim JM, et al. Unfolding protein response signaling is involved in development, maintenance, and regression of the corpus luteum during the bovine estrous cycle. Biochem Biophys Res Commun. 2013;441:344–50.

50. Lopes AS, Butler ST, Gilbert RO, Butler WR. Relationship of pre-ovulatory follicle size, estradiol concentrations and season to pregnancy outcome in dairy cows. Anim Reprod Sci. 2007;99:34–43.

51. Nishimoto H, Hamano S, Hill GA, Miyamoto A, Tetsuka M. Classification of bovine follicles based on the concentrations of steroids, glucose and lactate in follicular fluid and the status of accompanying follicles. J Reprod Dev. 2009;55:219–24.

52. Zhang Y, Yang L, Yang YJ, Liu XY, Jia JG, Qian JY, et al. Low-dose taurine upregulates taurine transporter expression in acute myocardial ischemia. Int J Mol Med. 2013;31:817–24.

53. Kuran M, Robinson JJ, Brown DS, McEvoy TG. Development, amino acid utilization and cell allocation in bovine embryos after in vitro production in contrasting culture systems. Reproduction. 2002;124:155–65.

54. Takahashi Y, Kanagawa H. Effects of glutamine, glycine and taurine on the development of in vitro fertilized bovine zygotes in a chemically defined medium. J Vet Med Sci. 1998;60:433–7.

55. Guyader-Joly C, Guérin P, Renard JP, Guillaud J, Ponchon S, Ménézo Y. Precursors of taurine in female genital tract: effects on developmental capacity of bovine embryo produced in vitro. Amino Acids. 1998;15:27–42.

56. Whitaker BD, Knight JW. Exogenous gamma-glutamyl cycle compounds supplemented to in vitro maturation medium influence in vitro fertilization, culture, and viability parameters of porcine oocytes and embryos. Theriogenology. 2004;62:311–22.

57. Ko TC, Beauchamp RD, Townsend CM, Thompson JC. Glutamine is essential for epidermal growth factor-stimulated intestinal cell proliferation. Surgery. 1993;114:147–54. discussion 153-144.

58. Mendes MI, Santos AS, Smith DE, Lino PR, Colaço HG, de Almeida IT, et al. Insights into the regulatory domain of cystathionine Beta-synthase: characterization of six variant proteins. Hum Mutat. 2014;35:1195–202.

59. Gupta N, Prasad PD, Ghamande S, Moore-Martin P, Herdman AV, Martindale RG, et al. Up-regulation of the amino acid transporter ATB(0,+) (SLC6A14) in carcinoma of the cervix. Gynecol Oncol. 2006;100:8–13.

60. Van Thuan N, Harayama H, Miyake M. Characteristics of preimplantational development of porcine parthenogenetic diploids relative to the existence of amino acids in vitro. Biol Reprod. 2002;67:1688–98.

61. Arai M, Yoshioka S, Tasaki Y, Okuda K. Remodeling of bovine endometrium throughout the estrous cycle. Anim Reprod Sci. 2013;142:1–9.

62. Reynolds LP, Kirsch JD, Kraft KC, Knutson DL, McClaflin WJ, Redmer DA. Time-course of the uterine response to estradiol-17beta in ovariectomized ewes: uterine growth and microvascular development. Biol Reprod. 1998;59:606–12.

63. Spencer TE, Johnson GA, Burghardt RC, Bazer FW. Progesterone and placental hormone actions on the uterus: insights from domestic animals. Biol Reprod. 2004;71:2–10.

64. Bettegowda A, Patel OV, Ireland JJ, Smith GW. Quantitative analysis of messenger RNA abundance for ribosomal protein L-15, cyclophilin-A, phosphoglycerokinase, beta-glucuronidase, glyceraldehyde 3-phosphate dehydrogenase, beta-actin, and histone H2A during bovine oocyte maturation and early embryogenesis in vitro. Mol Reprod Dev. 2006;73:267–78.

12

Cinnamicaldehyde regulates the expression of tight junction proteins and amino acid transporters in intestinal porcine epithelial cells

Kaiji Sun[1], Yan Lei[2], Renjie Wang[1,2], Zhenlong Wu[1,4*] and Guoyao Wu[1,3]

Abstract

Background: Cinnamicaldehyde (CA) is a key flavor compound in cinnamon essential oil possessing various bioactivities. Tight junction (TJ) proteins are vital for the maintenance of intestinal epithelial barrier function, transport, absorption and utilization of dietary amino acids and other nutrients. In this study, we tested the hypothesis that CA may regulate the expression of TJ proteins and amino acid transporters in intestinal porcine epithelial cells (IPEC-1) isolated from neonatal pigs.

Results: Compared with the control, cells incubated with 25 μmol/L CA had increased transepithelial electrical resistance (TEER) and decreased paracellular intestinal permeability. The beneficial effect of CA on mucosal barrier function was associated with enhanced protein abundance for claudin-4, zonula occludens (ZO)-1, ZO-2, and ZO-3. Immunofluorescence staining showed that 25 μmol/L CA promoted the localization of claudin-1 and claudin-3 to the plasma membrane without affecting the localization of other TJ proteins, including claudin-4, occludin, ZO-1, ZO-2, and ZO-3, compared with the control cells. Moreover, protein abundances for rBAT, xCT and LAT2 in IPEC-1 cells were enhanced by 25 μmol/L CA, while that for EAAT3 was not affected.

Conclusions: CA improves intestinal mucosal barrier function by regulating the distribution of claudin-1 and claudin-3 in enterocytes, as well as enhancing protein abundance for amino acid transporters rBAT, xCT and LAT2 in enterocytes. Supplementation with CA may provide an effective nutritional strategy to improve intestinal integrity and amino acid transport and absorption in piglets.

Keywords: Amino acid transporters, Barrier function, Cinnamicaldehyde, Intestinal epithelial cells, Tight junction proteins

Background

Cinnamicaldehyde (CA) is a key flavor compound in cinnamon essential oil extracted from the stem bark of *Cinnamomum cassia* in nature [1]. CA is widely used in the perfume, pharmacy, and food processing industries due to its antioxidant, anti-microbial, and anti-diabetic properties [2–6]. Moreover, it has been reported that CA has chemotherapeutic and anticancer effects through the inhibition of proliferation, inducing apoptosis, and blocking angiogenesis [7–9]. Importantly, CA is a safe flavor compound approved by FDA and the 'Flavor and Extract

Manufacturers' Association of the United States, suggesting the potential administration of this dietary factor may be achievable within an acceptable safety range for humans and animals [5]. The well documented antimicrobial properties and safety of CA have promoted its application to the nutrition of humans and animals. It has been reported that supplementation of essential oils, in which CA is the major component, increases the digestibility of crude protein in weaned pigs [10]. The beneficial effects of CA are associated with the enhanced secretion of digestive enzymes, improved nutrient digestion, and enhanced feed intake [11–14]. It remains unknown whether CA has any effect on intestinal barrier function, as well as nutrient transport and absorption in humans and pigs.

The epithelial barrier is formed by the apical plasma membrane and intercellular tight junction (TJ), which provides physical and functional barriers to prevent bacteria,

* Correspondence: bio2046@hotmail.com
[1]State Key Laboratory of Animal Nutrition, College of Animal Science and Technology, China Agricultural University, Beijing 100193, China
[4]Department of Animal Nutrition and Feed Science, China Agricultural University, Beijing 100193, China
Full list of author information is available at the end of the article

endotoxins and other harmful substances from entering the blood circulation, while also allowing for the absorption of nutrients [15]. Diverse physiological or pathological stimuli can regulate the intestinal mucosal-barrier function, which contributes to nutrient transport, absorption, and intracellular homeostasis [16–19]. Consistently, dysfunction of TJ proteins has been reported to be associated with increased paracellular permeability, and the development and progression of multiple intestinal disorders [20].

Although CA is known to improve the digestibility of dietary fiber, lipids, and crude protein, the underlying cellular and molecular mechanisms remain largely unknown. We have hypothesized that CA may up-regulate the expression of TJ proteins and amino acid transporters in intestinal epithelial cells, thus contributing to the intestinal barrier function in neonates. This hypothesis was tested in the present study using porcine intestinal epithelial cells (IPEC-1), isolated from neonatal pigs.

Methods

Reagents
Dulbecco's modified Eagle's F12 Ham medium (DMEM-F12) and fetal bovine serum (FBS) were purchased from Invitrogen (Carlsbad, CA, USA). Epidermal growth factor was a product of BD Biosciences (Carlsbad, CA, USA). Trypsin/EDTA was procured from Gibco (Carlsbad, CA, USA). Antibodies against occludin, claudin-1, claudin-3, claudin-4, zonula occludens (ZO)-1, ZO-2, and ZO-3 were products of Invitrogen (Carlsbad, CA, USA). Unless indicated, all other chemicals including CA were purchased from Sigma-Aldrich (St. Louis, MO, USA).

Cell culture
Intestinal porcine epithelial cell line 1 (IPEC-1) cells, which were isolated from the jejunum of newborn pigs without access to milk or any food [21], were then cultured in a DMEM-F12 medium supplemented with 5% FBS, insulin (5 µg/mL), transferrin (5 µg/mL), selenium (5 ng/mL), epidermal growth factor (5 µg/L), penicillin (50 µg/mL) and streptomycin (4 µg/mL) as previously described [22]. All cell cultures were carried out at 37 °C in a humidified incubator containing 5% CO_2.

Measurement of transepithelial electrical resistance (TEER)
The tightness of the TJ was assessed by measuring TEER as previously described [23]. Briefly, IPEC-1 Cells (5×10^4 cells per well) were seeded in culture transwells (the membrane area, 0.33 cm^2; pore size, 0.4 µm) which were placed in 24-well culture plates. Cells were incubated with 0, 12.5, or 25 µmol/L CA for the indicated time periods. TEER was determined every 12 h by using a Millicell ERS-2 Voltage-Ohm Meter (World Precision Instruments) equipped with a STX01 electrode as

described here [24]. All values are expressed as percentages of the basal level for the controls.

Monolayer paracellular permeability determination
Paracellular permeability was determined as previously described [25]. Briefly, IPEC-1 cells were seeded in culture transwells as for TEER determination. 1 mg/mL FITC-dextran (20 kDa) was added to the apical side of the monolayer and the flux of FITC-dextran was determined by serially sampling the basolateral compartment every 12 h. The concentration of FITC-dextran was measured using the SpectraMax M3 Multi-Mode Microplate Reader (Molecular Devices) with excitation and emission wavelengths of 490 and 520 nm, respectively. The permeability of monolayer cells was defined as the amount of FITC-dextran that was transported from the apical side into the basolateral chamber. FITC-dextran concentration was calculated by subtracting the fluorescence value of the FITC-free medium.

Western blot analysis
IPEC-1 cells treated with various concentrations of CA for 24 h were harvested for the analysis of the abundance of TJ proteins, as previously described [24]. Equal amounts of proteins (25 µg) were separated on SDS-PAGE gels, transferred to polyvinylidene difluoride membranes (Millipore), and then incubated with a primary antibody (1:2,000) overnight at 4 °C and then incubated with an appropriate secondary antibody (1:2,000) at 25 °C for 1 h. The blots were detected with the Image Quant LAS 4000 mini system (GE Healthcare Bio-sciences AB, Inc., Sweden) after incubation with the ECL plus system (Amersham Biosciences, Sweden). Chemifluorescence was quantified with the use of the Quantity One software (Bio-Rad Laboratories). All results were normalized to GAPDH and expressed as relative values to those of the control group.

Immunofluorescence assay
IPEC-1 cells treated with or without CA were fixed with 4% paraformaldehyde at 37 °C for 20 min, and then were incubated with a specific primary antibody against claudin-1, claudin-3, claudin-4, ZO-1, ZO-2 and ZO-3 for 16 h at 4 °C. Cells were washed three times with PBS, and then were incubated with an appropriate secondary antibody (1:100) for 1 h at 25 °C. Nuclei were stained by using Hoechst 33258 (1 µg/mL) for 10 min at 25 °C. The distribution of TJ proteins was visualized under a fluorescence microscope (Axio Vert. A1, Zeiss, Germany).

Statistical analysis
Values are expressed as mean ± SEM. Data was analyzed by one-way ANOVA and the Student-Newman-Keuls multiple comparisons test, using the SPSS statistical software

(SPSS for Windows, version 17.0). $P \leq 0.05$ were taken to indicate statistical significance.

Results

Effects of CA on barrier function in the IPEC-1 cell monolayer

As shown, incubation of cells with 25 μmol/L CA led to greater ($P < 0.05$) TEER at 36-48 h (Fig. 1a) when compared with controls. In contrast, no difference was observed between the cells treated with 12.5 μmol/L CA and the control cells at 12-48 h. Consistent with increased TEER, cells incubated with 25 μmol/L CA had reduced ($P < 0.05$) paracellular permeability, as indicated by FITC-dextran flux at 12-48 h (Fig. 1b) when compared with controls. Cells treated with 12.5 μmol/L CA had lowered

permeability ($P < 0.05$), compared with the control cells at 24-48 h. Although both 12.5 and 25 μmol/L CA treatment led to decreased permeability in enterocytes compared with the controls, cells treated with 25 μmol/L CA had lower permeability ($P < 0.05$) when compared with cells incubated with 12.5 μmol/L CA at 36-48 h.

Effects of CA on expression of TJ proteins in IPEC-1 cells

Compared with control cells, 25 μmol/L CA enhanced ($P < 0.05$) the abundance of proteins for claudin-4 (Fig. 2c) and ZO family proteins including ZO-1, ZO-2, and ZO-3 (Fig. 3). The protein abundance for ZO-2 (Fig. 3b), instead of other proteins, was enhanced by 12.5 μmol/L CA ($P < 0.05$) compared with that of the control. The protein abundances for claudin-1(Fig. 2a), claudin-3 (Fig. 2b), and occludin (Fig. 2d) were not affected ($P > 0.05$) by 12.5 or 25 μmol/L CA treatment.

Effects of CA on the intracellular distribution of TJ proteins in IPEC-1 cells

The cellular distributions of TJ proteins were assessed by an immunofluorescence microscope. Treatment with 25 μmol/L CA promoted the localization of claudin-1 and claudin-3 (Fig. 4a and b) to the plasma membrane without affecting the localization of other TJ proteins, including claudin-4, occludin, ZO-1, ZO-2, and ZO-3, compared to the control cells (Fig. 4). In contrast, 12.5 μmol/L CA had no effect on the localization of TJ proteins determined in our study, such as claudin-1, claudin-3, claudin-4, occludin, ZO-1, ZO-2, and ZO-3 (Fig. 4). It should be noted that most of the ZO-3 was located at the nucleus membrane which was not affected by CA exposure (Fig. 4g).

Effects of CA on the protein abundance for amino acid transporters

The active transport of amino acids is the major mechanism for their uptake into enterocytes [26–28]. We determined the protein abundance for the following amino acid transporters, EAAT3 (high-affinity glutamate transporter), LAT2 (arginine and leucine transporter), rBAT (basic amino acid transporter), and xCT (acidic amino acid transporter) in IPEC-1 cells by Western blot analysis. The protein abundance for rBAT (Fig. 5a) and LAT2 (Fig. 5c) in IPEC-1 cells were enhanced ($P < 0.05$) by both 12.5- and 25 μmol/L CA, compared with the control cells. In contrast, 25 μmol/L CA increased ($P < 0.05$) the protein abundance for xCT in the intestinal epithelial cells, but 12.5 μmol/L CA had no effect (Fig. 5b). The protein abundance for EAAT3 was not affected ($P > 0.05$) by either 12.5 or 25 μmol/L CA (Fig. 5d), compared with the control cells.

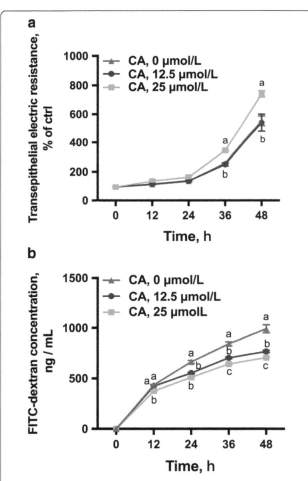

Fig. 1 Effects of CA on intestinal barrier function in IPEC-1 cells. Cells were cultured for 24 h in the absence or presence of 12.5- or 25 μmol/L CA. **a** TEER and **b** paracellular permeability were then determined. Values are expressed as means ± SEM, n = 6. Means at a time point without a common letter differ, $P < 0.05$. CA, Cinnamicaldehye; IPEC-1, intestinal porcine epithelial cell line 1; TEER, trans-epithelial electrical resistance

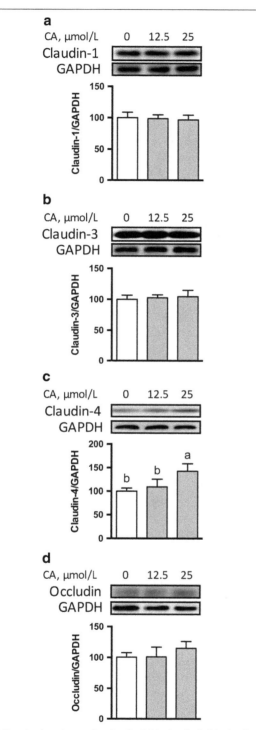

Fig. 2 Protein abundances for claudin-1 (**a**), claudin-3 (**b**), claudin-4 (**c**), and occludin (**d**) in IPEC-1 cells. IPEC-1 cells were cultured in the absence or presence of 12.5 or 25 μmol/L CA for 24 h. Cells were collected and protein abundances were analyzed. Values are expressed as means ± SEM, $n = 3$. Means without a common letter differ, $P < 0.05$. CA, Cinnamicaldehyde; IPEC-1, intestinal porcine epithelial cell line 1

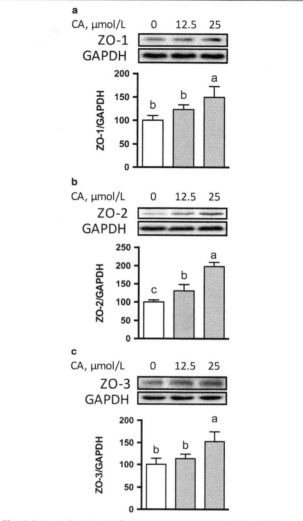

Fig. 3 Protein abundances for ZO-1 (**a**), ZO-2 (**b**), and ZO-3 (**c**) in IPEC-1 cells. Cells were cultured in the absence or presence of 12.5 or 25 μmol/L CA for 24 h. Cells were collected and protein abundances were analyzed. Values are expressed as means ± SEM, $n = 3$. Means without a common letter differ, $P < 0.05$. CA, Cinnamicaldehye; IPEC-1, intestinal porcine epithelial cell line 1; ZO, zonula occludens

Discussion

In the present study, we have shown that CA, a key flavor compound in cinnamon essential oil, promoted the intestinal mucosal-barrier function as indicated by increased TEER and decreased paracellular permeability. Western blot analysis revealed that cells treated with CA had enhanced protein abundances for TJ proteins, such as claudin-4 and scaffolding proteins. Moreover, the protein abundance for amino acid transporters, including rBAT, LAT2, and xCT, which are required for amino acid transport and absorption in enterocytes, were also enhanced by CA treatment.

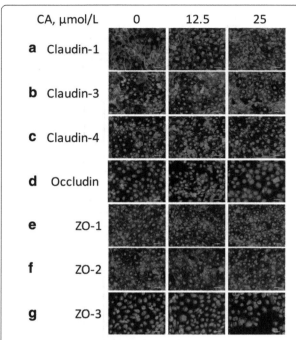

Fig. 4 The distributions of TJ proteins claudin-1 (**a**), claudin-3 (**b**), claudin-4 (**c**), occludin (**d**), ZO-1 (**e**), ZO-2 (**f**), and ZO-3 (**g**) in IPEC-1 cells. Cells were treated as in Fig. 3, and immunofluorescence staining was performed to identify the distributions of the proteins. CA, Cinnamicaldehyde; IPEC-1, intestinal epithelial porcine cell line 1. Scale bar, 50 μm

CA, a natural compound isolated from the stem bark of *Cinnamomum cassia*, is widely used in food processing and animal diets due to its antioxidant, antimicrobial, and anti-diabetic attributes [2–4]. Studies in pigs, a widely used animal model for various disorders in humans, have demonstrated that CA supplementation enhances nutrient digestibility in pigs [10]. It remains largely unknown whether CA supplementation can have any effect on intestinal barrier integrity, thereby improving nutrient transport, absorption, and intracellular homeostasis.

To test this hypothesis, we first measured TEER, an indicator of intestinal epithelial integrity and permeability of intestinal epithelium. Incubation of the enterocyte with CA led to increased TEER and decreased FITC-dextran flux in intestinal porcine monolayers, suggesting a beneficial effect of CA on barrier function. Epithelial barrier function and paracellular permeability are primarily determined by epithelial TJ proteins [29, 30]. Transmembrane proteins (e.g., the claudin family protein, occludin) and peripheral membrane proteins (e.g., ZO-1, ZO-2 and ZO-3) have been identified as critical components of TJ proteins [31, 32]. Disruption of epithelial TJ proteins has been reported to be associated with multiple intestinal disorders [29, 33]. Consistently, restoration of TJ proteins by nutrients or prebiotics

can improve mucosal barrier integrity and function in humans and animals [34]. We have recently found that dietary supplementation of glutamine prevented weanling stress-induced intestinal-mucosal barrier breakdown by augmenting TJ protein abundance [35], suggesting a functional role for amino acids in regulating mucosal barrier function.

In the present study, we found that CA regulates the protein abundance and cellular distributions of TJ proteins in intestinal cells. Specifically, the presence of 25 μmol/L CA led to enhanced protein abundances for claudin-4, ZO-1, ZO-2, and ZO-3, which are correlated well with augmented TEER values in IPEC-1 cells (Fig. 1). The claudin family proteins and ZO family proteins, play a crucial role in establishing cell–cell contacts and maintaining paracellular permeability [36, 37]. Recent studies have demonstrated that the reduction of claudin family proteins is strongly associated with intestinal barrier disruption in rodents [38, 39]. The regulatory effects of CA on the protein abundance of TJ suggest that supplementation with CA might be a preventive strategy to maintain the appropriate function of the intestinal-mucosal barrier. Another novel finding of our study is that CA treatment led to the distributions of claudin-1 and claudin-3 to the cellular plasma membrane (Fig. 4a and b) without affecting their protein abundances (Fig. 2a and b). Thus, CA regulates both the abundance and localization of TJ proteins in enterocytes. Considering that the disruption of TJ is caused by various stresses and pathogens in pigs [40], supplementation of CA may provide an effective nutritional strategy to alleviate mucosal barrier dysfunction. At present, the underlying mechanisms responsible for this effect remain incompletely understood [31]. More research involving our IPEC-1 cell model is required to answer this question.

In addition to providing physical and functional barriers to prevent the entry of bacteria, endotoxins, and other harmful substances from entering the blood circulation, appropriate amounts of TJ proteins maintains the integrity of the intestinal epithelium and, therefore, are also required for the absorption of nutrients [15]. Amino acids released from the hydrolysis of dietary proteins and peptides in the lumen of the small intestine are transported across cell membranes by a complex system of multiple amino acid transporters [26–28]. A number of transporters have been identified on the apical surface of the mammalian small intestine that are responsible for the intestinal absorption of amino acids [26, 41, 42]. The defective intestinal uptake of amino acids leads to alterations in plasma amino acids, growth retardation, and the Hartnup disorder [26]. We have

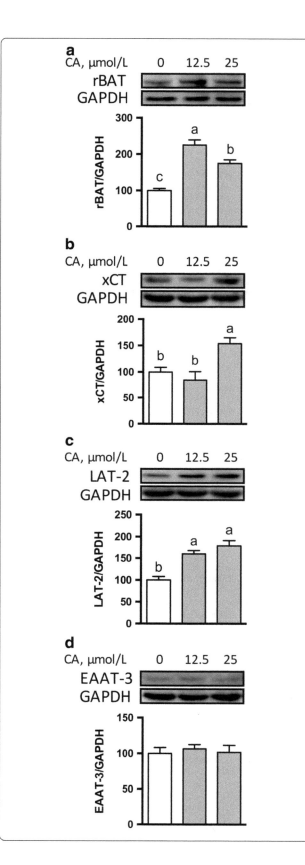

Fig. 5 Protein abundances for rBAT (**a**), xCT (**b**), LAT2 (**c**), and EAAT3 (**d**) in IPEC-1 cells. Cells were cultured in the absence or presence of 12.5 or 25 μmol/L CA for 24 h. Cells were collected and protein abundances for amino acid transporters were analyzed. Values are expressed as means ± SEM, $n = 3$. Means without a common letter differ, $P < 0.05$. CA, Cinnamicaldehyde; IPEC-1, intestinal porcine epithelial cell line 1; ZO, zonula occludens

found that CA increased protein abundances for amino acid transporters, including LAT2, rBAT, and xCT in porcine enterocytes. This is the first study showing that this flavor compound has the ability to up-regulate the expression of amino acid transporters in enterocytes. The enhanced protein abundance for amino acids transporters might promote the transport and absorption of amino acids, which, in turn, stimulates protein synthesis and contributes to the growth performance of pigs observed in previous studies [43, 44].

Conclusions

In summary, studies with porcine enterocytes have revealed that CA improved the intestinal epithelial barrier integrity, as indicated by increased TEER and decreased paracellular permeability. This beneficial effect of CA is accompanied by enhanced distribution of specific TJ proteins in intestinal epithelial cells. Importantly, the protein abundance for amino acid transporters was enhanced by CA. Further studies with animal model are needed to validate this beneficial effect of CA on intestinal barrier function observed in the present study. Supplementation with CA might be a potential nutritional strategy to improve the intestinal mucosal barrier function and nutrient absorption in neonatal piglets.

Abbreviations
CA: Cinnamic aldehyde; DMEM-F12: Dulbecco's modified Eagle's F12 Ham medium; EGF: Epidermal growth factor; FBS: Fetal bovine serum; HRP: Horseradish peroxidase; ITS: Insulin-transferrin-selenium; IPEC-1: Intestinal porcine epithelial cells-1; TEER: Trans-epithelial electrical resistance; TJ: Tight junction; ZO: Zonula occludens

Acknowledgements
Not applicable.

Funding
This work was supported the National Natural Science Foundation of China (31572410, 31572412, 31625025), the 111 Project (B16044), the Program for New Century Excellent Talents in University (NCET-12-0522), the Agriculture and Food Research Initiative Competitive Grant from the USDA National Institute of Food and Agriculture (No. 2014-6701521770), and Texas A&M AgriLife Research (H-8200).

Authors' contributions

ZLW and GW designed research; KJS, YL, and RJW conducted research; ZLW, ZL. Dai, and GW analyzed data; KJS, and ZLW wrote the paper. ZLW and GW had primary responsibility for final content. All authors read and approved the final manuscript.

Competing interests

The authors declare that they have no competing interests.

Author details

[1]State Key Laboratory of Animal Nutrition, College of Animal Science and Technology, China Agricultural University, Beijing 100193, China. [2]DadHank (Chengdu) Biotech Corp, Sichuan, China. [3]Department of Animal Science, Texas A&M University, College Station, TX 77843, USA. [4]Department of Animal Nutrition and Feed Science, China Agricultural University, Beijing 100193, China.

References

1. Bickers D, Calow P, Greim H, Hanifin JM, Rogers AE, Saurat JH, et al. A toxicologic and dermatologic assessment of cinnamyl alcohol, cinnamaldehyde and cinnamic acid when used as fragrance ingredients. Food Chem Toxicol. 2005;43(6):799–836.

2. Anderson RA, Broadhurst CL, Polansky MM, Schmidt WF, Khan A, Flanagan VP, et al. Isolation and characterization of polyphenol type-a polymers from cinnamon with insulin-like biological activity. J Agric Food Chem. 2004;52(1):65–70.

3. Matan N, Rimkeeree H, Mawson AJ, Chompreeda P, Haruthaithanasan V, Parker M. Antimicrobial activity of cinnamon and clove oils under modified atmosphere conditions. Int J Food Microbiol. 2006;107(2):180–5.

4. Singh G, Marimuthu P, de Heluani CS, Catalan CA. Antioxidant and biocidal activities of Carum Nigrum (seed) essential oil, oleoresin, and their selected components. J Agric Food Chem. 2006;54(1):174–81.

5. Dugoua JJ, Seely D, Perri D, Cooley K, Forelli T, Mills E, et al. From type 2 diabetes to antioxidant activity: a systematic review of the safety and efficacy of common and cassia cinnamon bark. Can J Physiol Pharmacol. 2007;85(9):837–47.

6. Wondrak GT, Villeneuve NF, Lamore SD, Bause AS, Jiang T, Zhang DD. The cinnamon-derived dietary factor cinnamic aldehyde activates the Nrf2-dependent antioxidant response in human epithelial colon cells. Molecules. 2010;15(5):3338–55.

7. Cabello CM, Bair WB 3rd, Lamore SD, Ley S, Bause AS, Azimian S, et al. The cinnamon-derived Michael acceptor cinnamic aldehyde impairs melanoma cell proliferation, invasiveness, and tumor growth. Free Radic Biol med. 2009;46(2):220–31.

8. Kim BH, Lee YG, Lee J, Lee JY, Cho JY. Regulatory effect of cinnamaldehyde on monocyte/macrophage-mediated inflammatory responses. Mediat Inflamm. 2010;2010:529359.

9. Bae WY, Choi JS, Kim JE, Jeong JW. Cinnamic aldehyde suppresses hypoxia-induced angiogenesis via inhibition of hypoxia-inducible factor-1alpha expression during tumor progression. Biochem Pharmacol. 2015; 98(1):41–50.

10. Jiang XR, Awati A, Agazzi A, Vitari F, Ferrari A, Bento H, et al. Effects of a blend of essential oils and an enzyme combination on nutrient digestibility, ileum histology and expression of inflammatory mediators in weaned piglets. Animal. 2015;9(3):417–26.

11. Jamroz D, Wiliczkiewicz A, Wertelecki T, Orda J, Skorupinska J. Use of active substances of plant origin in chicken diets based on maize and locally grown cereals. Brit Poultry Sci. 2005;46(4):485–93.

12. Lee KW, Everts H, Kappert HJ, Frehner M, Losa R, Beynen AC. Effects of dietary essential oil components on growth performance, digestive enzymes and lipid metabolism in female broiler chickens. Brit Poultry Sci. 2003;44(3):450–7.

13. Lee J, Finn CE, Wrolstad RE. Comparison of anthocyanin pigment and other phenolic compounds of Vaccinium membranaceum and Vaccinium ovatum native to the pacific northwest of north America.J Agr. Food Chem. 2004;52(23):7039–44.

14. Lee S, Han JM, Kim H, Kim E, Jeong TS, Lee WS, et al. Synthesis of cinnamic acid derivatives and their inhibitory effects on LDL-oxidation, acyl-CoA :

cholesterol acyltransferase-1 and-2 activity, and decrease of HDL-particle size. Bioorg med Chem Lett. 2004;14(18):4677–81.

15. Jacobi SK, Odle J. Nutritional factors influencing intestinal health of the neonate. Adv Nutr. 2012;3(5):687–96.

16. Wang B, Wu G, Zhou Z, Dai Z, Sun Y, Ji Y, et al. Glutamine and intestinal barrier function.Amino Acids. 2015;47(10):2143–54.

17. Camilleri M, Madsen K, Spiller R, Greenwood-Van Meerveld B, Verne GN. Intestinal barrier function in health and gastrointestinal disease. Neurogastroenterol Motil. 2012;24(6):503–12.

18. Ulluwishewa D, Anderson RC, McNabb WC, Moughan PJ, Wells JM, Roy NC. Regulation of tight junction permeability by intestinal bacteria and dietary components. J Nutr. 2011;141(5):769–76.

19. Nusrat A, Turner JR, Madara JL. Molecular physiology and pathophysiology of tight junctions. IV. Regulation of tight junctions by extracellular stimuli: nutrients, cytokines, and immune cells. Am J Physiol Gastrointest Liver Physiol. 2000;279(5):G851–7.

20. Turner JR, Buschmann MM, Romero-Calvo I, Sailer A, Shen L. The role of molecular remodeling in differential regulation of tight junction permeability. Semin Cell dev Biol. 2014;36:204–12.

21. Haynes TE, Li P, Li X, Shimotori K, Sato H, Flynn NE, et al. L-glutamine or L-alanyl-L-glutamine prevents oxidant- or endotoxin-induced death of neonatal enterocytes. Amino Acids. 2009;37(1):131–42.

22. Wang W, Wu Z, Lin G, Hu S, Wang B, Dai Z, et al. Glycine stimulates protein synthesis and inhibits oxidative stress in pig small intestinal epithelial cells. J Nutr. 2014;144(10):1540–8.

23. Li N, DeMarco VG, West CM, Neu J. Glutamine supports recovery from loss of transepithelial resistance and increase of permeability induced by media change in Caco-2 cells. J Nutr Biochem. 2003;14(7):401–8.

24. Jiao N, Wu Z, Ji Y, Wang B, Dai Z, Wu G. L-glutamate enhances barrier and Antioxidative functions in intestinal porcine epithelial cells. J Nutr. 2015; 145(10):2258–64.

25. Wang B, Wu Z, Ji Y, Sun K, Dai Z, Wu G. L-glutamine enhances tight junction integrity by activating CaMK Kinase 2-AMP-activated protein Kinase signaling in intestinal porcine epithelial cells. J Nutr. 2016;146(3):501–8.

26. Broer S. Amino acid transport across mammalian intestinal and renal epithelia. Physiol rev. 2008;88(1):249–86.

27. Buddington RK, Elnif J, Puchal-Gardiner AA, Sangild PT. Intestinal apical amino acid absorption during development of the pig. Am J Physiol Regul Integr Comp Physiol. 2001;280(1):R241–7.

28. Steffansen B, Nielsen CU, Brodin B, Eriksson AH, Andersen R, Frokjaer S. Intestinal solute carriers: an overview of trends and strategies for improving oral drug absorption. Eur J Pharm Sci. 2004;21(1):3–16.

29. Clayburgh DR, Shen L. Turner JR.a porous defense: the leaky epithelial barrier in intestinal disease. Lab Investig. 2004;84(3):282–91.

30. Schneeberger EE, Lynch RD. The tight junction: a multifunctional complex. Am J Physiol Cell Physiol. 2004;286(6):C1213–28.

31. Niessen CM. Tight junctions/adherens junctions: basic structure and function. J Invest Dermatol. 2007;127(11):2525–32.

32. Tsukita S, Furuse M, Itoh M. Multifunctional strands in tight junctions. Nat rev Mol Cell Biol. 2001;2(4):285–93.

33. Marchiando AM, Graham WV, Turner JR. Epithelial barriers in homeostasis and disease. Annu rev Pathol. 2010;5:119–44.

34. Turner JR. Intestinal mucosal barrier function in health and disease. Nat rev Immunol. 2009;9(11):799–809.

35. Wang H, Zhang C, Wu G, Sun Y, Wang B, He B, et al. Glutamine enhances tight junction protein expression and modulates corticotropin-releasing factor signaling in the jejunum of weanling piglets. J Nutr. 2015;145(1):25–31.

36. Markov AG, Veshnyakova A, Fromm M, Amasheh M, Amasheh S. Segmental expression of claudin proteins correlates with tight junction barrier properties in rat intestine. J Comp Physiol B. 2010;180(4):591–8.

37. Rahner C, Mitic LL, Anderson JM. Heterogeneity in expression and subcellular localization of claudins 2, 3, 4, and 5 in the rat liver, pancreas, and gut. Gastroenterology. 2001;120(2):411–22.

38. Lu Z, Ding L, Lu Q, Chen YH. Claudins in intestines: distribution and functional significance in health and diseases. Tissue Barriers. 2013;1(3): e24978.

39. Yuan B, Zhou S, Lu Y, Liu J, Jin X, Wan H, et al. Changes in the expression and distribution of Claudins, increased epithelial apoptosis, and a Mannan-binding Lectin-associated immune response lead to barrier dysfunction in Dextran sodium sulfate-induced rat colitis. Gut and Liver. 2015;9(6):734–40.

40. Diesing AK, Nossol C, Danicke S, Walk N, Post A, Kahlert S, et al. Vulnerability

of polarised intestinal porcine epithelial cells to mycotoxin deoxynivalenol depends on the route of application. PLoS One. 2011;6(2):e17472.

41. Christensen HN. Role of amino acid transport and countertransport in nutrition and metabolism. Physiol rev. 1990;70(1):43–77.

42. Dave MH, Schulz N, Zecevic M, Wagner CA, Verrey F. Expression of heteromeric amino acid transporters along the murine intestine. J Physiol. 2004;558(Pt 2):597–610.

43. Morales A, Buenabad L, Castillo G, Arce N, Araiza BA, Htoo JK, et al. Low-protein amino acid-supplemented diets for growing pigs: effect on expression of amino acid transporters, serum concentration, performance, and carcass composition. J Anim Sci. 2015;93(5):2154–64.

44. Sun YL, Wu ZL, Li W, Zhang C, Sun KJ, Ji Y, et al. Dietary L-leucine supplementation enhances intestinal development in suckling piglets. Amino Acids. 2015;47(8):1517–25.

13

The impact of synbiotic administration through in ovo technology on the microstructure of a broiler chicken small intestine tissue on the 1st and 42nd day of rearing

A. Sobolewska[1*], J. Bogucka[1], A. Dankowiakowska[1], G. Elminowska-Wenda[2], K. Stadnicka[1] and M. Bednarczyk[1]

Abstract

Background: Application the innovative method which is in ovo technology provides a means of modulating the immune system at early embryonic stages. The aim of study was to determine influence of the in ovo stimulation, on d 12 of incubation, with synbiotics (synbiotic 1- *L. salivarius* IBB3154 + Bi[2]tos, Clasado Ltd. and the synbiotic 2 - *L. plantarum* IBB3036 + lupin RFOs) on the microstructure of duodenum, jejunum and ileum in the 1st and 42nd day of rearing.

Results: On the 1st day of chickens life, in the duodenum of both experimental groups (SYN1 and SYN2), a significantly higher and wider intestinal villi as well as a significantly larger absorbent surface of these villi were found in comparison with the Control group ($P \leq 0.01$). On the 42nd day of rearing the beneficial effect of synbiotic 1 was reflected by the numerically higher villi (no statistical differences) with a larger surface ($P \leq 0.01$) in the duodenum in the SYN1 group compare to the Control group. In the jejunum on the 1st day of life, in the SYN1 group, significantly higher villi than in the Control group, with a simultaneous decrease in the depth of crypts ($P \leq 0.01$), and also the largest width of villi and their absorbent area ($P \leq 0.01$) in comparison to the other groups were found. On the 42nd day of life, in the jejunum, an increase in the height of the villi whilst reducing the crypt depth in the SYN2 group was found ($P \leq 0.01$). In turn, in the SYN1 group, there were significantly more neutral goblet cells observed compared with the control group ($P \leq 0.05$). In the ileum of 1-day-old chickens, the widest villi ($P \leq 0.05$) and the deepest crypts ($P \leq 0.01$) were found in the SYN2 group. In the same group, there was also the least amount of neutral goblet cells in comparison to the other groups ($P \leq 0.05$).

Conclusions: We observed that synbiotic 1 and 2 beneficially affected the examined characteristics on the 1st and 42nd day of life. The obtained results allow us to conclude that the use of synbiotics significantly affect gut structure which should contribute to improvement in nutrient absorption by the gut.

Keywords: Broiler chicken, In ovo, Small intestine, Synbiotics

* Correspondence: sobolewska@utp.edu.pl
[1]Department of Animal Biochemistry and Biotechnology, University of Science and Technology in Bydgoszcz, Mazowiecka 28, 85-084 Bydgoszcz, Poland
Full list of author information is available at the end of the article

Background

Synbiotics consist of a combination of synergistically interacting probiotics and prebiotics which can be aimed at improving the resistance and stability of health-promoting organisms in the gut of birds by providing a substrate for fermentation [1, 2]. Thus, synbiotics are factors that modulate the immune system of birds by acting on the bacterial flora of their gastrointestinal tract [3]. To ensure the greatest protection of the immune system for newly hatched chickens, external supplementation with bioactive substances, such as synbiotics, should occur as early as possible. Application of an innovative method, such as in ovo technology, provides a means of modulating the immune system at early embryonic stages. This technology involves the introduction - on the appropriate day of embryonic development of birds - of the particular substance in solution form to the air chamber of eggs or directly into a developing embryo [4]. Of course, the effectiveness of the injection and the level of use of the injected bioactive substance by the avian embryo depend on various factors [5]. These can be the chemical and physical features of the injected substances, its dose and the egg surface where the injection was performed (i.e. the embryo, the amnion, the allantois, the air chamber egg or the yolk sac). A thorough understanding of the various stages of embryonic development in birds allows the optimal time of injection to be defined [6, 7]. According to Villaluenga et al. [8], the optimal time of prebiotic injection is on the 12th day of embryonic development. In comparison with injections on d 1, 8 and 17, a significantly higher number of Bifidobacteria was observed in the colon of two-day-old chickens. A similar result was obtained by Pilarski et al. [9], who studied the effect of alpha-galactosides (RFOs) administered on the 12th day of egg incubation on selected traits of chickens.

This is mainly due to the fact that on this day of bird embryonic development, allantochorion is already fully developed and vascularised. The embryo is surrounded by the amniotic fluid that remains in contact with the embryonic gastrointestinal tract, which permits the transport of substances from the air chamber into the intestine [10]. Thus, a highly vascularised allantochorion enables efficient transport of bioactive substances given by injection in ovo on the 12th day of egg incubation between the air chamber of the eggs and the digestive tract of chickens. Upon hatching, the in ovo modulated profile of the gut microflora has an influence on the good condition of health of a chick, eliminating the need for antibiotics. We presume that this beneficial condition of the GI (gastrointestinal) tract would be reflected in the morphology of the intestines and might be maintained throughout the life of the chicken.

The prebiotics used in this study as components of the synbiotics are: commercially developed non-digestive transgalacto - oligosaccharides (Bi²tos, Clasado Ltd.) and raffinose family oligosaccharides (RFOs), which was obtained from lupin seeds as a white powder.

In previous projects carried out by our team, we evaluated the impact of bioactive substances in the form of pre-, pro- and synbiotics injected on the 12th day of incubation into the air chamber of the egg on production traits and the macro- and microstructure of the small intestine. The best composition of these substances and their optimal dose was selected. On the basis of these studies, we found the most beneficial effect of synbiotics on the above-mentioned parameters [11, 12].

The aim of study was to determine the influence of the in ovo stimulation with synbiotics on the microstructure of the duodenum, jejunum and ileum on the 1st and 42nd day of rearing.

Methods

The experiment was conducted on hatching eggs of the Cobb 500 FF line incubated in commercial hatchery conditions (Drobex - Agro Ltd., Solec Kujawski, Poland) in Petersime incubators and on 1- and 42-day-old broilers. On d 12 of incubation, the eggs were candled, and the infertile ones or those containing dead embryos were discarded. Eggs containing living embryos were randomly divided into 3 groups. Bioactive substances were administered in an amount of 0.2 mL into the air chamber of the egg on the 12th day of embryonic development by the in ovo technique. The hole in the shell of the egg was sealed with the use of a special automatic system [13]. 5000 eggs were injected. The SYN1 group received synbiotic $L.$ $salivarius$ IBB3154 + Bi²tos, Clasado Ltd. (2 mg of Bi²tos prebiotic $+10^5$ bacteria/egg), and the SYN2 group received symbiotic $L.$ $plantarum$ IBB3036 + lupin RFOs (2 mg of RFO prebiotic $+10^5$ bacteria/egg). The Control group was injected with physiological saline 0.9% NaCl.

Animals

The rearing experiment was conducted on the experimental farm of the PIAST company in Olszowa according to the technological recommendations and lasted for 42 d. All groups were fed and watered ad libitum. Commercial diets were used according to the age of the chickens: 1-10 d - starter, 10–20 d - grower, 20-41 d – finisher (Table 1).

Histomorphological examination

The material for the morphological and histological assays (approx. 2 cm) of the small intestine (duodenum, jejunum, ileum) were collected from 1- and 42-day-old chickens. Before slaughter, 680 chickens from each group were weighed, and their mean body weight was calculated. Subsequently, 15 birds per group, with a body weight similar to the mean for the group, were selected. Directly after slaughter, the small intestine was extracted,

Table 1 Composition of premix for starter, grower and finisher diets for chickens

Vitamin/ Element	Declared name	Inclusion rate Additive Code	kg/t feed Units	Starter diet Quantity per tonne of finished feed	Quantity of compound per tonne of finished feed	Grower diet Quantity per tonne of finished feed	Quantity of compound per tonne of finished feed	Finisher diet Quantity per tonne of finished feed	Quantity of compound per tonne of finished feed
Vitamin A	Retinyl acetate	E672	MIU	13.00	-	10.00	-	10.00	-
Vitamin D_3	Cholecalciferol	E671	MIU	5.00	-	5.00	-	5.00	-
Vitamin E	alpha Tocopherol	3a700	g	80.00	-	50.00	-	50.00	-
Vitamin K	Vitamin K	-	g	3.00	-	3.00	-	3.00	-
Vitamin B_1	Vitamin B_1	-	g	3.00	-	2.00	-	2.00	-
Vitamin B_2	Vitamin B_2	-	g	9.00	-	8.00	-	6.00	-
Vitamin B_6	Vitamin B_6	3a831	g	4.00	-	3.00	-	3.00	-
Vitamin B_{12}	Vitamin B_{12}	-	mg	20.00	-	15.00	-	15.00	-
Biotin	Biotin	-	g	0.15	-	0.12	-	0.12	-
Cal-D-Pan	Calcium Pantothenate	-	g	15.00	-	12.00	-	10.00	-
Nicotinic acid	Nicotinic acid	-	g	60.00	-	50.00	-	50.00	-
Folic	Folic acid	-	g	2.00	-	2.00	-	1.50	-
Choline	Choline chloride	-	g	0.50	-	0.40	-	0.35	-
Methio-nine	Methionine	3.1.1	g	3405.00	-	3018.00	-	2514.00	-
Threonine	Threonine	3.3.1	g	745.00	-	726.00	-	361.00	-
Lysine	Lysine	3.2.3	g	2812.00	-	2831.00	-	1779.00	-
Iodine	Calcium iodate	E2	g	1.00	1.59	1.00	1.59	1.00	1.59
Selenium	Sodium selenite	E8	g	0.35	7.78	0.35	7.78	0.35	7.78
Iron	Ferrous sulphate	E1	g	40.00	133.33	40.00	133.33	40.00	133.33
Molybdenum	Sodium molybdate	E7	g	0.50	1.27	0.50	1.27	0.50	1.27
Manganese	Manganous oxide	E5	g	100.00	161.29	100.00	161.29	100.00	161.29
Copper	Cupric sulphate	E4	g	15.00	60.00	15.00	60.00	15.00	60.00
Zinc	Zinc oxide	E6	g	100.00	138.89	100.00	138.89	100.00	138.89

and all sections of the small intestine was excised, measured and weighed. Samples were taken from the midpoint of the duodenum, from the midpoint of the jejunum between the point of entry of the bile duct and Meckel's diverticulum and the midpoint of the ileum between Meckel's diverticulum and the ileocecal junction. A total of 270 samples were collected (3 groups × 3 sections × 15 birds × 2 repetitions). The individual sections of the intestine were flushed with 0.9% saline and then fixed in 4% $CaCO_3$ buffered formalin. The fixed samples were dehydrated, cleared and permeated with paraffin in

a tissue processor (Thermo Shandon, Chadwick Road, Astmoor, Runcorn, Cheshire, United Kingdom), and subsequently embedded in paraffin blocks using an embedding system (Medite, Burgdorf, Germany). Thus, the formed blocks were cut into 10 μm-thick sections using a rotary microtome (Thermo Shandon, Chadwick Road, Astmoor, Runcorn, Cheshire, United Kingdom), which in turn were placed on microscope slides coated with egg protein with the addition of glycerine.

Before staining, the preparations were deparaffinised and hydrated. They were subsequently subjected to PAS

(Periodic acid–Schiff) staining with Schiff's reagent to conduct morphometric analyses of the small intestine and to stain and count neutral goblet cells. Measurements were made using a DELTA EVOLUTION 300 microscope equipped with a digital camera by ToupCam™. Measurements included: height and width of intestinal villi and intestinal crypt depth, and the number of neutral goblet cells in the intestine were calculated using the MultiScan-Base v. 18.03 (Computer Scanning Systems II, Warsaw, Poland). In order to measure the height of the villi, ten villi per bird were randomly selected from a cross section. The length was measured from the tip of the villus to its base at the crypt-villus outlet. The perimeter length of the villus was measured to calculate the surface area using the formula cited by Sakamoto et al. [14]. The depth of intestinal crypts was defined as the invagination depth between neighbouring intestinal villi and was measured between 10 villi [15]. The number of PAS - positive cells was calculated per 1 mm^2 of the intestinal villi surface area.

The data were analysed by means of a one-way analysis of variance using STATISTICA 10.0 PL. The arithmetic mean ($\overline{x}$) and the standard error of the mean (SEM) were calculated. Significant differences between the groups were tested using Duncan's Multiple Range Test.

Results

Synbiotic *L. plantarum* IBB3036 + lupin RFO (SYN2 group) significantly increased the body weight of chickens on the 1st day of life ($P \leq 0.05$) (Table 2), while significantly higher feed intake was found in both injected groups compared to the Control group ($P \leq 0.05$) (Table 3). FCE (Feed Conversion Ratio) did not significantly increase after synbiotics throughout the entire rearing period (Table 3). In all study groups, we also observed a very low mortality rate, which was below 2% [16].

While macroscopically evaluating the small intestines of the chickens, we found that the use of synbiotic 2 reduced the length ($P \leq 0.05$ and $P \leq 0.01$) and weight ($P \leq 0.01$) of the duodenum on the 1st day of life of the chickens in relation to the other groups (Table 4). However, it did not affect the microstructure of the analysed section at this

Table 2 Effect of synbiotics treatment injected in ovo on d 12 of incubation on the body weight of broiler chickens

Item	Day	
	1	42
Body weight, g		
Control	40.7[b]	3127
SYN1	40.3[b]	3146
SYN2	41.6[a]	3111
SEM	0.1803	9.7890

[a-b]Difference ($P \leq 0.05$) between treatments (vertical), Means ± SEM representing 680 birds per group. SYN 1 - *L. salivarius* IBB3154 + Bi^2tos, Clasado Ltd., SYN 2 - *L. plantarum* IBB3036 + lupin RFOs

Table 3 Effect of synbiotics treatment injected in ovo on d 12 of incubation on the feed intake and on the FCE (the efficiency of feed conversion) of broiler chickens

Item	Day	
	1-10	1-41
Feed intake, g		
Control	247[b]	4930
SYN1	254[a]	4940
SYN2	258[a]	4898
SEM	1.3291	24.4097
FCE, g/g		
Control	1.21	1.60
SYN1	1.24	1.59
SYN2	1.29	1.60
SEM	0.0146	0.0067

[a-b]Difference ($P \leq 0.05$) between treatments (vertical), Means ± SEM representing 680 birds per group. SYN 1 - *L. salivarius* IBB3154 + Bi^2tos, Clasado Ltd., SYN 2 - *L. plantarum* IBB3036 + lupin RFOs

age, because in the duodenum, in both experimental groups (SYN1 and SYN2), a significantly higher and wider intestinal villi as well as a significantly larger absorbent surface of these villi were found in comparison with the Control group ($P \leq 0.01$) (Table 5). Simultaneously, in the

Table 4 Effect of synbiotics treatment injected in ovo on d 12 of incubation on the length and on the weight of intestine of broiler chickens

Item			Day	
			1	42
Length of intestine, cm	Duodenum	Control	9.60 ± 0.24[Aa]	28.67 ± 0.73[b]
		SYN 1	9.03 ± 0.24[a]	31.37 ± 0.59[a]
		SYN 2	8.28 ± 0.19[Bb]	30.10 ± 0.96[ab]
	Jejunum	Control	17.20 ± 0.65	64.00 ± 1.93[b]
		SYN 1	16.60 ± 0.65	70.20 ± 2.34[a]
		SYN 2	17.27 ± 0.49	68.50 ± 1.86[ab]
	Ileum	Control	15.60 ± 0.41	60.80 ± 1.73[b]
		SYN 1	14.77 ± 0.55	68.30 ± 2.03[a]
		SYN 2	14.54 ± 0.47	66.47 ± 2.15[a]
Weight of intestine, g	Duodenum	Control	0.44 ± 0.01[a]	17.44 ± 0.36[B]
		SYN 1	0.41 ± 0.01[ab]	19.38 ± 0.39[A]
		SYN 2	0.38 ± 0.02[b]	17.22 ± 0.45[B]
	Jejunum	Control	0.47 ± 0.03	45.31 ± 1.39[AB]
		SYN 1	0.43 ± 0.02	48.12 ± 2.19[A]
		SYN 2	0.44 ± 0.01	40.61 ± 1.72[B]
	Ileum	Control	0.36 ± 0.02	39.25 ± 1.37
		SYN 1	0.32 ± 0.02	41.45 ± 2.65
		SYN 2	0.32 ± 0.01	35.98 ± 1.86

[a-b]Difference ($P \leq 0.05$), [A-B]difference ($P \leq 0.01$) between treatments (vertical), Means ± SEM representing 15 birds. SYN 1 - *L. salivarius* IBB3154 + Bi^2tos, Clasado Ltd., SYN 2 - *L. plantarum* IBB3036 + lupin RFOs

Table 5 Effect of synbiotics injected in ovo on the duodenum morphology of chickens at 1st and 42nd day of age

Item	Day	
	1	42
Villus height, μm		
Control	529.96 ± 19.96[B]	1889.79 ± 39.79
SYN1	603.47 ± 12.39[A]	1940.95 ± 36.10
SYN2	594.37 ± 17.07[A]	1880.52 ± 30.88
Villus width, μm		
Control	62.88 ± 1.62[B]	188.59 ± 5.13[B]
SYN1	75.72 ± 1.23[A]	216.69 ± 8.05[A]
SYN2	73.96 ± 1.71[A]	202.13 ± 5.45[AB]
Villus surface area, μm^2		
Control	106,267 ± 6158.5[B]	1123,441 ± 39,725.1[B]
SYN1	144,041 ± 4238.0[A]	1323,359 ± 58,298.1[A]
SYN2	139,322 ± 5970.8[A]	1187,036 ± 32,976.4[AB]
Crypt depth, μm		
Control	49.95 ± 2.13[AB]	136.82 ± 3.54[B]
SYN1	45.81 ± 1.89[B]	189.31 ± 4.26[A]
SYN2	53.97 ± 1.27[A]	138.47 ± 2.30[B]
No. of neutral goblet cells/1 mm^2		
Control	808 ± 49.08[A]	134 ± 8.78
SYN1	526 ± 21.30[B]	116 ± 4.85
SYN2	456 ± 14.28[C]	133 ± 3.63

[A-C]Difference ($P \leq 0.01$) between treatments (vertical), Means ± SEM representing 15 birds. SYN 1 - L. salivarius IBB3154 + Bi²tos, Clasado Ltd., SYN 2 - L. plantarum IBB3036 + lupin RFOs

Table 6 Effect of synbiotics injected in ovo on the jejunum morphology of chickens at 1st and 42nd d of age

Item	Day	
	1	42
Villus height, μm		
Control	337.88 ± 11.01[B]	1334.68 ± 46.19[AB]
SYN1	383.86 ± 10.01[A]	1190.20 ± 50.03[B]
SYN2	352.47 ± 10.80[AB]	1517.95 ± 54.52[A]
Villus width, μm		
Control	51.64 ± 1.06[b]	170.44 ± 7.44
SYN1	56.42 ± 1.30[a]	178.89 ± 8.10
SYN2	52.55 ± 1.46[b]	172.66 ± 5.66
Villus surface area, μm^2		
Control	55,022.6 ± 2410.6[B]	724,927.0 ± 50,976.2
SYN1	67,746.0 ± 2309.5[A]	688,644.6 ± 51,191.4
SYN2	58,459.9 ± 2489.5[B]	814,432.8 ± 30,911.1
Crypt depth, μm		
Control	46.92 ± 0.85[A]	150.87 ± 7.65[A]
SYN1	37.59 ± 0.93[B]	125.41 ± 2.87[B]
SYN2	48.00 ± 0.97[A]	112.88 ± 3.95[B]
No. of neutral goblet cells/1mm^2		
Control	743 ± 42.21	153 ± 20.30[b]
SYN1	778 ± 32.57	232 ± 26.24[a]
SYN2	681 ± 22.49	185 ± 6.49[ab]

[a-b]Difference ($P \leq 0.05$), [A-B]difference ($P \leq 0.01$) between treatments (vertical), Means ± SEM representing 15 birds. SYN 1 - L. salivarius IBB3154 + Bi²tos, Clasado Ltd., SYN 2 - L. plantarum IBB3036 + lupin RFOs

SYN1 group,a significant decrease in the depth of crypts relative to the SYN2 group was reported. In the same group, a reduction in the number of goblet cells on the surface of 1 mm^2 of the villi was observed, while the villi were significantly larger compared to the Control group ($P \leq 0.01$) (Table 5). In turn, on the 42nd day of rearing after treatment of synbiotic 1 (SYN 1 group),a greater weight of the duodenum ($P \leq 0.01$) than in the other groups and an increased length of this section ($P \leq 0.05$) compare to the Control group were observed (Table 4). This beneficial effect was reflected by the numerically higher villi (no statistical differences) with a larger surface ($P \leq 0.01$) in the duodenum in the SYN1 group compare to the Control group (Table 5).

In the jejunum on the 1st day of life, similar to the duodenum, in the SYN1 group, significantly higher villi than in the Control group, with a simultaneous decrease in the depth of crypts ($P \leq 0.01$), and also the largest width of villi and their absorbent area ($P \leq 0.01$) in comparison to the other groups were found (Table 6). On the 42nd day of life, in the jejunum, an increase of the villi height whilst reducing the crypt depth in the SYN2 group was found ($P \leq 0.01$). In turn, in the SYN1 group, there were

significantly more neutral goblet cells observed compared with the control group ($P \leq 0.05$), (Table 6).

In the ileum of 1-day-old chickens, the widest villi ($P \leq 0.05$) and the deepest crypts ($P \leq 0.01$) were found in the SYN2 group. In the same group, there was also the least amount of neutral goblet cells in comparison to the other groups ($P \leq 0.05$) (Table 7). Despite the lack of a significant effect of the applied synbiotics on the length and weight of the analysed sections of the small intestine on the 1st day of life, on the 42nd day of rearing, the length of the ileum ($P \leq 0.05$) was significantly greater in both treatment groups compared to the Control group (Table 4). In both experimental groups, a deepening crypt depth was observed ($P \leq 0.01$) (Table 7). Other parameters, such as villus height, villus width and villus surface area, were similar in all the groups.

Discussion

Studying the impact of injected synbiotics on the body weight of chickens throughout their rearing, we found a positive effect of synbiotic 2 on this parameter just in 1-day-old chickens. Opposite, in the research conducted by Bogucka et al. [11], no significant effect of any bioactive

Table 7 Effect of synbiotics injected in ovo on the ileum morphology of chickens at 1st and 42nd d of age

Item	Day	
	1	42
Villus height, µm		
Control	295.04 ± 9.16	933.31 ± 43.10
SYN1	298.67 ± 10.86	847.78 ± 50.01
SYN2	289.81 ± 9.09	949.05 ± 38.70
Villus width, µm		
Control	46.42 ± 1.29[b]	162.69 ± 5.58
SYN1	46.62 ± 1.34[b]	166.18 ± 4.45
SYN2	50.85 ± 1.66[a]	161.32 ± 5.30
Villus surface area, µm^2		
Control	43,547.4 ± 2124.7	471,197.5 ± 20,933.6
SYN1	44,138.3 ± 2318.3	445,293.0 ± 27,184.3
SYN2	45,834.9 ± 1735.9	484,522.9 ± 26,619.7
Crypt depth, µm		
Control	46.14 ± 0.77[B]	91.28 ± 2.59[B]
SYN1	38.55 ± 0.85[C]	116.81 ± 3.18[A]
SYN2	50.14 ± 0.98[A]	117.39 ± 5.53[A]
No. of neutral goblet cells/1mm^2		
Control	974 ± 50.59[b]	263 ± 26.05
SYN1	1129 ± 61.42[a]	358 ± 40.66
SYN2	678 ± 19.60[c]	285 ± 37.21

[a-c]Difference ($P \leq 0.05$), [A-C]difference ($P \leq 0.01$) between treatments (vertical), Means ± SEM representing 15 birds. SYN 1 - *L. salivarius* IBB3154 + Bi^{2}tos, Clasado Ltd., SYN 2 - *L. plantarum* IBB3036 + lupin RFOs

substance injected in ovo on the 12th day of incubation on the body weight of 1-day-old chickens was shown.

Our study showed a differential effect of the applied synbiotics (synbiotic 1 - *L. salivarius* IBB3154 + galacto-oligosaccharides and synbiotic 2 - *L. plantarum* IBB3036 + Raffinose Family Oligosaccharides) on the microstructure of individual sections of a broiler chicken small intestine. We found a positive effect for both synbiotics given in ovo on the height, width and the absorbent surface of duodenum villi in comparison to the Control group on the 1st day of life of the chickens. In turn, on the 42nd day of the life of the chickens, only synbiotic 1 demonstrated a positive impact in comparison to the Control group on the villi width and villi surface area. A similar effect of synbiotics on the width of the villi was observed by Bogucka et al. [12], however, it did not reflect on the absorbent surface of the intestine. Additionally, in birds from the same group at the end of rearing, the deepest crypts were found. The positive effect of in ovo injection of the synbiotic composed of Bi^{2}tos and *Lactococcus lactis* subsp. *cremoris* IBB SC1 on the height of duodenum villi on the 1st day of life of chickens was also demonstrated in our previous studies [11]. According to Pluske et al. [17],

longer villi and their greater absorbent surface area translate into better utilisation of feed, and thereby improve the health of the birds. Deeper crypts, in turn, indicate rapid tissue regeneration processes to permit the renewal of villi to normal sloughing or inflammation due to the presence of pathogens or their toxins [18]. Awad et al. [19], studying the effect of synbiotic supplementation, which is a combination of *Enterococcus faecium* probiotic and a prebiotic derived from chicory rich in inulin and immuno-modulatory substances derived from sea algae, did not demonstrate significantly higher villi and significantly deeper crypts in the duodenum of 35-day-old broiler chickens. Similar results were obtained by Awad et al. in their further study in 2009 [20].

In the jejunum of 1-day-old chickens, a beneficial impact of synbiotic 1 on the microstructure was demonstrated, but this wasn't maintained for 42 d. Similar results in relation to the heights of villi after in ovo administration of bioactive substances (prebiotic: inulin, synbiotics: inulin + *Lactococcus lactis* spp. *lactis*, Bi^{2}tos + *Lactococcus lactis* spp. *cremoris*) in 1-day-old chickens were obtained by Bogucka et al. [11]. At the end of rearing in the group, in which was applied synbiotic 2, significantly higher intestinal villi were stated in comparing to the SYN1 group and the biggest surface area of these villi (in this case no significant differences) which may have an impact on better absorption of nutrients. In our previous studies [12], there were no significant effects of both prebiotics and synbiotics on the height of intestinal villi in 35-day-old broiler chickens.

Analysing the crypts depth of the jejunum in examined groups of birds their significant shortening was stated in both experimental groups (SYN1 and SYN2). The crypt depth is one of the indicators of the health and functional status of the intestine in chickens, and their size can be a measure of the intensity of intestinal epithelial cell renewal processes [21, 22]. Xu et al. [23] indicate that "The crypt can be regarded as the villus factory, and a large crypt indicates fast tissue turnover and a high demand for new tissue" so in our study the decreased crypt depth may indicates an efficient tissue turnover and good condition of the gut. Different results were obtained after in ovo injection of a synbiotic containing the prebiotics inulin and Bi^{2}tos. Bioactives had a significant impact on the deepening intestinal crypts of 35-day-old broiler chickens [12]. Rehman et al. [18] examining the effect of the bioactive substance - inulin - on the jejunum histomorphology on the 35th day of the rearing of chickens and found significantly longer villi ($P \leq 0.05$) and significantly deeper crypts ($P \leq 0.01$) as a result of this prebiotic supplementation compared to the Control group.

The ileum is characterised by much shorter villi and less absorbent surface of these structures compared to the duodenum and jejunum, which may explain the reduction of the absorption of nutrients at this stage. Evidence of this was the fact that most of the digested food substances had already been absorbed in the upper parts of the small intestine, i.e. the duodenum and jejunum [24]. Injection of the applied synbiotics slightly affected the microstructure of ileum differently compared to the duodenum and jejunum. In 1-day-old chickens, synbiotic *L. plantarum* IBB3036 + lupin RFOs (SYN2 group) significantly increased villi width and crypt depth relative to the other groups. In the studies of Bogucka et al. [11], a synbiotic containing prebiotic Bi^2tos significantly contributed to the shortening of the villi in 1-day-old chicks, whereas on the 42nd day of rearing, the width of the villi was already significantly lower in both experimental groups than in the Control group. In this section, at the end of rearing, a reduction in the absorbent surface of the villi was found in the SYN1 group in comparison to the Control group while prolonging the crypt depth in both experimental groups. A reduction of the absorbent surface area along with a deepening of the crypts was observed in the group of 35-day-old chickens treated in ovo with a synbiotic containing prebiotic Bi^2tos and probiotic bacteria *Lactococcus lactis* spp. *Cremoris* in the study of Bogucka et al. [12]. Different results were obtained by Awad et al. [19, 20], which showed a significant extension of the villi and decreased crypt depth in the ileum after application of the synbiotic, although the substance was administered as a feed additive.

The mucus layer in the small intestine is secreted by goblet cells, which permits the excretion of gastric contents and form a protective barrier against mechanical and chemical injuries (ingested food, microorganisms, pathogens) [25, 26]. The percentage of goblet cells in the duodenum is the lowest and increases towards the large intestine throughout the rearing period. This is because a greater number of microbial organisms is present in the proximal colon [27]. This was also confirmed by our findings. On the 1st day after hatching - in groups injected with synbiotics – a significantly lower number of neutral goblet cells on the surface of 1 mm^2 of duodenum villi was found. At the end of rearing, in the SYN1 group, there were significantly more neutral goblet cells in the jejunum compared to the other groups. In our previous studies [12], the significant effect of the symbiotic (galactooligosaccharides + *L. lactis* subsp. *cremoris* 477) on an increase in the number of goblet cells in the same segment of the intestine was also shown. However, only a several-fold increase of goblet cells may indicate an infection of the intestinal pathogens [28]. According to Langhout et al. [29] and Sharma et al. [30], the effect for increasing the number of neutral goblet cells both in the jejunum and in the ileum may be due to delayed feed intake by the animals. This results in a reduction in the surface of intestinal absorption and an increase in the number of these cells, which was also confirmed in our results.

The above mentioned research and the results of our study present the different effects of synbiotics, depending on their composition and depending on the anatomical structure of GI tract. These two synbiotics: synbiotic 1 (*L. salivarius* IBB3154 + Bi^2tos, Clasado Ltd.) and synbiotic 2 (*L. plantarum* IBB3036 + lupin RFOs) designed based on our in vitro results [16] presented the two types of synergism, i.e., between the prebiotic and probiotic components (synbiotic 1) and synergism between the two independent bioactive compounds and the host (synbiotic 2). In this last case RFO is less efficiently used by the probiotic bacteria, and it remains available to other indigenous stains of intestinal microbiota. In this situation, the prebiotic has a positive influence on the host organism through improvement of microbial balance in the intestines.

However, the positive effect of injected synbiotics was mainly indicated in the duodenum and jejunum.

Conclusions/implications

Summarising our research focused on the evaluation of the impact of synbiotics given in ovo on the production traits and histomorphology of a broiler chicken small intestine, we observed that synbiotic *L. salivarius* IBB3154 + Bi^2tos, Clasado Ltd. and synbiotic *L. plantarum* IBB3036 + lupin RFOs beneficially affected the examined characteristics on the 1st and 42nd day of life. The obtained results allow us to conclude that the use of synbiotics significantly affect gut structure which should contribute to improvement in nutrient absorption by the gut.

Abbreviations
FCR: Feed conversion ratio; PAS: Periodic acid–Schiff; RFO: Raffinose family of oligosaccharides; SEM: The standard error of the mean

Acknowledgments
The research leading to these results has received funding from the European Union's Seventh Framework Programme for research, technological development and demonstration under grant agreement n°311794. It was co-financed from funds for science of the Polish Ministry of Science and Education allocated to an international project ECO FCE in the years 2013–2017. The analyses in this work was done with equipment granted in the project "Implementation of the second phase of the Regional Center of Innovation" co-financed by the European Regional Development Fund under the Regional Operational Programme of Kujawsko-Pomorskie for the years 2007–2013.

Funding
This project was funded from the European Union's Seventh Framework Programme for research, technological development and demonstration under grant agreement n°311,794. It was co-financed from funds for science of the Polish Ministry of Science and Education allocated to an international project ECO FCE in the years 2013–2017.

Authors' contributions

AS, JB, AD data and results interpretation, performed the statistical analysis, histological analysis, drafted the manuscript. JB histological study coordination, aided in data interpretation. GEW data and results interpretation. KS participated in study design and coordination. MB designed research. All authors read and approved the final manuscript.

Competing interests

The authors declare that they have no competing interests.

Author details

[1]Department of Animal Biochemistry and Biotechnology, University of Science and Technology in Bydgoszcz, Mazowiecka 28, 85-084 Bydgoszcz, Poland. [2]Department of Normal Anatomy, Ludwik Rydygier Collegium Medicum in Bydgoszcz, Nicolaus Copernicus University in Toruń, Karłowicza 24, 85-092 Bydgoszcz, Poland.

References

1. Yang Y, Iji PA, Choct M. Dietary modulation of gut microflora in broiler chickens: a review of the role of six kinds of alternatives to in-feed antibiotics. Worlds Poult Sci J. 2009;65:97–114.
2. Adil S, Magray SN. Impact and manipulation of gut microflora in poultry: a review. J Anim Vet Adv. 2012;11:873–7.
3. Slawinska A, Siwek M, Zylinska J, Bardowski J, Brzezinska J, Gulewicz KA, et al. Influence of synbiotics delivered in ovo on immune organ development and structure. Folia Biol (Krakow). 2014;62:277–85.
4. Ciesiolka D, Gulewicz P, Martinez-Villaluenga C, Pilarski R, Bednarczyk M, Gulewicz K. Products and biopreparations from alkaloid-rich lupin in animal nutrition and ecological agriculture. Folia Biol (Krakow). 2005;53:59–66.
5. Johnston PA, Liu H, O'Connel T, Phelps P, Bland M, Tyczkowski J, et al. Applications in ovo technology. Poult Sci. 1997;76:165–78.
6. Williams C. In ovo vaccination for disease prevention. Int Poult Prod. 2007; 15(8):7–9.
7. Williams CJ. In ovo vaccination and chick quality. Int Hatch Prac. 2011;19(2):7–13.
8. Villaluenga CM, Wardenska M, Pilarski R, Bednarczyk M, Gulewicz K. Utilization of the chicken embryo model for assessment of biological activity of different oligosaccharides. Folia Biol (Krakow). 2004;52:135–42.
9. Pilarski R, Bednarczyk M, Lisowski M, Rutkowski A, Bernacki Z, Wardenska M, et al. Assessment of the effect of α-galactosides injected during embryogenesis on selected chicken traits. Folia Biol (Krakow). 2005;53:13–20.
10. Jankowski J. Hodowla i użytkowanie drobiu. Warszawa: PWRiL; 2012.
11. Bogucka J, Dankowiakowska A, Elminowska - Wenda G, Sobolewska A, Szczerba A, Bednarczyk M. Effect of prebiotics and synbiotics delivered in ovo on the broiler small intestine histomorphology during the first days after hatching. Folia Biol (Krakow). 2016; doi:10.3409/fb64_3.131.
12. Bogucka J, Dankowiakowska A, Elminowska - Wenda G, Sobolewska A, Jankowski J, Szpinda M, et al. Performance and small intestine morphology and ultrastructure of male broilers injected in ovo with bioactive substances. Ann Anim Sci. 2016b;DOI: 10.1515/aoas-2016-0048.
13. Bednarczyk M, Urbanowski M, Gulewicz P, Kasperczyk K, Maiorano G. Field and in vitro study on prebiotic effect of raffinose family oligosaccharides in chickens. Bull Vet Inst Pulawy. 2011;55:465–9.
14. Sakamoto K, Hirose H, Onizuka A, Hayashi M, Futamura N, Kawamura Y, et al. Quantitative study of changes in intestinal morphology and mucus gel on total parenteral nutrition in rats. J Surg Res. 2000;94:99–106.
15. Uni Z, Platin R, Sklan D. Cell proliferation in chicken intestinal epithelium occurs both in the crypt and along the villus. J Comp Physiol B. 1998;168:241–7.
16. Dunislawska A, Slawinska A, Stadnicka K, Bednarczyk M, Gulewicz P, Jozefiak D, et al. Synbiotics for broiler chickens—in vitro design and evaluation of the influence on host and selected microbiota populations following in ovo delivery. PLoS One. 2017;12(1):e0168587. doi:10.1371/journal.pone.0168587.
17. Pluske JR, Thompson MJ, Atwood CS, Bird PH, Williams IH, Hartmann PE. Maintenance of villus height and crypt depth, and enhancement of disaccharide digestion and monosaccharide absorption, in piglets fed on cows' whole milk after weaning. Br J Nutr. 1996;76:409–22.
18. Rehman H, Rosenkranz C, Böhm J, Zentek J. Dietary inulin affects the morphology but not the sodium-dependent glucose and glutamine transport in the jejunum of broilers. Poult Sci. 2007;86:118–22.
19. Awad W, Ghareeb K, Böhm J. Intestinal structure and function of broiler chickens on diets supplemented with a symbiotic containing Enterococccus faecium and oligosaccharides. Int J Mol Sci. 2008;9:2205–16.
20. Awad WA, Ghareeb K, Abdel - Raheem S, Böhm J. Effects of dietary inclusion of probiotic and synbiotic on growth performance, organ weights, and intestinal histomorphology of broiler chickens. Poult Sci. 2009;88:49–56.
21. Fan Y, Croom J, Christensen V, Black B, Bird A, Daniel L, et al. Jejunal glucose uptake and oxygen consumption in turkey poults selected for rapid growth. Poult Sci. 1997;76:1738–45.
22. Samanya M, Yamauchi KE. Histological alterations of intestinal villi in chickens fed dried Bacillus subtilis Var. natto. Comp Biochem Physiol A Mol Integr Physiol. 2002;133:95–104.
23. Xu ZR, Hu CH, Xia MS, Zhan XA, Wang MQ. Effects of dietary fructooligosaccharide on digestive enzyme activities, intestinal microflora and morphology of male broilers. Poult Sci. 2003;82(6):1030–6.
24. Yamauchi K, Incharoen T, Yamauchi K. The relationship between intestinal histology and function as shown by compensatory enlargement of remnant villi after midgut resection in chickens. Anat Rec. 2010;293:2071–79.
25. Kim JJ, Khan WI. Goblet cells and mucins: role in innate defense in enteric infections. Pathogens. 2013;2:55–70.
26. Hollingsworth MA, Swanson BJ. Mucins in cancer: protection and control of the cell surface. Nat Rev Cancer. 2004;4:45–60.
27. Karam SM. Lineage commitment and maturation of epithelial cells in the gut. Front Biosci. 1999;4:D286–98.
28. Kim YS, Ho SB. Intestinal goblet cells and mucins in health and disease: recent insights and progress. Curr Gastroenterol Rep. 2010;12:319–30.
29. Langhout DJ, Schutte JB, Van Leeuwen PV, Wiebenga J, Tamminga S. Effect of dietary high- and low- methylated citrus pectin on the activity of the ileal microflora and morphology of the small intestinal wall of broiler chicks. Br Poult Sci. 1999;40:340–7.
30. Sharma R, Schumacher U. Morphometric analysis of intestinal mucins under different dietary conditions and gut flora in rats. Dig Dis Sci. 1995;40:2532–9.

14

Regulatory elements and transcriptional control of chicken *vasa* homologue (*CVH*) promoter in chicken primordial germ cells

So Dam Jin[1], Bo Ram Lee[1], Young Sun Hwang[1], Hong Jo Lee[1], Jong Seop Rim[1] and Jae Yong Han[1,2*]

Abstract

Background: Primordial germ cells (PGCs), the precursors of functional gametes, have distinct characteristics and exhibit several unique molecular mechanisms to maintain pluripotency and germness in comparison to somatic cells. They express germ cell-specific RNA binding proteins (RBPs) by modulating tissue-specific *cis-* and *trans-*regulatory elements. Studies on gene structures of chicken *vasa* homologue (*CVH*), a chicken RNA binding protein, involved in temporal and spatial regulation are thus important not only for understanding the molecular mechanisms that regulate germ cell fate, but also for practical applications of primordial germ cells. However, very limited studies are available on regulatory elements that control germ cell-specific expression in chicken. Therefore, we investigated the intricate regulatory mechanism(s) that governs transcriptional control of *CVH*.

Results: We constructed green fluorescence protein (GFP) or luciferase reporter vectors containing the various 5′ flanking regions of *CVH* gene. From the 5′ deletion and fragmented assays in chicken PGCs, we have identified a *CVH* promoter that locates at −316 to +275 base pair fragment with the highest luciferase activity. Additionally, we confirmed for the first time that the 5′ untranslated region (UTR) containing intron 1 is required for promoter activity of the *CVH* gene in chicken PGCs. Furthermore, using a transcription factor binding prediction, transcriptome analysis and siRNA-mediated knockdown, we have identified that a set of transcription factors play a role in the PGC-specific *CVH* gene expression.

Conclusions: These results demonstrate that *cis-*elements and transcription factors localizing in the 5′ flanking region including the 5′ UTR and an intron are important for transcriptional regulation of the *CVH* gene in chicken PGCs. Finally, this information will contribute to research studies in areas of reproductive biology, constructing of germ cell-specific synthetic promoter for tracing primordial germ cells as well as understanding the transcriptional regulation for maintaining germness in PGCs.

Keywords: Chicken, Chicken *vasa* homologue, Primordial germ cell, Regulatory element, siRNA-mediated knockdown

Background

Primordial germ cells (PGCs) that emerge during early embryogenesis undergo a series of developmental events, such as specification, migration, and differentiation, to produce a new organism in the next generation [1, 2]. They express RNA binding proteins (RBPs) by modulating tissue-specific *cis-* and *trans-*regulatory elements and have specialized genetic programs distinct from those of other somatic cells for maintaining their unique characteristics [3–5]. Significant efforts have been made to elucidate the detailed molecular mechanisms regulating transcriptional control in germ cells [6–8]. At the transcriptional level, certain genes are effectively silenced, whereas other genes are exclusively expressed to maintain the levels of germline-expressed gene products [9, 10].

In chicken, PGCs separate from the epiblast in the blastoderm at Eyal-Giladi and Kochav stage X, which consist of 40,000 to 60,000 undifferentiated embryonic cells, and translocate into the hypoblast area of the pellucida [11, 12]. During gastrulation, they circulate through the vascular system and finally settle in the gonadal anlagen. After the arrival of PGCs, these cells

* Correspondence: jaehan@snu.ac.kr
[1]Department of Agricultural Biotechnology, Research Institute of Agriculture and Life Sciences, College of Agriculture and Life Sciences, Seoul National University, Seoul 08826, South Korea
[2]Institute for Biomedical Sciences, Shinshu University, Minamiminowa, Nagano 399-4598, Japan

continue to proliferate until they enter meiosis. This development of the PGC lineage is a highly complex process that is controlled by the coordinated action of many key factors, such as the expression and regulation of germline-specific genes [13, 14].

Evolutionarily conserved germ cell-specific *vasa* has been characterized in germ cells in several organisms, including chicken [15–17], zebrafish [18], mouse [19], and human [20]. Several studies have demonstrated that *vasa* plays critical roles in germ cell specification, supporting germ line development, translational control of transcribed genes, and RNA processes involving the biosynthesis of PIWI-interacting RNAs (piRNAs) in germ cells at the post-transcriptional level [21–26]. However, the intricate regulatory mechanism(s) that governs transcriptional control of *vasa* expression during chicken germline development has yet to be investigated in detail.

Understanding the cellular and molecular mechanisms that regulate germ cell-specific gene expression during PGC development is critical for the practical use of genetic modifications and germ-cell biology. In the current study, to characterize the promoter of chicken *vasa* homologue (*CVH*) for inducing germ cell-specific gene expression, we conducted 5′ deletion and fragment assays using both enhanced green fluorescent protein (eGFP) and NanoLuc luciferase expression vector. Furthermore, we investigated the predicted putative binding of transcription factors (TFs) on the promoter for *CVH*. Finally, we demonstrated that the transcriptional control of *CVH* expression through *cis*-elements and TFs is important for germ cell-specific gene expression in chicken PGCs.

Methods

Experimental designs, animals and animal care

This study was designed with the aim of investigating the *cis*- and *trans*-regulatory elements for modulating the transcription of *CVH* gene in chicken PGCs through dual luciferase assay and transcriptome analysis. The care and experimental use of chickens were approved by the Institute of Laboratory Animal Resources, Seoul National University (SNU-150827-1). The chickens were maintained in accordance with a standard management program at the University Animal Farm, Seoul National University, Korea. The procedures for animal management, reproduction, and embryo manipulation adhered to the standard operating protocols of our laboratory.

Construction of eGFP and NanoLuc luciferase expression vectors controlled by CVH promoters of different sizes

For construction of eGFP expression vectors, the 5′ flanking regions of the *CVH* gene (NM_204708.2) were amplified using genomic DNA extracted from adult chicken blood, and subsequently inserted into the pGEM T easy vector (Promega, Madison, WI, USA). Primer sets were used to clone fragments of the *CVH* promoter of different sizes (Table 1). The eGFP coding sequence and polyadenylated (Poly-A) tail were inserted into the clone vectors including *CVH* promoter using the restriction enzymes *SpeI* and *NdeI*. For the construction of NanoLuc luciferase expression vectors, different lengths of the 5′ upstream region of the *CVH* gene were inserted between the *KpnI* and *XhoI* sites of pNL1.2 vectors (Promega).

Culture of chicken PGCs and DF-1

Chicken PGCs were cultured in accordance with our standard procedure [2]. Briefly, PGCs from White Leghorn embryonic gonads at 6 days old (Hamburger-Hamilton stage 28) were maintained in knockout Dulbecco's Modified Eagle's Medium (DMEM) (Gibco, Grand Island, NY, USA) supplemented with 20% fetal bovine serum (FBS) (Hyclone, South Logan, UT, USA), 2% chicken serum (Sigma-Aldrich, St. Louis, MO, USA), 1× nucleosides (Millipore, Billerica, MA, USA), 2 mmol/L L-glutamine (Gibco), 1× nonessential amino acids (Gibco), β-mercaptoethanol (Gibco), 1 mmol/L sodium pyruvate (Gibco), and 1× antibiotic-antimycotic (Gibco). Human basic fibroblast growth factor (bFGF) (Koma Biotech, Seoul, Korea) at 10 ng/mL was used for PGC self-renewal. The cultured PGCs were subcultured onto mitomycin-inactivated mouse embryonic fibroblasts at 5- to 6-day intervals by gentle pipetting without any enzyme treatment. For DF-1, the cells were maintained in DMEM with high glucose (Hyclone), 10% FBS, and 1× antibiotic-antimycotic. Cultured cells were grown at 37 °C in a 5% CO_2 incubator.

Table 1 List of primer sequences used for cloning of the *CVH* promoter

Primer sets	Primer sequence (5′ → 3′)
CVH −1,575 bp_F	GACACAGCTTTCCCACGTGAG
CVH −1,231 bp_F	TGGCCACGTGCTATCATATTAGT
CVH −625 bp_F	CTCTGATCATGCCTGCAGCC
CVH −316 bp_F	CAGGACAGGCCTAGGGACAGA
CVH −227 bp_F	AGCATAAACAGGGAAAGCGC
CVH −135 bp_F	GCGCCACCTTCTCACCCC
CVH +25 bp_F	GCTATTTGGAGCGGAGAGTGAAA
CVH promoter_R	AGCGAATGCCAGCAGCC
CVH −135/+222_R	CGCCCTGACGCCACCAT
CVH −135/+162_R	AGCACGCACTGCCCTTGC
eGFP poly A _F	ACTAGTCCGCGGATGGTGAGCAAG
eGFP poly A_R	CATATGGACGTCTCCCCAGCATGCC

In vitro transfection

In vitro transfection was performed using Lipofectamine 2000 in accordance with the manufacturer's instructions (Invitrogen, Carlsbad, CA, USA). For expression analysis of eGFP, the constructed *CVH* promoter vector (1 μg) and 2 μL of Lipofectamine 2000 were separately diluted with 50 μL of Opti-MEM I reduced serum medium (Invitrogen) and incubated at room temperature for 5 min. Liposome-DNA solutions were then mixed and incubated at room temperature for 20 min to form the lipid-DNA complex. Liposome-DNA complex solution was added to 2.5×10^5 cultured PGCs in 500 μL of PGC culture medium. Transfected cells were incubated for 24 h without feeders. After incubation, cells were analyzed using a fluorescence microscope.

Luciferase reporter assay

Nano-Glo Dual-Luciferase Reporter Assay System (Promega) was used to measure the *CVH* promoter activities. The prepared cells were seeded in a 96-well plate and co-transfected with pGL4.53 firefly luciferase (Fluc) and pNL1.2 (NlucP/CVH RE) NanoLuc luciferase (Nluc) plasmid using Lipofectamine 2000 (Invitrogen). The transfected cells were then lysed with lysis buffer with Fluc substrate and incubated on an orbital shaker for 3 min. Fluc signals were then quenched, followed by reaction with Nluc substrate. The signals in arbitrary unit (AU) from both Nluc and Fluc were measured using a luminometer (Glomax-Multi-Detection System; Promega). The promoter activities were calculated by the ratio of the respective AU values of Nluc/Fluc. pNL1.2, an empty vector, was used as a negative control.

Prediction of putative transcriptional binding elements by in silico sequence analysis

The 591-base pair (bp) fragment (−316/+275) of the *CVH* promoter that had the highest activity was analyzed for TF binding sites. Such sites were predicted by MatInspector, a Genomatix program (http://www.genomatix.de) using TRANSFAC matrices (vertebrate matrix; core similarity 1.0 and matrix similarity 0.8) PROMO, which uses version 8.3 TRANSFAC (http://alggen.lsi.upc.es/cgi-bin/promo_v3/promo/promoinit.cgi?dirDB=TF_8.3) and TFBIND, which uses weight matrix in the database TRANSFAC R.3.4 (http://tfbind.hgc.jp).

Small interfering RNA (siRNA) transfection in chicken PGCs

Chicken PGCs were seeded at a density of 2.5×10^5 per well of a 12-well plate in 1 mL of medium. Then, the cells were transfected with each siRNA (50 pmol/L) using RNAiMAX (Invitrogen). Negative control siRNA with no complementary sequence in the chicken genome was used as a control. The sequence of each siRNA is listed in Table 2. After transfection for 48 h, total RNA

Table 2 List of small interfering RNA sequences used for knockdown analysis

Target genes	siRNA sequence (5′ → 3′)	
	Sense	Antisense
EP300-#1	GAGUUCUCCUCACUACGAA	UUCGUAGUGAGGAGAACUC
EP300-#2	GAUGAAUGCUGGCAUGAAU	AUUCAUGCCAGCAUUCAUC
GABPA-#1	GAGCAAGGUAUGUGUCUGU	ACAGACACAUACCUUGCUC
GABPA-#2	CACAAGAAGUCAACCAUCA	UGAUGGUUGACUUCUUGUG
HSF2-#1	GUGUUGGAUGAACAGAGAU	AUCUCUGUUCAUCCAACAC
HSF2-#2	CAGAACUGAGAGCAAAACA	UGUUUUGCUCUCAGUUCUG
NFYA-#1	UCAGACAGCUUACAGACUA	UAGUCUGUAAGCUGUCUGA
NFYA-#2	CAGUACACAGCCAACAGUA	UACUGUUGGCUGUGUACUG
SP3-#1	CAGUACAGUGCUGGCAUCA	UGAUGCCAGCACUGUACUG
SP3-#2	CUGGUAAUAUAGUACAGA	AUCUGUACUAUAUUACCAG
ZNF143-#1	GCAGCGUUUCAUAGCACCU	AGGUGCUAUGAAACGCUGC
ZNF143-#2	GAGCUUGAAAACUCCUAGU	ACUAGGAGUUUUCAAGCUC

was extracted using TRIzol reagent (Invitrogen). The knockdown efficiency of predicted TFs and their effects on the expression of germ cell-related genes including *CVH*, *cDAZL*, *CIWI*, and *CDH* were measured using quantitative reverse transcription-polymerase chain reaction (RT-PCR).

RNA isolation and quantitative RT-PCR

Total RNA of siRNA-treated PGCs was extracted using TRIzol reagent (Life Technologies, Carlsbad, CA, USA), in accordance with the manufacturer's protocol. One microgram of each RNA was reverse-transcribed with the Superscript III First-strand Synthesis System (Invitrogen). The cDNA was diluted fourfold and used as a template for quantitative real-time PCR, which was performed using the Step One Plus real-time PCR system (Applied Biosystems) with EvaGreen (Biotium, Hayward, CA, USA). Each test sample was investigated in triplicate. Then, the relative gene expression of individual samples was calculated after normalization with glyceraldehyde 3-phosphate dehydrogenase (*GAPDH*) expression as an endogenous control [14, 27]. The primer pairs, which were designed using NCBI Primer-BLAST (https://www.ncbi.nlm.nih.gov/tools/primer-blast/index.cgi?LINK_LO C = BlastHome), used for the detection of cDNAs are listed in Table 3.

Statistical analysis

All data are expressed as means ± standard deviation from three independent experiments. One-way analysis of variance with Bonferroni correction was used to calculate the significance of differences between experimental groups. GraphPad Prism software (ver. 5.0; GraphPad Software, La Jolla, CA, USA) was used to evaluate the data. $P < 0.05$; was considered statistically significant.

Table 3 List of primer sequences used for quantitative reverse transcription polymerase chain reaction

				Primer sequence (5′ → 3′)		
No.	Gene Symbol	Description	Accession No.	Forward	Reverse	Product Size, bp
1	EP300	E1A binding protein p300	XM_004937710.1	AGCTGCAGATGGAGGAGAAATC	ACGGTAAAGTGCCTCCAGTG	242
2	GABPA	GA binding protein transcription factor, alpha subunit 60 kDa	NM_001007858.1	TGAACAGGTGACACGATGGG	GGGACTCGCTGGAAGAAGTC	225
3	HSF2	heat shock transcription factor 2	NM_001167764.1	CCAGTTATCACCTGGAGCC	CCAACAAGTCCTCTCGACCC	242
4	NFYA	nuclear transcription factor Y, alpha	NM_001006325.1	TCAGCCACCTGTGAAGACAC	TCGAACTGGGCTTTCACCTC	231
5	SP3	Sp3 transcription factor	NM_204603.1	GGCAAAAGGTTCACTCGCAG	GTGTTGTTCCTCCCGCAGTA	229
6	ZNF143	zinc finger protein 143	XM_004941377.1	GAAGCGGCACATCCTTACCT	CCCTGACTTCCACAGCGATT	216

Results

Identification of the minimal promoter region for transcription of the CVH gene

To investigate the gene promoter for the expression of CVH mRNA, we constructed eGFP expression vectors containing different sizes of the CVH promoter by 5′ deletion, spanning a 1,850-bp region from the 5′ flanking region to the 5′ untranslated region (UTR) (Fig. 1a). Subsequently, we tested whether the differently sized CVH promoters can induce the expression of eGFP in cultured chicken PGCs. As shown in Fig. 1b, the following eGFP reporters were associated with the expression of green fluorescence in chicken PGCs: 1,850-bp fragment (−1,575/+275), 1,506-bp fragment (−1,231/+275), 900-bp

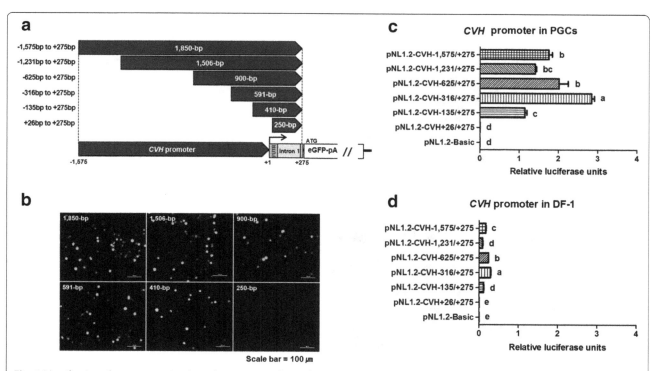

Fig. 1 Identification of promoter region for inducing germ cell-specific gene expression in the chicken *vasa* homologue (*CVH*) promoter through 5′ deletion assays. **a** Schematic diagram of the constructed enhanced green fluorescent protein (eGFP) expression vectors with CVH promoters of different sizes. By 5′ deletion assays, six constructs including differently sized 5′ flanking sequences containing the 5′ untranslated region (UTR) were randomly designed. eGFP expression vector of a 250-bp fragment of the CVH promoter containing only the 5′ UTR. **b** Twenty-four hours after transfection, the expression of eGFP under the control of the differently sized promoters in cultured chicken primordial germ cells (PGCs) was monitored by microscopy. Each fragment used for driving eGFP expression was ligated into the NanoLuc luciferase expression vector (pNL1.2-Basic) to measure promoter activity. Dual luciferase assay of CVH promoter activity in PGCs (**c**) and DF-1 (**d**). NanoLuc luciferase expression levels were normalized to the luciferase activity of internal firefly control and are expressed as relative luciferase units. Scale bar = 100 μm. Different letters (a–e) indicate significant differences ($P < 0.05$)

fragment (−625/+275), 591-bp fragment (−316/+275), and 410-bp fragment (−135/+275); however, the smallest fragment (+26/+275) did not induce this expression. To evaluate the promoter activity further, we performed the dual luciferase reporter assay using the same fragments of the *CVH* promoter in chicken PGCs and DF-1. Consistent with the findings for eGFP expression, the luciferase reporters containing the promoter region of *CVH* presented strong enzyme activity, but we could not detect enzyme activity from the smallest 250-bp fragment and pNL1.2-basic, an empty vector (Fig. 1c). Notably, compared with DF-1 fibroblast cells, chicken PGCs generally presented at least 10 times higher luciferase activities (Fig. 1c and d). Collectively, these results suggest that the minimal promoter region of the *CVH* gene is located at −135 to +275 bp, which includes the 5′ UTR, and plays an important role in the transcription of this gene in chicken PGCs.

Investigation of the *cis*-regulatory elements of the *CVH* gene

For further investigation of the potential transcriptional *cis*-elements in the *CVH* promoter, we performed 5′ and 3′ fragmentation assays using the 591-bp fragment

(−316/+275) that presented the highest luciferase reporter activity, as well as a 410-bp fragment (−135/+275) (Fig. 2a). First, we confirmed the eGFP expression with the designed fragments of the *CVH* promoter in chicken PGCs. Among six fragment constructs, the 591-bp fragment (−316 /+275), 502-bp fragment (−227/+275), and 410-bp fragment (−135/+275) were associated with the strong expression of green fluorescence in chicken PGCs compared with the 357-bp fragment (−135/+222) and the 297-bp fragment (−135/+162). These latter two fragments (357-bp and 297-bp fragments) still showed minimal promoter activity, while the 250-bp fragment (+26/+275) showed none (Fig. 2b). We also conducted a dual luciferase reporter assay using NanoLuc luciferase expression vectors to compare the *CVH* promoter activity in chicken PGCs and DF-1. As shown in Fig. 2c, deletion of the 92-bp fragment between −227/+275 bp and −135/+275 bp resulted in a dramatic decrease in luciferase activity. These results suggest that a positive transcriptional *cis*-element is located in this region. Furthermore, partial deletion of the 5′ UTR including intron 1 (−135/+222 bp and −135/+162 bp) also produced a dramatic change in promoter activity (Fig. 2c). Interestingly, all

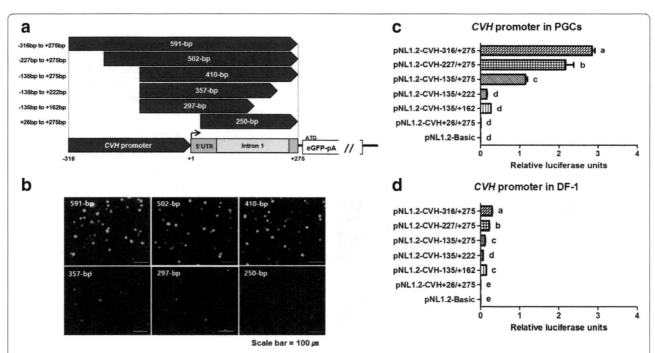

Fig. 2 Identification of *CVH* promoter regions through 5′ and 3′ fragmentation assays. **a** Schematic diagram of six fragmented constructs of the *CVH* promoter. Constructs were designed for expression analysis using an eGFP expression vector. 5′ and 3′ random deletion assays were conducted for a 591-bp fragment (−316/+275) and a 410-bp fragment (−135/+275), respectively. For 3′ deletion for a 410-bp fragment (−135/+275), two constructs for a 357-bp fragment (−135/+222) and a 297-bp fragment (−135/+162) contained part of the intron region. **b** Expression analysis of eGFP under different fragments of the *CVH* promoter in cultured chicken PGCs. Twenty-four hours after transfection, the expression of eGFP was observed by fluorescence microscopy. Each fragment used for driving eGFP expression was ligated into the NanoLuc luciferase expression vector (pNL1.2-Basic) for measuring promoter activity. The expression of NanoLuc luciferase was measured using Nano-Glo-dual luciferase assays in PGCs (**c**) and DF-1 (**d**). Promoter activity was measured as the ratio of NanoLuc luciferase expression levels to internal firefly control and is expressed as relative luciferase units. Scale bar = 100 μm. Different letters (a–e) indicate significant differences (*P* < 0.05)

tested fragments showed higher luciferase activity in PGCs than in DF-1 fibroblasts (Fig. 2d). Collectively, these results indicate that the PGC-specific gene expression requires at least a 410-bp sequence of the 5′ upstream region of the *CVH* gene along with the 5′ UTR including intron 1.

Prediction and selection of the TFs involved in transcriptional control of the *CVH* promoter in chicken PGCs

Based on the findings of the *CVH* promoter activity mentioned above, we predicted TFs that have binding sites in the 591-bp fragment (−316/+275) of the *CVH* promoter using three software programs (PROMO, TFBIND, and MatInspector). Additionally, we attempted to clarify the TFs that were more highly expressed in chicken PGCs than in other cell types, such as Stage X blastodermal cells, gonadal stromal cells (GSCs), and chicken embryonic fibroblasts (CEFs) using previously obtained transcriptome data (Fig. 3a) [3, 4, 13]. From these analyses, we identified six TFs (*EP300*, *GABPA*, *HSF2*, *NFYA*, *SP3*, and *ZNF143*) that were expressed at

significantly higher levels in PGCs and have putative binding sites in the 591-bp fragment (−316/+275) of the *CVH* promoter. To summarize our findings, we marked the consensus sequences and positions of the predicted TFs in sequences of the *CVH* promoter including TATA-box sequence and transcription start codon in Fig. 3b and c.

Predicted TFs affecting the transcriptional activity of germ cell-specific RBPs

To confirm the expression of the selected TFs in chicken PGCs, we conducted quantitative RT-PCR using the RNA samples prepared from various cells/tissues (PGCs, Stage X, CEFs, DF-1, and GSCs). The results showed that the expression of five TFs is highly PGC-specific, with the exception being *GABPA*, which is expressed in both PGCs and Stage X equally (Fig. 4). These results indicate that these TFs may be involved in transcriptional control of the *CVH* promoter by directly interacting with it in chicken PGCs. We further examined whether these TFs affect the transcription of germ cell-specific RBPs

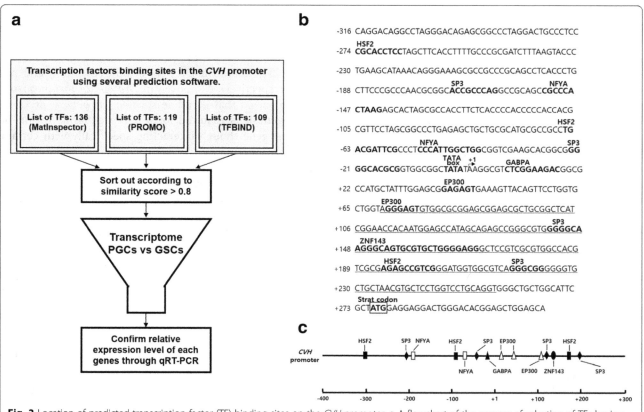

Fig. 3 Location of predicted transcription factor (TF) binding sites on the *CVH* promoter. **a** A flowchart of the process of selection of TFs having putative binding sites in the 591-bp fragment (−316/+275) of the *CVH* promoter. The input is sets of TFs predicted by several prediction software programs (MatInspector, PROMO, and TFBIND). TFs with a similarity score below 0.8 were then removed. The putative TFs were extracted depending on significant expression in chicken PGCs, from the transcriptome data. **b** Nucleotide sequences of the 5′ flanking region of the 591-bp fragment (−316/+275) of the *CVH* promoter. Selected TFs were marked on the sequences of the *CVH* promoter region. **c** Schematic diagrams of the locations of these predicted factors in the promoter region. +1 indicates the transcriptional initiation site and boxed ATG indicates the translational start site. The bold sequences show predicted binding sites of the TFs. Underline indicates the intron region of the *CVH* gene. GSCs, gonadal stromal cells

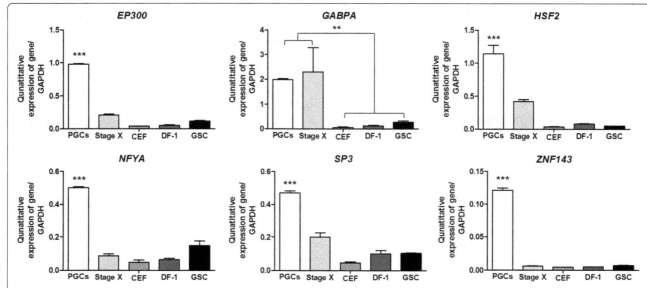

Fig. 4 Quantitative expression analysis of predicted TFs in various cell types. By quantitative reverse transcription-polymerase chain reaction (qRT-PCR) analysis, predicted TFs were analyzed with the prepared PGCs, Stage X blastoderm, chicken embryonic fibroblasts, DF-1, and GSCs. Error bars indicate the standard deviation of triplicate analysis. Significant differences are indicated as ***$P < 0.001$ and **$P < 0.01$

(*CVH*, *cDAZL*, *CIWI*, and *CDH*) in chicken PGCs using a siRNA-mediated knockdown assay. As shown in Fig. 5, in the samples with the highest knockdown efficiency of *HSF2*, *NFYA*, *SP3*, and *ZNF143* mRNA expression in chicken PGCs, the expression of RBP mRNA was significantly reduced, suggesting that these TFs function in regulating transcriptional control of the *CVH* promoter, and other PGC-specific RBPs, such as *cDAZL*, *CIWI*, and *CDH*, while *EP300* and *GABPA* remain unaffected. Taken together, these results suggest that these TFs (*HSF2*, *NFYA*, *SP3*, and *ZNF143*) play a role in the transcription of PGC-specific RBPs through direct binding to 5′ upstream promoter regions.

Discussion

The results of the current study suggest that the promoter region of the *CVH* gene, which extends from −316 to +275 bp and contains the 5′ UTR and intron 1, can control the transcription of the *CVH* gene in chicken PGCs. They also suggest that significantly upregulated TFs such as *HSF2*, *NFYA*, *SP3*, and *ZNF143* in chicken PGCs play a role in expression of the *CVH* gene by directly interacting with putative binding sites of the *CVH* gene promoter.

VASA, an evolutionarily conserved RBP that promotes translational control of germ cell-specific genes, is expressed specifically in germ cells during germline development [28]. Several reports have shown that *vasa* play a critical role in the formation of the germplasm and gametogenesis in invertebrates such as *Caenorhabditis elegans* and *Drosophila melanogaster* [29, 30]. In addition, VASA expression in germ cells is essential for

their survival and proliferation [31, 32]. In transgenic animals, 5.1-kb, 4.7-kb, 2.4-kb, 5.6-kb, 8-kb and 4.3-kb of *vasa* promoter have been used for germ cell specific expression of reporter genes in medaka [33], rainbow trout [34], zebrafish [35], mice [36], cows [37] and pig [38], respectively. Moreover, in *Drosophila melanogaster*, it is reported that germline specific *vas* gene expression in oogenesis is required for a 40-bp genomic region of the *vas* gene though interacting specifically with certain ovarian protein [39]. In addition, in the malaria mosquito, *vasa*-like gene is specifically expressed in both the male and female gonads in adult mosquitoes and is characterized the regulatory regions that are the entire 5′UTR and only 380-bp of upstream sequence for the specific germline expression in the GSCs of both sexes [40]. Although studies on the transcriptional control of *CVH* for temporal and spatial regulation hold great promise for practical applications, regarding using a germ cell-specific promoter for tracing germ cells as well as understanding the molecular network of transcriptional regulation behind their unique characteristics, very limited information is available on the regulatory elements involved in transcriptional control of the *CVH* gene in chicken.

Previous study showed that the *CVH* gene requires a 5′ flanking region of 1,555-bp for higher induction of specific expression in germ cells at the transcriptional level [16]. However, as shown in Fig. 1, we described that the highest promoter activity region of the *CVH* gene, which is a 591-bp fragment (−316/+275) containing the 5′ UTR, is sufficient for the induction of specific expression in chicken PGCs, as determined by a 5′ deletion

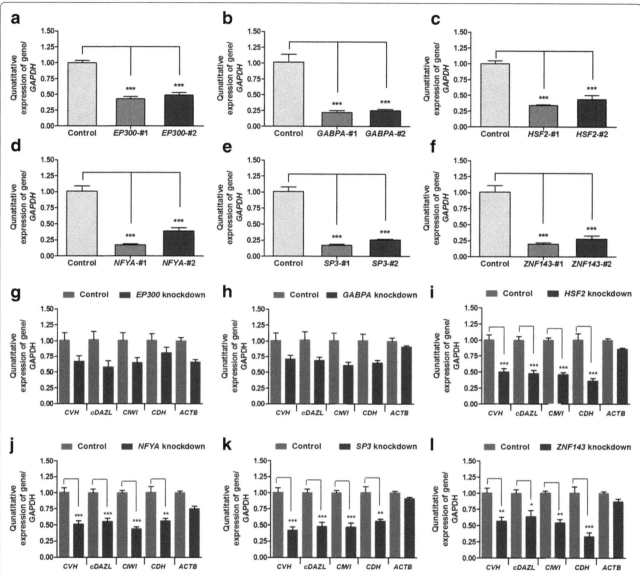

Fig. 5 Relative gene expression analysis after knockdown of predicted TFs in cultured PGCs. **a–f** Confirmation of siRNA knockdown efficiency in vitro in chicken PGCs for the predicted TFs as determined by qRT-PCR. **g–l** Relative expression analysis of germ cell-related RNA binding proteins (*CVH*, *cDAZL*, *CIWI*, and *CDH*) in cultured PGCs after treatment of each siRNA. *ACTB* (*beta-actin*) was used as a control for silencing the specificity of the knockdown probes. qPCR was conducted in triplicate, with data being normalized to the expression of *GAPDH* as a control. Significant differences between control and treatment groups are indicated as ***$P < 0.001$, **$P < 0.01$, and *$P < 0.05$. Error bars indicate the standard error of triplicate analyses

assay. Additionally, our findings demonstrate for the first time that the 5′ UTR containing intron 1 is required for promoter activity of the *CVH* gene in chicken PGCs, as determined through 5′ and 3′ fragmentation assays (Fig. 2). With regard to the roles of introns in transcriptional control in diverse organisms, several reports have shown that introns play a pivotal role in controlling transcription, including that of germline-specific genes, and act as enhancers to control gene expression [41–47]. In search for regulatory elements of *vasa* promoter in medaka, Li et al. demonstrated that the

first intron plays an important role in the VAS activity from total of 11 regions identified within the 5.1-kb *vasa* promoter [47], and subsequently found that the first 35-bp of exon 1 of *vasa* gene is sufficient to increase transcriptional activity as a enhancer [48]. Therefore, it seems likely that the 5′ UTR containing intron 1 of the *CVH* gene would be valuable for constructing a germ cell-specific *CVH* promoter vector for the practical utilization of genetic resources.

Transcriptional control is required for regulatory elements such as specialized promoter sequences and promoter recognition *trans*-acting factors [49]. But, there are few reports on transcriptional regulators of *vasa* gene. A previous study revealed that Mitf acts as transcriptional activator of germ cell-specific genes encoded RNA-binding proteins such as *vasa*, *dazl* and *dnd* in medaka spermatogonial cell line [50]. Therefore, we investigated whether predicted TFs that have putative binding sites in the 591-bp fragment (−316/+275) of the *CVH* promoter can directly regulate the expression of chicken *CVH*. Using our previous transcriptome analysis, we identified TFs that were more highly expressed in chicken PGCs than in other cell types. The integrated approaches used in this study were complementary for finding novel TFs with putative binding sites in the *CVH* promoter. However, we could not find Mitf in the analyzed *CVH* promoter above mentioned. Finally, we selected six TFs (*EP300*, *GABPA*, *HSF2*, *NFYA*, *SP3*, and *ZNF143*) that have putative binding sites in the 591-bp fragment (−316/+275) of the *CVH* promoter through a series of experiments. In Fig. 3, all TFs are marked for consensus sequences and positions in the sequence of the *CVH* promoter, including the TATA-box sequence for transcriptional initiation and the start codon. We also validated their expression levels using quantitative RT-PCR. Based on the results, five TFs (*EP300*, *HSF2*, *NFYA*, *SP3*, and *ZNF143*) were significantly expressed in PGCs compared with their levels in other samples, while *GABPA* were significantly expressed in PGCs compared with CEF, DF-1, and GSC, but showed no significant difference in expression compared with that at Stage X (Fig. 4). We further examined whether these TFs affect the transcription of germ cell-specific RBPs (*CVH*, *cDAZL*, *CIWI*, and *CDH*) in chicken PGCs through siRNA-mediated knockdown.

With regard to significant expression and functions in germ cells, it has been reported that heat shock factor 2 (HSF2) plays a role during embryonic development and under stress conditions, prevents the formation of damaged gametes, and ensures the integrity of the reproductive process [51]. In addition, knockout mouse models have shown that HSF2 is involved in oogenesis and spermatogenesis and strongly expressed in PGCs [52, 53]. As a general transcription activator, NF-Y binds strongly at CCAAT motifs and consists of NF-YA, −YB, and -YC subunits [54]. In *C. elegans*, mutations in *nfya-1* affect the development of germ cells and also reduce the number of sperm [55]. Moreover, NF-Y is greatly affected by the CCAAT motif in terms of its transcriptional activity regarding *Miwi* and *CIWI* gene [56, 57]. Additionally, SP3 is one of the Sp family of TFs, which is characterized by three conserved zinc fingers [58], and positively or negatively controls the transcriptional activity of numerous genes through binding to the GC box in *cis*-regulatory elements [59]. Importantly, it has been

reported that regulation of *Nanog* gene expression is required for Sp1 and Sp3 expression, besides Oct4 and Sox2, in mouse [60]. Zinc finger protein 143 (ZNF143) was first identified in *Xenopus* [61], and most ZNF143 binding sites are disturbed in promoters associated with CpG islands near the transcription start site in the mammalian genome [62]. *Znf143* is particularly expressed in the mouse ICM [63] and its expression has been implicated in the regulation of mammalian embryonic stem cell survival and renewal [64]. It was also proven that ZNF143, interacting with Oct4, governs Nanog expression through direct binding to the *Nanog* proximal promoter [65]. In addition, ZNF143 has recently been identified as a new factor connecting promoters and distal regulatory elements as an insulator function for lineage-specific gene expression [66]. With regard to transcriptional regulator of Mouse Vasa Homologue (MVH), *Znf143* preferred histone H3 lysine 27 acetylation (H3K27ac)-marked regions associated with the early genes including *Kit*, *Prdm1*, and *Sox2* rather than the late genes such as *Mvh*, *Piwil1*, *Piwil2*, *Tdrd7*, and *Tdrd9* in germ cells [67]. However, our results revealed that *HSF2*, *NFYA*, *SP3*, and *ZNF143* would be expected to function as transcriptional regulators in chicken PGCs. Collectively, our results demonstrate for the first time that these TFs are involved in promoter activity of germ cell-specific RBPs in chicken PGCs; however, it remains to be determined whether these TFs directly act on each gene promoter during chicken PGC development.

Conclusion

In conclusion, we have identified the promoter region of the *CVH* gene for PGC-specific gene expression and found TFs such as *HSF2*, *NFYA*, *SP3*, and *ZNF143* associated with transcriptional control of the *CVH* gene in chicken PGCs. This information should aid a wide range of studies in constructing germ cell-specific synthetic promoters for tracing germ cells using transgenesis, as well as our understanding of the transcriptional regulation that maintains germness in PGCs.

Abbreviations

ACTB: Beta-actin; bFGF: Basic fibroblast growth factor; cDAZL: Chicken DAZL; CDH: Chicken dead end homologue; CEF: Chicken embryonic fibroblast; CIWI: Chicken PIWI-like protein 1; CVH: Chicken vasa homolog; DAZL: Deleted in azoospermia-like; DMEM: Dulbecco's modified eagle medium; eGFP: Enhanced green fluorescent protein; EP300: E1A binding protein p300; FBS: Fetal bovine serum; GABPA: GA binding protein transcription factor, alpha; GADPH: Glyceraldehyde 3-phosphate dehydrogenase; GSC: Gonadal stromal cell; H3K27ac: Histone H3 lysine 27 acetylation; HSF2: Heat shock transcription factor 2; ICM: Inner cell mass; NFYA: Nuclear transcription factor Y, alpha; PGC: Primordial germ cell; piRNAs: PIWI-interacting RNAs; PIWI: P-element induced wimpy testis; qRT-PCR: Quantitative reverse transcription-polymerase chain reaction; RBP: RNA binding protein; siRNA: Small interfering RNA; SP3: Stimulating protein 3; TF: Transcription factor; UTR: Untranslated region; ZNF143: Zinc finger protein 143

Acknowledgements

Not applicable.

Funding

This work was supported by a National Research Foundation of Korea (NRF) grant funded by the Korea government (MSIP) (No. 2015R1A3A2033826).

Authors' contributions

HJY participated in study design and coordination. JSD participated in the design of the study, carried out the experiments, statistical analysis and wrote the first draft of the manuscript. LBR participated in oversaw manuscript preparation. JSD, HYS and LHJ were involved in data interpretation. RJS participated in writing the final versions of the manuscript. All authors have read and approved the final manuscript.

Competing interests

The authors declare that they have no competing interests.

References

1. Han JY. Germ cells and transgenesis in chickens. Comp Immunol Microbiol Infect Dis. 2009;32:61–80.
2. Park TS, Han JY. Piggybac transposition into primordial germ cells is an efficient tool for transgenesis in chickens. Proc Natl Acad Sci U S A. 2012;109:9337–41.
3. Kim H, Park TS, Lee WK, Moon S, Kim JN, Shin JH, et al. MPSS profiling of embryonic gonad and primordial germ cells in chicken. Physiol Genomics. 2007;29:253–9.
4. Han JY, Park TS, Kim JN, Kim MA, Lim D, Lim JM, et al. Gene expression profiling of chicken primordial germ cell ESTs. BMC Genomics. 2006;7:220.
5. Donovan PJ. The germ cell–the mother of all stem cells. Int J Dev Biol. 1998;42:1043–50.
6. Zheng YH, Rengaraj D, Choi JW, Park KJ, Lee SI, Han JY. Expression pattern of meiosis associated SYCP family members during germline development in chickens. Reproduction. 2009;138:483–92.
7. Rengaraj D, Zheng YH, Kang KS, Park KJ, Lee BR, Lee SI, et al. Conserved expression pattern of chicken DAZL in primordial germ cells and germ-line cells. Theriogenology. 2010;74:765–76.
8. Kim TH, Yun TW, Rengaraj D, Lee SI, Lim SM, Seo HW, et al. Conserved functional characteristics of the PIWI family members in chicken germ cell lineage. Theriogenology. 2012;78:1948–59.
9. Reinke V. Germline genomics. WormBook. 2006. doi:10.1895/wormbook.1.74.1.
10. Seydoux G, Braun RE. Pathway to totipotency: lessons from germ cells. Cell. 2006;127:891–904.
11. Hamburger V, Hamilton HL. A series of normal stages in the development of the chick embryo. J Morphol. 1951;88:49–92.
12. Ginsburg M, Eyal-Giladi H. Temporal and spatial aspects of the gradual migration of primordial germ cells from the epiblast into the germinal crescent in the avian embryo. Development. 1986;95:53–71.
13. Lee SI, Lee BR, Hwang YS, Lee HC, Rengaraj D, Song G, et al. MicroRNA-mediated posttranscriptional regulation is required for maintaining undifferentiated properties of blastoderm and primordial germ cells in chickens. Proc Natl Acad Sci U S A. 2011;108:10426–31.
14. Lee BR, Kim H, Park TS, Moon S, Cho S, Park T, et al. A set of stage-specific gene transcripts identified in EK stage X and HH stage 3 chick embryos. BMC Dev Biol. 2007;7:1.
15. Tsunekawa N, Naito M, Sakai Y, Nishida T, Noce T. Isolation of chicken vasa homolog gene and tracing the origin of primordial germ cells. Development. 2000;127:2741–50.
16. Minematsu T, Harumi T, Naito M. Germ cell-specific expression of GFP gene induced by chicken vasa homologue (Cvh) promoter in early chicken embryos. Mol Reprod Dev. 2008;75:1515–22.
17. Lavial F, Acloque H, Bachelard E, Nieto MA, Samarut J, Pain B. Ectopic expression of Cvh (Chicken Vasa homologue) mediates the reprogramming of chicken embryonic stem cells to a germ cell fate. Dev Biol. 2009;330:73–82.
18. Yoon C, Kawakami K, Hopkins N. Zebrafish vasa homologue RNA is localized to the cleavage planes of 2-and 4-cell-stage embryos and is expressed in the primordial germ cells. Development. 1997;124:3157–65.
19. Fujiwara Y, Komiya T, Kawabata H, Sato M, Fujimoto H, Furusawa M, et al. Isolation of a DEAD-family protein gene that encodes a murine homolog of Drosophila vasa and its specific expression in germ cell lineage. Proc Natl Acad Sci U S A. 1994;91:12258–62.
20. Castrillon DH, Quade BJ, Wang T, Quigley C, Crum CP. The human VASA gene is specifically expressed in the germ cell lineage. Proc Natl Acad Sci U S A. 2000;97:9585–90.
21. Styhler S, Nakamura A, Swan A, Suter B, Lasko P. Vasa is required for GURKEN accumulation in the oocyte, and is involved in oocyte differentiation and germline cyst development. Development. 1998;125: 1569–78.
22. Carrera P, Johnstone O, Nakamura A, Casanova J, Jäckle H, Lasko P. VASA mediates translation through interaction with a Drosophila yIF2 homolog. Mol Cell. 2000;5:181–7.
23. Raz E. The function and regulation of vasa-like genes in germ-cell development. Genome Biol. 2000;1:1017.
24. Noce T, Okamoto-Ito S, Tsunekawa N. Vasa homolog genes in mammalian germ cell development. Cell Struct Funct. 2001;26:131–6.
25. Liu N, Han H, Lasko P. Vasa promotes Drosophila germline stem cell differentiation by activating mei-P26 translation by directly interacting with a (U)-rich motif in its 3′ UTR. Genes Dev. 2009;23:2742–52.
26. Xiol J, Spinelli P, Laussmann MA, Homolka D, Yang Z, Cora E, et al. RNA clamping by Vasa assembles a piRNA amplifier complex on transposon transcripts. Cell. 2014;157:1698–711.
27. Livak KJ, Schmittgen TD. Analysis of relative gene expression data using real-time quantitative PCR and the 2− ΔΔCT method. Methods. 2001;25: 402–8.
28. Gustafson EA, Wessel GM. Vasa genes: emerging roles in the germ line and in multipotent cells. Bioessays. 2010;32:626–37.
29. Spike C, Meyer N, Racen E, Orsborn A, Kirchner J, Kuznicki K, et al. Genetic analysis of the Caenorhabditis elegans GLH family of P-granule proteins. Genetics. 2008;178:1973–87.
30. Illmensee K, Mahowald AP. Transplantation of posterior polar plasm in Drosophila. Induction of germ cells at the anterior pole of the egg. Proc Natl Acad Sci U S A. 1974;71:1016–20.
31. Parvinen M. The chromatoid body in spermatogenesis. Int J Androl. 2005;28: 189–201.
32. Medrano JV, Ramathal C, Nguyen HN, Simon C, Reijo Pera RA. Divergent RNA-binding proteins, DAZL and VASA, induce meiotic progression in human germ cells derived in vitro. Stem Cells. 2012;30:441–51.
33. Tanaka M, Kinoshita M, Kobayashi D, Nagahama Y. Establishment of medaka (Oryzias latipes) transgenic lines with the expression of green fluorescent protein fluorescence exclusively in germ cells: a useful model to monitor germ cells in a live vertebrate. Proc Natl Acad Sci U S A. 2001;98:2544–9.
34. Yoshizaki G, Takeuchi Y, Tominaga H, Kobayashi T, Takeuchi T. Visualization of primordial germ cells in transgenic rainbow trout carrying green fluorescent protein gene driven by vasa promoter. Fish Sci. 2002;68:1067–70.
35. Krøvel AV, Olsen LC. Expression of a vas: EGFP transgene in primordial germ cells of the zebrafish. Mech Dev. 2002;116:141–50.
36. Gallardo T, Shirley L, John GB, Castrillon DH. Generation of a germ cell-specific mouse transgenic Cre line. Vasa-Cre Genesis. 2007;45:413–7.
37. Luo H, Zhou Y, Li Y, Li Q. Splice variants and promoter methylation status of the Bovine Vasa Homology (Bvh) gene may be involved in bull spermatogenesis. BMC Genet. 2013;14:1.
38. Song Y, Lai L, Li L, Huang Y, Wang A, Tang X, et al. Germ cell-specific expression of Cre recombinase using the VASA promoter in the pig. FEBS Open Bio. 2016;6:50–5.
39. Sano H, Nakamura A, Kobayashi S. Identification of a transcriptional regulatory region for germline-specific expression of vasa gene in Drosophila melanogaster. Mech Dev. 2002;112:129–39.
40. Papathanos PA, Windbichler N, Menichelli M, Burt A, Crisanti A. The vasa regulatory region mediates germline expression and maternal transmission of proteins in the malaria mosquito Anopheles gambiae: a versatile tool for genetic control strategies. BMC Mol Biol. 2009;10:65.
41. Kawamoto T, Makino K, Niwa H, Sugiyama H, Kimura S, Amemura M, et al. Identification of the human beta-actin enhancer and its binding factor. Mol Cell Biol. 1988;8:267–72.
42. Liu Z, Moav B, Faras A, Guise K, Kapuscinski A, Hackett P. Functional analysis of elements affecting expression of the beta-actin gene of carp. Mol Cell Biol. 1990;10:3432–40.

43. Tomaras GD, Foster DA, Burrer CM, Taffet SM. ETS transcription factors regulate an enhancer activity in the third intron of TNF-alpha. J Leukoc Biol. 1999;66:183–93.

44. Henkel G, Weiss DL, McCoy R, Deloughery T, Tara D, Brown MA. A DNase I-hypersensitive site in the second intron of the murine IL-4 gene defines a mast cell-specific enhancer. J Immunol. 1992;149:3239–46.

45. Wong TT, Tesfamichael A, Collodi P. Identification of promoter elements responsible for gonad-specific expression of zebrafish Deadend and its application to ovarian germ cell derivation. Int J Dev Biol. 2013;57:767–72.

46. Mohapatra C, Barman HK. Identification of promoter within the first intron of Plzf gene expressed in carp spermatogonial stem cells. Mol Biol Rep. 2014;41:6433–40.

47. Li M, Guan G, Hong N, Hong Y. Multiple regulatory regions control the transcription of medaka germ gene vasa. Biochimie. 2013;95:850–7.

48. Li M, Zhao H, Wei J, Zhang J, Hong Y. Medaka vasa gene has an exonic enhancer for germline expression. Gene. 2015;555:403–8.

49. DeJong J. Basic mechanisms for the control of germ cell gene expression. Gene. 2006;366:39–50.

50. Zhao H, Li M, Purwanti YI, Liu R, Chen T, Li Z, et al. Mitf is a transcriptional activator of medaka germ genes in culture. Biochimie. 2012;94:759–67.

51. Abane R, Mezger V. Roles of heat shock factors in gametogenesis and development. FEBS J. 2010;277:4150–72.

52. Sarge KD, Park-Sarge OK, Kirby JD, Mayo KE, Morimoto RI. Expresion of heat shock factor 2 in mouse testis potential role as a regulator of heat-shock protein gene expression during spermatogenesis. Biol Reprod. 1994;50:1334–43.

53. Kallio M, Chang Y, Manuel M, Alastalo TP, Rallu M, Gitton Y, et al. Brain abnormalities, defective meiotic chromosome synapsis and female subfertility in HSF2 null mice. EMBO J. 2002;21:2591–601.

54. Mantovani R. The molecular biology of the CCAAT-binding factor NF-Y. Gene. 1999;1:15–27.

55. Deng H, Sun Y, Zhang Y, Luo X, Hou W, Yan L, et al. Transcription factorNFY globally represses the expression of the C. elegans Hox gene Abdominal-B homolog egl-5. Dev Biol. 2007;308:583–92.

56. Hou Y, Yuan J, Zhou X, Fu X, Cheng H, Zhou R. DNA Demethylation and USF Regulate the Meiosis-Specific Expression of the Mouse Miwi. PLoS Genet. 2012;8:e1002716.

57. Sohn YA, Lee SI, Choi HJ, Kim HJ, Kim KH, Park TS, et al. The CCAAT element in the CIWI promoter regulates transcriptional initiation in chicken primordial germ cells. Mol Reprod Dev. 2014;81:871–82.

58. Philipsen S, Suske G. A tale of three fingers: the family of mammalian Sp/XKLF transcription factors. Nucleic Acids Res. 1999;27:2991–3000.

59. Suske G. The Sp-family of transcription factors. Gene. 1999;238:291–300.

60. Wu DY, Yao Z. Functional analysis of two Sp1/Sp3 binding sites in murine Nanog gene promoter. Cell Res. 2006;16:319–22.

61. Myslinski E, Krol A, Carbon P. ZNF76 and ZNF143 are two human homologsof the transcriptional activator Staf. J Biol Chem. 1998;273:21998–2006.

62. Myslinski E, Gerard MA, Krol A, Carbon P. A genome scale location analysisof human Staf/ZNF143-binding sites suggests a widespread role for human Staf/ZNF143 in mammalian promoters. J Biol Chem. 2006;281:39953–62.

63. Yoshikawa T, Piao Y, Zhong J, Matoba R, Carter MG, Wang Y, et al. High-throughput screen for genes predominantly expressed in the ICM of mouse blastocysts by whole mount in situ hybridization. Gene Expr Patterns. 2006;6:213–24.

64. Chia NY, Chan YS, Feng B, Lu X, Orlov YL, Moreau D, et al. A genome-wide RNAi screen reveals determinants of human embryonic stem cell identity. Nature. 2010;468:316–20.

65. Chen X, Fang F, Liou YC, Ng HH. Zfp143 regulates Nanog through modulation of Oct4 binding. Stem Cells. 2008;26:2759–67.

66. Bailey SD, Zhang X, Desai K, Aid M, Corradin O, Cowper-Sal R, et al. ZNF143 provides sequence specificity to secure chromatin interactions at gene promoters. Nat Commun. 2015;2:6186.

67. Ng J-H, Kumar V, Muratani M, Kraus P, Yeo J-C, Yaw L-P, et al. In vivo epigenomic profiling of germ cells reveals germ cell molecular signatures. Dev Cell. 2013;24:324–33.

15

The regulation of IMF deposition in pectoralis major of fast- and slow- growing chickens at hatching

Lu Liu[1,2†], Huanxian Cui[1,2†], Ruiqi Fu[1,2†], Maiqing Zheng[1,2], Ranran Liu[1,2], Guiping Zhao[1,2] and Jie Wen[1,2*] (iD)

Abstract

Background: The lipid from egg yolk is largely consumed in supplying the energy for embryonic growth until hatching. The remaining lipid in the yolk sac is transported into the hatchling's tissues. The gene expression profiles of fast- and slow-growing chickens, Arbor Acres (AA) and Beijing-You (BJY), were determined to identify global differentially expressed genes and enriched pathways related to lipid metabolism in the pectoralis major at hatching.

Results: Between these two breeds, the absolute and weight-specific amounts of total yolk energy (TYE) and intramuscular fat (IMF) content in pectoralis major of fast-growing chickens were significantly higher ($P < 0.01$, $P < 0.01$, $P < 0.05$, respectively) than those of the slow-growing breed. IMF content and u-TYE were significantly related ($r = 0.9047$, $P < 0.01$). Microarray analysis revealed that gene transcripts related to lipogenesis, including *PPARG, RBP7, LPL, FABP4, THRSP, ACACA, ACSS*1, *DGAT2*, and *GK*, were significantly more abundant in breast muscle of fast-growing chickens than in slow-growing chickens. Conversely, the abundance of transcripts of genes involved in fatty acid degradation and glycometabolism, including *ACAT1, ACOX2, ACOX3, CPT1A, CPT2, DAK, APOO, FUT9, GCNT1,* and *B4GALT3*, was significantly lower in fast-growing chickens. The results further indicated that the PPAR signaling pathway was directly involved in fat deposition in pectoralis major, and other upstream pathways (Hedgehog, TGF-beta, and cytokine–cytokine receptor interaction signaling pathways) play roles in its regulation of the expression of related genes.

Conclusions: Additional energy from the yolk sac is transported and deposited as IMF in the pectoralis major of chickens at hatching. Genes and pathways related to lipid metabolism (such as PPAR, Hedgehog, TGF-beta, and cytokine–cytokine receptor interaction signaling pathways) promote the deposition of IMF in the pectoralis major of fast-growing chickens compared with those that grow more slowly. These findings provide new insights into the molecular mechanisms underlying lipid metabolism and deposition in hatchling chickens.

Keywords: Chicken, Gene expression, Intramuscular fat deposition, Pathway, Pectoralis major, Yolk at birth

Background

Lipid is one of the main nutrients in chicken yolk, and has an important role in fueling the embryonic development of chickens [1, 2]. When the glucose is fully consumed, lipids in the yolk are used predominantly at early embryonic developmental stages [3], and more than 90% of the embryos energy comes from fatty acid oxidation [4]. Differences in the usage of lipids from the yolk sac influence the different growth patterns seen in chicken embryogenesis [5].

Two mechanisms exist for consuming lipids from the yolk sac: absorption through blood circulation or absorption through the small intestine. A total of 80% of the lipids in the yolk sac are consumed by embryo growth by the final week of the prehatching stage [4, 6], and the remainder is transported into the embryonic abdominal cavity and deposited into nearby tissues after hatching [7, 8]. This results in a rapid increase in the

* Correspondence: Jiewen@iascaas.net.cn
†Equal contributors
[1]Institute of Animal Sciences, Chinese Academy of Agricultural Sciences, Beijing 100193, China
[2]State Key Laboratory of Animal Nutrition, Beijing 100193, China

content of intramuscular fat (IMF), which coexists with muscle tissue in breast and thigh.

A few studies on chicken lipid metabolism before and after hatching have been reported [9, 10], but there is still a lack of systematic research on the molecular regulation of IMF deposition from the yolk sac at hatching. In this study, with the aim of identifying global differentially expressed genes (DEGs) and pathways related to lipid metabolism in chicken breast associated with fast- and slow-growing breeds at hatching, a comparative analysis of the gene expression levels between fast- (Arbor Acres, AA; a commercial fast-growing broiler) and slow-growing chickens (Beijing-You, BJY; a slow-growing Chinese breed) was performed using microarray technology.

Methods

Animals and sample collection

Six AA and six BJY chickens (half male and half female) on the day of hatching were used in this study. Individuals within each breed had the same genetic background. Animal experiments were approved by the Science Research Department (in charge of animal welfare issues) at the Institute of Animal Sciences, Chinese Academy of Agricultural Sciences (CAAS), Beijing, China.

After birds had been weighed and their live weight had been recorded, they were sacrificed and the pectoralis major and yolk sac were excised. The pectoralis major samples were stored at –80 °C or –20 °C for RNA isolation and the measurement of IMF content. The yolk sac samples were stored at –20 °C for the measurement of total yolk energy content (TYE).

Measurement of biochemical indexes

TYE was determined with a Parr 1281 bomb calorimeter (Parr Instrument Co., Moline, IL, USA). IMF content in pectoralis major was measured by the Soxhlet method [11], using anhydrous ether as the solvent, and is expressed as a percentage of dry tissue weight.

Total RNA preparation, microarray hybridization, and analysis of DEGs

Total RNA samples from six AA and six BJY chickens were isolated individually and pooled for microarray analysis with equal amounts (1 μg) from every sample. Microarray hybridization was performed by Shanghai Biotechnology Corporation (Shanghai, China) using an Agilent Chicken Gene microarray (ID: 015068). Array scanning and data extraction were accomplished following standard protocols. The normalized fluorescence intensity values from each dye-swapped experiment were averaged separately, after which averaged sample and reference fluorescence values were $\log_2$-transformed for each probe. The expression value of each probe set was normalized and calibrated using the RMA method.

DEGs were screened and genes were considered to be differentially expressed only when the relative abundance fold change between the two breeds exceeded 2.

Quantitative real-time PCR (qPCR)

Individual RNA samples from all chickens were used. All PCR primers were designed at or just outside exon/exon junctions to avoid the amplification of residual genomic DNA, and specificity was determined using BLASTN (Table 1).

qPCR analysis was performed after a reverse transcription reaction, as previously described [12]. cDNA was prepared with 2.0 μg of total RNA of each sample, in accordance with the manufacturer's instructions. For qPCR, each PCR mixture with a volume of 25 μL contained 12.5 μL of 2 × iQ™ SYBR Green Supermix, 0.5 μL (10 mmol/L) of each primer, and 1 μL of cDNA. Mixtures were incubated in an iCycler iQ Real-time Detection system (Bio-Rad, Hercules, CA, USA) programmed to conduct 40 cycles (95 °C for 15 s and 65 °C for 35 s). Quantitation of the transcripts was performed using a standard curve with 10-fold serial dilutions of cDNA. A melting curve was constructed to ensure that only a single PCR product was amplified. Samples were assayed in triplicate with standard deviations of threshold cycle (CT) values not exceeding 0.5, and each experiment was repeated at least twice. Negative (without template) control reactions were performed for each sample.

Gene ontology (GO) enrichment analysis and visualization

GO enrichment analysis was performed to identify the gene function classes and categories corresponding to the DEGs using the GOEAST software toolkit. The significance level for GO term enrichment was set at a false discovery rate (FDR) adjusted to less than 0.5 and a P-value of less than 0.05, by the Yekutieli method.

Kyoto encyclopedia of genes and genomes (KEGG) pathway analysis

KEGG pathway [13, 14] information was also used in the analysis. A ProbeName for each category was first mapped to an NCBI Entrez gene ID according to the Agilent Chicken microarray annotation file, and then each was mapped to an appropriate KEGG gene ID according to the KEGG gene cross-reference file. Pathways that were significantly enriched with DEGs were identified using a hypergeometric test from the R package ($P < 0.1$, FDR-adjusted). Pathways with fewer than three known chicken genes were discarded.

Statistical analyses

The significance of differences between groups was evaluated using Student's t-test. $P < 0.05$ (*) or <0.01 (**) was considered significant. Data are presented as mean ± SEM.

Table 1 The specific primers for qPCR in this study

Gene	Sequence	Product size	Accession NO.
THRSP	F:5'-ATCAAGCCCGTGGTGGAGC-3' R:5'-CTTTGGTGTTTTTGGTGAGGTCG-3'	184 bp	NM_213577
ACACA	F:5'-AACCTGCTAAACCCCTGG-3' R:5'-AGTCCCAAATCCGAAAGG-3'	175 bp	NM_205505
ACSS1	F:5'-TGGGAGATGTTACCACAC-3' R:5'-GCAGAATACACCAAGAGAG-3'	181 bp	XM_415011
PPARG	F:5'-TAAAGTCCTTCCCGCTGACCAAA-3' R:5'-AAATTCTGTAATCTCCTGCACTGCCTC-3'	230 bp	NM_001001460
LPL	F: 5'-AGGAGAAGAGGCAGCAATA-3' R:5'-AAAGCCAGCAGCAGATAAG-3'	222 bp	AB016987
FABP4	F:5'-GGGGTTTGCTACCAGGAAGATG-3' R:5'-CATTCCACCAGCAGGTTCCC-3'	276 bp	NM_204290
RBP7	F: 5'-TTCCATCCATACCACAAGCACA-3' R:5'-AGTGAGTCCAGCCCCTGTTCTT-3'	179 bp	XM_417606
DGAT2	F: 5'-AATGGGTCCTCACGTTCC-3' R:5'-TGGTGGTCAGCAGGTTGT-3'	237 bp	XM_419374
GK	F: 5'-TATGGCTGCTACTTTGTGC-3' R:5'-GTATCCCGCAGTCCTTGT-3'	187 bp	XM_416788
ACAT1	F: 5'-CTCCAGCAAGACAGGCAGT-3' R:5'-CACCAGCAACCATTACATCC-3'	150 bp	XM_417162
ACOX2	F: 5'-TATGTAAGGCGTGGGTCA-3' R:5'-TATGTAAGGCGTGGGTCA-3'	198 bp	XM_414406
ACOX3	F: 5'-ACATCTGGCTGTGCTCTATC-3' R:5'-ACTCCCCGCTAGCTTTAC-3'	179 bp	XM_420814
CPT1A	F: 5'-AGACGGACACTGCAAAGGAG-3' R:5'-AGCCCCTTCCCAAAAACA-3'	174 bp	NM_001012898
CPT2	F: 5'-GGGTCGTGTTGGGCTGTT-3' R:5'-AAAGAGGTTTCTGGGCGTTC-3'	168 bp	XM_001234342
DAK	F:5'-AGAGGAGGAAGGAATTGACCTC-3' R:5'-GTCGAAGACCACATGGCTGT-3'	272 bp	NM_001079500
APOO	F: 5'-CTGCCTTCTGCCTCAGGAAA-3' R:5'-CAATGCTGATCCTGCAACGG-3'	162 bp	XM_015272548
FUT9	F: 5'-TGAAATGTGTAGCTGCGTGGA-3' R:5'-AGACGTCTCCGAATTGCTTGT-3'	141 bp	NM_001079502
GCNT1	F: 5'-ACCAAGATACTGGAGGGCGA-3' R:5'-CTCACTGCTGAGAGGTTCCA-3'	174 bp	XM_003643022
B4GALT3	F:5'-TCCTCCTGCACGATGTGAAC-3' R:5'-TCGCCCCAGTATGTGTTTGG-3'	202 bp	XM_416564

Results

Fast-growing chickens had higher levels of TYE and fat deposition at hatching than did slow-growing chickens

Data on live weight (LW), IMF content, absolute TYE amount, and LW-specific TYE amount (u-TYE) in the two breeds are plotted in Fig. 1a–d. The content of IMF in the pectoralis major of AA chickens (2.57%) was significantly higher ($P < 0.05$) than that (2.14%) of BJY chickens. Similarly, the LW, TYE, and LW-specific u-TYE amounts were also significantly higher ($P < 0.01$, $P < 0.01$, $P < 0.05$) in AA chickens (40.46 g, 57.92 kJ, 1.43 kJ/g) than in BJY ones (31.39 g, 33.58 kJ, 1.07 kJ/g).

As shown in Fig. 2a, the correlation between IMF and u-TYE was $r = 0.9047$ ($P < 0.01$). There was more fat

deposition in the pectoralis major of AA chickens than in BJY chickens on the day of hatching, which was suggested to have occurred because more energy had been supplied from the yolk sac in the former group.

Higher expression of genes related to lipid biosynthesis in muscle of fast-growing than slow-growing chickens

Using an Agilent Chicken Gene microarray, a total of 787 known DEGs, 364 upregulated and 423 downregulated ones, were found in the pectoralis major of AA chickens at hatching, compared with their levels in BJY chickens (Additional file 1). Based on these 787 known DEGs, GO analysis was performed and GO terms enriched ($P < 0.05$) for biological processes were

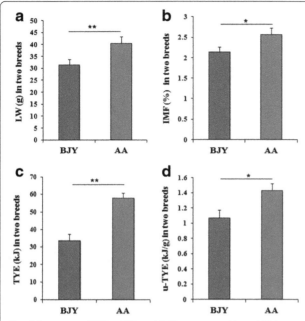

Fig. 1 Summary of TYE, u-TYE, and IMF content in AA and BJY chickens on the day of hatching. Means within the same panel indicate significant differences between the two breeds ($P < 0.01$ or $P < 0.05$). Data are presented as mean ± SEM ($n = 6$)

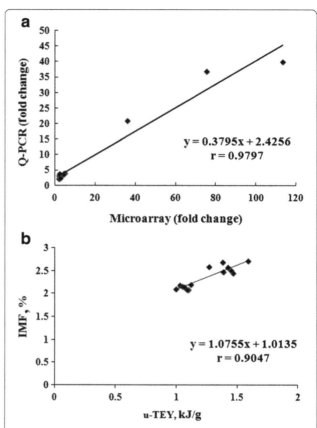

Fig. 2 The correlation analysis by Spearman rank correlation in fast- (AA) and slow-growing (BJY) chickens. The high correlation coefficient ($r = 0.9047$) indicates that the IMF content correlated strongly with u-TYE in the two breeds ($n = 12$). The very high correlation coefficient ($r = 0.9797$) indicates that the qPCR fold changes of the two breeds correlated strongly with the microarray data ($n = 15$)

selected, as presented in Additional file 2. Based on GO-term analysis, 44 known DEGs related to lipid metabolism were screened. Compared with their levels in BJY chicken, 25 upregulated and 19 downregulated DEGs related to lipid metabolism were identified in AA chickens (Additional file 3), which are involved in many biological pathways: fatty acid biosynthesis, preadipocyte differentiation, triglyceride biosynthesis, and fatty acid degradation.

From these 44 DEGs, 15 representative ones were selected to validate the microarray results by qPCR, and the correlation of the fold changes of the two breeds between these two sets of results (Fig. 2b) was $r = 0.9797$ ($P < 0.01$), indicating extremely strong correspondence for all 15 genes. It was also found that the relative expression of 9 of the 15 genes related to fatty acid biosynthesis (*THRSP, ACACA, ACSS1*) (Fig. 3a), preadipocyte differentiation (*PPARG, LPL, FABP4, RBP7*) (Fig. 3b), and triglyceride biosynthesis (*DGAT2, GK*) (Fig. 3c) was significantly upregulated ($P < 0.05$ or $P < 0.01$) in AA compared with the level in BJY chickens, consistent with the differences in lipid deposition. These results suggested that these genes are responsible for the greater IMF deposition in fast-growing chickens than in slow-growing ones.

Lower expression of genes related to fatty acid degradation and glycometabolism in fast-growing than in slow-growing chickens

The expression levels of 10 genes (among 44 genes) related to fatty acid degradation (*ACAT1, CPT1A, CPT2, DAK,*

ACOX2, ACOX3) and glycometabolism (*APOO, FUT9, GCNT1, B4GALT3*) were significantly lower ($P < 0.05$ or $P < 0.01$) in AA than in BJY chickens. Verification of these microarray results was obtained by qPCR, which showed that the expression of these 10 genes, related to fatty acid degradation (*ACAT1, CPT1A, CPT2, DAK, ACOX2, ACOX3*) (Fig. 4a) and glycometabolism (*APOO, FUT9, GCNT1, B4GALT3*) (Fig. 4b), was significantly lower ($P < 0.05$ or $P < 0.01$) in AA chickens than in BJY chickens.

The difference in expression level of these genes between the two breeds was consistent with these genes possibly contributing to greater deposition of IMF in the pectoralis major of AA chickens than that in BJY chickens at hatching.

PPAR and other related signaling pathways regulate differences of expression in related DEGs between the two breeds at birth

A KEGG pathway analysis was performed on the 787 known DEGs to investigate the regulation of lipid metabolism in the pectoralis major at hatching. Seventeen

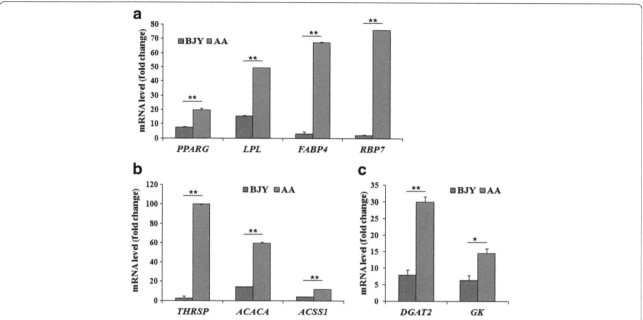

Fig. 3 The expression levels of DEGs related to lipid biosynthesis determined by qPCR in fast- (AA) and slow-growing (BJY) chickens. These genes are all involved in fatty acid biosynthesis, preadipocyte differentiation, or triglyceride biosynthesis. All of these DEGs were significantly ($P < 0.01$ or $P < 0.05$) more highly expressed in AA chickens than in BJY chickens. Data are presented as mean ± SEM ($n = 6$)

KEGG pathways were identified in AA and BJY chickens (Additional file 4). Based on the 44 known DEGs related to lipid metabolism, another pathway analysis was performed and 7 KEGG pathways were identified in the two breeds (Additional file 5); then, five common enriched pathways (PPAR signaling, fatty acid metabolism, Hedgehog signaling, TGF-beta signaling, and cytokine–cytokine receptor interactions) were identified by the two KEGG pathway analysis methods.

Discussion

Lipid is an essential energy source and cell membrane component in animals and, in chickens, is mainly deposited in abdominal, subcutaneous, and muscle adipose

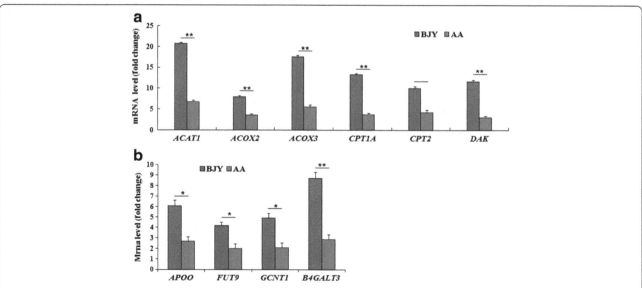

Fig. 4 The expression levels of DEGs related to fatty acid degradation or glycometabolism determined by qPCR in fast- (AA) and slow-growing (BJY) chickens. These genes are all involved in fatty acid degradation or glycometabolism. Each of these DEGs was significantly ($P < 0.01$ or $P < 0.05$) downregulated in AA chickens compared with its level in BJY chickens. Data are presented as mean ± SEM ($n = 6$)

tissues. Although global gene expression research has been performed on abdominal adipose tissues [15–17], systematic research on the molecular regulation of IMF deposition in chicken pectoralis major at hatching has not. In this study, gene expression profiling was used to identify global DEGs and the pathways related to lipid metabolism in the breast muscle of fast- and slow-growing chickens at hatching.

cDNA array analysis

RNA samples were pooled from individuals (*n* = 6) of each breed, as such a pooling strategy can dramatically improve accuracy when only one array is available for each biological condition [18]. Seventeen genes that are well known to be related to lipid metabolism were selected, and nearly 100 qPCR tests were performed to confirm the microarray results. Overall, 2.16% of the total DEGs (38.64% of the DEGs related to lipid metabolism) were verified as being present. As shown in Fig. 2b, the fold changes in gene expression strongly corresponded (*r* = 0.9797, *P* < 0.01) between the qPCR and microarray analyses. Despite the microarray analysis being performed once for each breed, the data exhibited high reliability and persuasive results were ensured because of the RNA pooling strategy and the high degree of verification.

DEGs related to lipid metabolism in chicken pectoralis major at hatching

The yolk is the sole source of energy during the embryonic stages and at hatching. The remaining yolk energy supply upon hatching is absorbed and transported to

tissues for deposition. The TYE and deposition of IMF in the pectoralis major in AA chickens were significantly higher (*P* < 0.01, *P* < 0.05) than those in BJY chickens at hatching, and were correlated in both breeds.

The energy remaining in birds at hatching is recovered from the yolk and stored (deposited) as IMF after transport, uptake, and re-esterification; a series of genes regulate these processes. To reveal the molecular regulation of IMF deposition in chicken pectoralis major at hatching, DEGs related to lipid metabolism were identified in the fast- and slow-growing chicken breeds.

These DEGs include *Spot 14* (encoded by *THRSP*) [19, 20], *ACACA* [21], and *ACSS1* [22, 23] play important roles in lipid metabolism by accelerating fatty acid biosynthesis. *PPARG*, *RBP7*, *LPL*, and *FABP4*, which positively regulate the process of preadipocyte differentiation [24, 25]. Similarly, *DGAT2* and *GK* promote esterification [26, 27]. All of these nine genes had significantly higher expression in the pectoralis major of AA chickens than in BJY chickens at hatching. Conversely, the expression of several genes was significantly lower in AA chickens than in BJY chickens. Among these, *ACAT1*, *ACOX2*, *ACOX3*, *CPT1A*, *CPT2*, and *DAK* positively regulate different steps of fatty acid oxidation [28–31], and *FUT9*, *GCNT1*, *APOO*, and *B4GALT3* all positively regulate energy use in maintaining metabolic balance between carbohydrates and lipids [32–36].

All of these differences in gene expression were either positively or negatively correlated with IMF content. This suggests that these genes play a role in regulating IMF deposition in the pectoralis major of chickens at hatching.

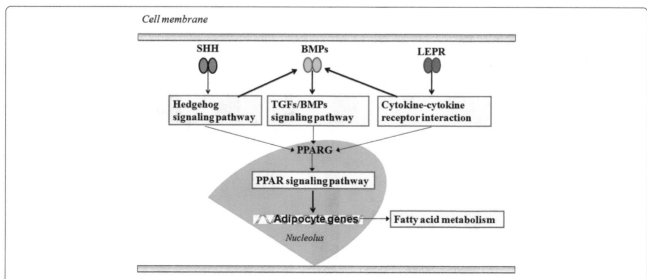

Fig. 5 Lipid metabolism regulatory network proposed for the breast of female chickens at hatching, based on significant DEGs and KEGG pathway analysis. The network involves Hedgehog, TGF-beta, and cytokine–cytokine receptor interaction signaling pathways, through *SHH*, *BMP*, and *LEPR* molecular interactions. These three pathways potentially regulate lipid metabolism via the PPAR signaling pathway

Signaling pathways related to lipid metabolism in breast muscle at hatching

GO-term analysis was used to explore the function of the DEGs, and KEGG pathway analysis was used to explore the regulatory networks underlying IMF content. As expected, several well-known pathways relating to lipid metabolism were identified, including PPAR signaling, fatty acid metabolism, Hedgehog, TGF-beta, and cytokine–cytokine receptor interactions [37–40].

The PPAR signaling pathway is known to play an important role in regulating lipid metabolism [40, 41]. Many DEGs identified here are involved in PPAR signaling pathways, including *ACOX2, ACOX3, CPT1A, CPT2, CYP8B, DBI, FABP4, GK, LPL,* and *PPARG*. Several DEGs, including those of the *BMP* family and receptors (*BMP2, BMP5, BMP7, BMPR1B*), *LEPR*, and *SHH* participate in the Hedgehog, TGF-beta, and cytokine–cytokine receptor interaction signaling pathways. The expression levels of *bone morphogenetic proteins 2* (*BMP2*), *bone morphogenetic proteins 5* (*BMP5*), and *bone morphogenetic proteins 7* (*BMP7*), and *BMP receptor 1B* (*BMPR1B*), *sonic Hedgehog homolog* (*SHH*), and *leptin receptor* (*LEPR*) were all significantly different between the two breeds (Additional file 3). Previous studies have shown that Hedgehog, TGF-beta, and cytokine–cytokine receptor interaction signaling pathways can regulate lipid metabolism [37–39] through the PPAR signaling pathway. This is consistent with the present results from KEGG pathway analysis. The cytokine–cytokine receptor interaction signaling pathway can regulate cell differentiation through the *LEPR* and *TGF* families (*TGF-beta* and *BMP*). Therefore, the cytokine–cytokine receptor interaction signaling pathway may share a similar role to the Hedgehog and TGF-beta signaling pathways as an upstream regulator of the PPAR signaling pathway in lipid metabolism.

Several DEGs, including the *BMP* family and its receptors, participated in more than one of the Hedgehog, TGF-beta, and cytokine–cytokine receptor interaction signaling pathways in the present study, so it is suggested that all of the Hedgehog, TGF-beta, and cytokine–cytokine receptor interaction signaling pathways play roles in the upstream regulation of the PPAR signaling pathway in lipid metabolism. These results suggest that these pathways form a network, along with others related to lipid metabolism, to influence IMF deposition in the chicken pectoralis major at hatching (Fig. 5). The KEGG pathway analysis suggests that lipid metabolism in chicken pectoralis major at hatching is regulated both directly by genes encoding participating enzymes and indirectly via signaling pathways.

Conclusion

In summary, residual sources of energy from the yolk sac are transported to be deposited as IMF in chickens at hatching. Genes and pathways related to lipid metabolism (such as PPAR, Hedgehog, TGF-beta, and cytokine–cytokine receptor interaction signaling pathways) account for greater IMF deposition in the pectoralis major of fast-growing chickens (AA) than that in a slow-growing breed (BJY). These findings provide new insights into the molecular mechanisms underlying lipid metabolism in chickens at hatching.

Additional files

Additional file 1: Annotation and changing of 787 DEGs in pectoralis major of AA and BJY chickens at hatching.

Additional file 2: The enriched GO terms among the 787 DEGs in both AA and BYJ chickens.

Additional file 3: A total of 44 known DEGs related to lipid metabolism in AA and BJY chickens.

Additional file 4: The enriched KEGG pathways based on 787 known DEGs in AA and BJY chickens.

Additional file 5: The enriched KEGG pathways based on 44 known DEGs related to lipid metabolism in AA and BJY chickens.

Abbreviations
AA: Arbor acres chicken; ACACA: Acetyl-coenzyme A carboxylase alpha; ACAT1: Acetyl-coenzyme A acetyltransferase 1; ACOX2: Acyl-coenzyme A oxidase 2; ACOX3: Acyl-coenzyme A oxidase 3; ACSS1: Acyl-CoA synthetase short-chain family member 1; APOO: Apolipoprotein O; B4GALT3: UDP-Gal:betaGlcNAc beta 1,4- galactosyl transferase, polypeptide 3; BJY: Beijing you chicken; CPT1A: Carnitine palmitoyltransferase IA; CPT2: Carnitine palmitoyltransferase II; DAK: Dihydroxyacetone kinase 2 homolog; DEGs: Differentially expressed genes; DGAT2: Diacylglycerol O-acyltransferase homolog 2; FABP4: Fatty acid binding protein 4; FDR: False discovery rate; FUT9: Fucosyltransferase 9; GCNT1: N-glucosaminyl (N-acetyl) transferase 1; GK: Glycerol kinase; GO: Gene ontology; KEGG: Kyoto encyclopedia of genes and genomes; LPL: Lipoprotein lipase; LW: Live weight; PPARG: Peroxisome proliferator-activated receptor gamma; qPCR: Quantitative real time PCR; RBP7: Retinol binding protein 7; THRSP: Thyroid hormone responsive; TYE: Total yolk energy; u-TYE: LW-specific TYE amount

Acknowledgements
The authors would like to thank W. Bruce Currie (Emeritus Professor, Cornell University) for his contributions to the preparation of the manuscript.

Funding
The research was supported by grants from the National Natural Science Foundation of China (31372305), the Agricultural Science and Technology Innovation Program (ASTIP-IAS04), and the Earmarked Fund for Modern Agro-industry Technology Research System (CARS-42).

Authors' contributions
JW designed the study and was in charge of the overall project. LL, HC, and RF contributed to the design and performance of the study, the interpretation of data, and writing of the manuscript. MZ contributed to the design of the study and assisted in animal handling. GZ contributed to the design of the study, interpretation of data, and writing of the manuscript. RL contributed to reviewing the manuscript. All authors submitted comments on drafts, and read and approved the final manuscript.

Competing interests
The authors declare that they have no competing interests.

References

1. Reidy TR, Atkinson JL, Leeson S. Size and components of poult yolk sacs. Poult Sci. 1998;77(5):639–43.
2. Noy Y, Sklan D. Energy utilization in newly hatched chicks. Poult Sci. 1999; 78(12):1750–6.
3. Ding ST, Lilburn MS. Characterization of changes in yolk sac and liver lipids during embryonic and early posthatch development of turkey poults. Poult Sci. 1996;75(4):478–83.
4. Noble RC, Cocchi M. Lipid metabolism and the neonatal chicken. Prog Lipid Res. 1990;29(2):107–40.
5. Sato M, Tachibana T, Furuse M. Heat production and lipid metabolism in broiler and layerchickens during embryonic development. Comp Biochem Physiol A Mol Integr Physiol. 2006;143(3):382–8.
6. Sklan D. Fat and carbohydrate use in posthatch chicks. Poult Sci. 2003; 82(1):117–22.
7. Murakami H, Akiba Y, Horiguchi M. Growth and utilization of nutrients in newly- hatched chick with or without removal of residual yolk. Growth Dev Aging. 1992;56(2):75–84.
8. Yadgary L, Wong EA, Uni Z. Temporal transcriptome analysis of the chicken embryo yolk sac. BMC Genomics. 2014;15:690.
9. Huang JX, Luo XG, Lu L, Liu B. Effects of age and strain on yolk sac utilization and leptin levels in newly hatched broilers. Poult Sci. 2008;87(12): 2647–52.
10. Yadgary L, Cahaner A, Kedar O, Uni Z. Yolk sac nutrient composition and fat uptake in late-term embryos in eggs from young and old broiler breeder hens. Poult Sci. 2010;89(11):2441–52.
11. AOAC. Fat or Ether Extract in Meat. Official Methods of Analysis. Washington: Academic; 1990. p. 931–48.
12. Cui HX, Zhao SM, Cheng ML, Guo L, Ye RQ, Liu WQ, et al. Cloning and expression levels of genes relating to the ovulation rate of the Yunling black goat. Biol Reprod. 2009;80(2):219–26.
13. Kanehisa M, Goto S, Hattori M, Aoki-Kinoshita KF, Itoh M, Kawashima S, et al. From genomics to chemical genomics: new developments in KEGG. Nucleic Acids Res. 2006;34:D354–7.
14. Kanehisa M, Araki M, Goto S, Hattori M, Hirakawa M, Itoh M, et al. KEGG for linking genomes to life and the environment. Nucleic Acids Res. 2008;36:D480–4.
15. Wang HB, Li H, Wang QG, Zhang XY, Wang SZ, Wang YX, et al. Profiling of chicken adipose tissue gene expression by genome array. BMC Genomics. 2007;27(8):193–207.
16. Zhang H, Wang SZ, Wang ZP, Da Y, Wang N, Hu XX, et al. A genome-wide scan of selective sweeps in two broiler chicken lines divergently selected for abdominal fat content. BMC Genomics. 2012;13(1):704.
17. Larkina TA, Sazanova AL, Fomichev KA, Oiu B, Sazanova AA, MaLWski T, et al. Expression profiling of candidate genes for abdominal fat mass in domestic chicken Gallus gallus. Genetika. 2011;47(8):1140–4.
18. Kendziorski C, Irizarry RA, Chen KS, Haag JD, Gould MN. On the utility of pooling biological samples in microarray experiments. Proc Natl Acad Sci U S A. 2005;102(12):4252–7.
19. Chou WY, Cheng YS, Ho CL, Liu ST, Liu PY, Kuo CC, et al. Human spot14 protein interacts physically and functionally with the thyroid receptor. Biochem Biophys Res Commun. 2007;357(1):133–8.
20. Zhan K, Hou ZC, Li HF, Xu GY, Zhao R, Yang N. Molecular cloning and expression of the duplicated thyroid hormone responsive Spot 14 genes in ducks. Poult Sci. 2006;85(10):1746–54.
21. Barber MC, Price NT, Travers MT. Structure and regulation of acetyl-CoA carboxylase genes of metazoa. Biochim Biophys Acta. 2005;1733(1):1–28.
22. Suzuki H, Kawarabayasi Y, Kondo J, Abe T, Nishikawa K, Kimura S, et al. Structure and regulation of rat long-chain acyl-CoA synthetase. J Biol Chem. 1990;265(15):8681–5.
23. Evans RM, Barish GD, Wang YX. PPARs and the complex journey to obesity. Nat Med. 2004;10(4):355–61.
24. Nagy L, Tontonoz P, Alvarez JG, Chen H, Evans RM. Oxidized LDL regulates macrophage gene expression through ligand activation of PPARgamma. Cell. 1998;93(2):229–40.
25. Zizola CF, Schwartz GJ, Vogel S. Cellular retinol-binding protein type III is a PPARγ target gene and plays a role in lipid metabolism. Am J Physiol Endocrinol Metab. 2008;295(6):E1358–68.
26. Lardizabal KD, Mai JT, Wagner NW, Wyrick A, Voelker T, Hawkins DJ. DGAT2 is a new diacylglycerol acyltrans-ferase gene family. Purification, cloning, and expression in insect cells of two polypeptides from Mortierella raman-

niana with diacylglycerol acyltransferase activity. J Biol Chem. 2001;276(42): 38862–9.
27. Feese MD, Faber HR, Bystron CE, Pettigrew DW, Remington SJ. Glycerol kinase from Escherichia coli and an Ala65->Thr mutant: the crystal structures reveal conformational changes with implications for allosteric regulation. Structure. 1998;6(11):1407–18.
28. Yu C, Chen J, Lin S, Liu J, Chang CC, Chang TY. Human acyl-CoA: cholesterol acyltransferase-1 is a homotetrameric enzyme in intact cells and in vitro. J Biol Chem. 1999;274(51):36139–45.
29. Chang TY, Li BL, Chang CC, Urano Y. Acyl-coenzyme A: cholesterol acyltransferases. Am J Physiol Endocrinol Metab. 2009;297(1):E1–9.
30. Longo N. Amat di San Filippo C, Pasquali M. Disorders of carnitine transport and the carnitine cycle. Am J Med Genet C Semin Med Genet. 2006;142C(2): 77–85.
31. Rinaldo P, Matern D, Bennett MJ. Fatty acid oxidation disorders. Annu Rev Physiol. 2002;64:477–502.
32. Kaneko M, Kudo T, Iwasaki H, Ikehara Y, Nishihara S, Nakagawa S, et al. α 1,3-fucosyltransferase IX (Fuc-TIX) is very highly conserved between human and mouse; molecular cloning, characterization and tissue distribution of human Fuc-TIX. FEBS Lett. 1999;452(3):237–42.
33. Varki A, Cummings RD, Esko JD, Freeze HH, Stanley P, Bertozzi C, et al. Essentials of Glycobiology. 2nd ed. NY: Cold Spring Harbor Laboratory Press; 2009.
34. Galvan M, Tsuboi S, Fukuda M, Baum LG. Expression of a specific glycosyltransferase enzyme regulates T cell death mediated by galectin-1. J Biol Chem. 2000;275(22):16730–7.
35. Lamant M, Smih F, Harmancey R, Philip-Couderc P, Pathak A, Roncalli J, et al. ApoO, a novel apolipoprotein, is an original glycoprotein up-regulated by diabetes in human heart. J Biol Chem. 2006;281(47):36289–302.
36. Almeida R, Amado M, David L, Levery SB, Holmes EH, Merkx G, et al. A family of human b4-galactosyl- transferases. Cloning and expression of two novel UDP-galactose: beta-N- acetylglucosamine beta1,4-galactosyltrans-ferases, b4Gal- T2 and beta4Gal-T3. J Biol Chem. 1997;272(51):31979–91.
37. Eaton S. Multiple roles for lipids in the Hedgehog signalling pathway. Nat Rev Mol Cell Biol. 2008;9(6):437–45.
38. Bhatia B, Hsieh H, Kenney AM, Nahlé Z. Mitogenic Sonic hedgehog signaling drives E2F1-dependent lipogenesis in progenitor cells and Medulloblastoma. Oncogene. 2011;30(4):410–22.
39. McNairn AJ, Doucet Y, Demaude J, Brusadelli M, Gordon CB, Uribe-Rivera A, et al. TGFbeta signaling regulates lipogenesis in human sebaceous glands cells. BMC Dermatol. 2013;13(1):2.
40. Szatmari I, Töröcsik D, Agostini M, Nagy T, Gurnell M, Barta E, et al. PPARγ regulates the function of human dendritic cells primarily by altering lipid metabolism. Blood. 2007;110(9):3271–80.
41. Gerhold DL, Liu F, Jiang G, Li Z, Xu J, Lu M, et al. Gene Expression Profile of Adipocyte Differentiation and Its Regulation by Peroxisome Proliferator-Activated Receptor-gamma Agonists. Endocrinology. 2002;143(6):2106–18.

16

Systematic analysis of feeding behaviors and their effects on feed efficiency in Pekin ducks

Feng Zhu[1], Yahui Gao[1], Fangbin Lin[1], Jinping Hao[2], Fangxi Yang[2] and Zhuocheng Hou[1*]

Abstract

Background: Feeding behavior study is important for animal husbandry and production. However, few studies were conducted on the feeding behavior and their relationship with feeding efficiency in Pekin ducks. In order to investigate the feeding behavior and their relationship with feed efficiency and other economic traits in Pekin ducks, we selected 358 male Pekin ducks and recorded feeding information between 3 to 6 wk of age using automatic electronic feeders, and compared the feeding behavior under different residual feed intake (RFI) levels.

Results: We observed that total feed time, daily feed intake and feed intake per meal had strong positive correlations with feed efficiency traits; moreover, strong correlation between feed intake per meal and body weight was found ($R=0.32, 0.36$). Daily feeding rate meal and meal duration had weak correlations with feed efficiency ($R=0.14\sim0.15$). The phenotypic correlation of between-meal pauses, with feed efficiency was not observed. When daily changes were analyzed, high RFI ducks had the highest feed consumption over all times, and obvious differences in daily visits were found among different RFI level animals during the middle period; these differences were magnified with age, but there was no difference in daily meal number. Moreover, our data indicate that high RFI birds mainly take their meals at the edge of the population enclosure, where they are more susceptible to environmental interference.

Conclusions: Overall, this study suggests that the general feeding behaviors can be accurately measured using automatic electronic feeders and certain feeding behaviors in Pekin ducks are associated with improved feed efficiency.

Keywords: Economic traits, Feeding behavior, Feed efficiency, Pekin duck

Background

Feed efficiency is a major economic trait of domestic animals. For most poultry, feed efficiency has been selected in variety of ways, such as feed conversion ratio (FCR) and residual feed intake (RFI) [1, 2]. Improvements in feed efficiency have been achieved by selecting for RFI [3, 4], and some studies have shown that FCR has the genetic potential to improve feed efficiency in Pekin ducks and mule ducks [5, 6]. New developed automatic recording machine helps us to account large number of individuals at the same time in breeding. In practice, we found ducks needs adaptation time at the beginning of using feeding machine. Adaptive learning ability was observed to be different in Pekin ducks

at beginning of using machine. So learning animal feeding behavior would be important for animal husbandry, productions system design and animal welfare [7, 8].

It is widely recognized that behavior is an important aspect of the physiological status of animals [9]. Feeding behavior may reflect animal meal habit, as a potential predictor of feed efficiency [10]. Moreover, feeding behavior can also be used as an indicator for reflecting animal health situation. Early studies that utilized behavioral data depended on human observation or photographic information acquired by camera equipment. With the development of electronic feeders, individual feeding information can be collected automatically and measured accurately [11]. Recent studies observed the meal intervals showed considerable variations in Pekin ducks [12]. In broilers, feeding behaviors were found to be related with feed efficiency in different selected lines [13]. However, the general feeding behavior pattern is not clear and whether these

* Correspondence: zchou@cau.edu.cn
[1]National Engineering Laboratory for Animal Breeding and MOA Key Laboratory of Animal Genetics and Breeding, Department of Animal Genetics and Breeding, China Agricultural University, Beijing 100193, China
Full list of author information is available at the end of the article

feeding behavior affect feed efficiency in Pekin ducks is also unknown.

In order to investigate the relationship between feeding behavior and feed efficiency in Pekin ducks, this study selected 358 male Pekin ducks and recorded feeding information between 3 to 6 wk of age using electronic feeders. Our results provide basic feeding behavior data in Pekin ducks and found feeding behavior are strongly correlated with feed efficiency in Pekin ducks.

Methods
Birds and housing
All ducks were reared in the same house with the same feeding and lighting program. Each house hold 100 feeding machines. Ten machines were grouped together as one pen with barrier for duck management. We randomly selected 3 batches of 3 wk. Male Pekin ducks obtained from the Beijing Golden Star Company (358 birds in total). Each of the 3 batches (batch one: n = 126; batch two: n = 119; batch three: n = 113) was measured from 3 wk of age until 6 wk of age. The intervals between batches was 8 d and measurement was same for each batch. Ducks in each batch were fed ad libitum for 25 d in a single pen. The pen area was 96 m^2 and contained 10 automatic feeders and a water tank. Illumination was 10 lx for 24 h during the entire period. The pen design is illustrated in Additional file 1: Figure S1.

Feeding behavior data acquisition
We designed duck feeding machine which use RFID as sensor to record each duck. The design of duck feeding machine is similar with Bley and Bessei [12]. Feeding behaviors (entering time, exiting time, feed consumption) and live body were recorded for each feeding for each duck from 19 d to 42 d. Only one duck can enter the machine for each feeding event adjusting barrier according to duck body size. The machine has been extensively used in the duck breeding practices [14]. Each feeding machine was spaced at 0.5 m intervals, and the feeder ID was set from 1 to 10 in linear. The first machine was nearest the gate and the last one was against the barrier. Tiny electronic tags were permanently fastened to the upper neck of each duck until slaughtering, which couldn't interfere with animal activity. Upon entering a feeder, each duck's electronic label ID was recognized by the machine. The date and time at beginning and end of each visit was recorded. The central server collected the raw data from each feeder and calculated feed intake and meal duration. At the end of feeding, the ducks with very small body size, without measurement or unusual feeding behavior were removed before analysis of feeding behavior. Any visit that could not be assigned to a specific duck was removed from the data set before analysis. During the whole measuring period, 0.001% of missing record events were observed and these records were removed. Only the ducks with complete records were used for further analysis. Additionally, the records that individual weight were far outweighed other records on the same day were removed, which may be caused by interference from other ducks.

Traits measurement
The body weight (BW) of each bird was automatically measured individually from the beginning to the end of feeding at each visit. Daily BW was calculated as the median of the BW measurement over the entire 24 h period of each day. To calculate feed efficiency, 19 d and 42 d BW was measured manually as beginning and end weight for the test period respectively. The BWG (body weight gain) and feed conversion ratio (FCR) were computed over the entire feeding period. The RFI for the feeding test period was calculated as the residual of the linear regression of the feed intake during the test. The regression equations shown below:

$$Feed\ Intake = -1.1720 + 1.8204 \times BWG$$
$$+ 17.9369 \times \left(BW_{19}^{0.75}\right) + RFI$$

Based on the distribution of RFI, three subsets were selected for later analysis. Ducks in the bottom 10% of RFI were defined as the low-RFI level, from 45% to 55% were defined as the middle-RFI level and the top 10% RFI individuals were selected as the high-RFI group. Each subgroup had 33 ducks, and these ducks were randomly chosen from the combined data.

Every visit was treated as a valid record if feed intake and meal duration were greater than zero. The intervals between visits were calculated to estimate the meal criterion (Additional file 2: Figure S2). The meal criterion was estimated using the method 1 developed by Howie et al. [15]. For each duck, visits were defined as part of a meal if the interval between visits was lower than the meal criterion. Feeding behavior traits were calculated daily and over the test period for each duck: feed intake, feeding rates, average intervals between meals, number of visits and number of meals. The distribution of records was also analyzed to investigate the feeding tendency of ducks.

Because feeding patterns of animals are affected by the environment and their growth needs, the data has been divided into three stages, based on previous analysis. The first three days were considered as a period of adaptation, the fourth day to the tenth day after the commencement of the study was defined as the pre-growth period, and the eleventh to twentieth day was classified as the growth period.

Data analysis
Data management and statistical analyses were performed using R software. The variance analysis for different RFI

levels was performed based on mixed models using R package *VCA*. The following fixed effects, for which $P < 0.05$, were retained:

$$y_{ijk} = \mu + batch_i + level_j + e_{ij}$$

In formation, $level_j$ was the RFI levels, and the $batch_i$ was the batch effect. E_{ij} was the residual of the model. Additionally interaction effect between batch and RFI level was not significant in the data.

Results

Descriptive statistics and meal interval estimation

A total of 105,785 visits were recorded by the feeders. 28 ducks' label were missing before slaughter. After filtering unqualified data, data from a total of 330 ducks were used in further analysis. We compared the batch effect for feed efficiency and feeding behavior. No significant batch effects were observed among three batches. In the following analysis, all three batch data were combined together for description analysis. The *R*-square correlation coefficient of regression of estimated feed intake was 0.42. The descriptive statistics of the measured traits are summarized in Table 1. During the test period, the average FCR was 2.7, average RFI was null by construction, with an SD of 0.6 kg. Moreover the average total feeding time during the test period was 320.6 min; the average number of meals per day was 8.6, and the average number of visits per day were 13.6. The average daily feed intake was 0.324 kg/d. The meal criterion was estimated to be 1,132 s. The frequency

distribution of $\log_e$ - transformed intervals is illustrated in Additional file 2: Figure S2.

Phenotypic correlation

Table 2 shows the phenotypic correlation between feeding behavior and general traits. Total feeding time (TFT), daily feed intake (DFI), feed intake per meal (MFI) daily feeding rate (DFR) and meal duration (MD) had significant positive correlation with feed efficiency traits; moreover, significant correlation between feed intake per meal and body weights was found ($R = 0.32, 0.36$, respectively). However, for between-meal pauses (BMP), no strong phenotypic correlation was found with general traits ($P > 0.05$).

Comparison among different RFI level

Variance analysis for different RFI level is presented in Table 3. Unsurprisingly TFT, DFI and MFI was significantly different among different RFI levels ($P < 0.05$), because these feeding behavior traits are highly phenotypically correlated with feed efficiency (Table 2). For other feeding behavior traits, our results showed that there were slight differences among different RFI level ducks, but these were not significant.

Phenotypic correlations between feeding behaviors and other traits in different RFI levels group were illustrated in Table 4. Unexpectedly, body weight and body weight gain was positively correlated with DFI in different levels ducks. DFI and MD were observed to be significantly correlated with feeding efficiency in high RFI ducks ($P < 0.05$). Moreover, significant correlation between TFT and feed efficiency was found in both low and high

Table 1 Descriptive statistics of measured traits for Pekin ducks

Items		All (n = 358)			
		Mean	S.D.	Min	Max
General traits	BW 19, kg	1.1	0.09	0.74	1.4
	BW 42, kg	3.8	0.2	3.1	4.4
	BWG, kg	2.7	0.2	1.8	3.3
	Feed Intake, kg	7.2	0.8	3.9	9.4
	RFI, kg	0.0	0.6	−3.9	2.1
	FCR	2.7	0.3	1.3	3.4
Feeding behavior traits	Total Feed Time, min	320.6	72.9	204.9	616.6
	Daily feed intake, kg/d	0.324	0.037	0.234	0.669
	Meal Feed Intake, g	33.8	22.8	1.0	350.0
	Visit duration, s	76.6	20.0	31.0	149.3
	Meal duration, s	159.2	162.4	79.9	316.8
	Average interval between meals, min	161.8	123	91.0	309.5
	Daily Feeding Rate, g/min	20.3	3.7	14.0	34.1
	Number of meals per day	8.6	1.4	2	14.4
	Number of visits per day	13.6	3.1	8.9	28.8

BW 19: average body eight at d 19; BW 42: average body eight at d 42; RFI, residual feed intake; FCR, feed conversion ratio. S.D., standard deviation

Table 2 Phenotypic correlation between feeding and other traits

	Total feed time	Daily feed intake	Daily feeding rate	Between-meal pause	Meal feed intake	Meal duration	BW 19	BW 42	BWG
BW 19	0.08	0.33**	0.10	0.14*	0.36**	0.15**	1.00	0.44**	−0.07
BW 42	0.22**	0.62**	0.07	−0.13*	0.32**	0.1	0.36**	1.00	0.90**
BWG	0.20**	0.44**	0.03	−0.19**	0.11*	0.04	−0.07	0.9**	1.00
RFI	0.23**	0.58**	0.14*	−0.07	0.23**	0.12*	0.03	0.01	0.00
FCR	0.16**	0.45**	0.14*	0.06	0.27**	0.15**	0.44**	−0.13*	−0.34**
FI	0.31**	0.84**	0.15*	−0.12*	0.33**	0.16**	0.31**	0.65**	0.55**

BW 19: average body eight at d 19; BW 42: average body eight at d 42; RFI, residual feed intake; FCR, feed conversion ratio.; FI, feed intake; "**"P value < 0.01; "*" P value < 0.05

RFI level ducks. For DFR, only correlation with RFI was observed in low RFI level.

Changes in feeding behavior over time

To further understand the relationship between RFI and feeding behavior, changes in feeding behavior over time were investigated and are illustrated in Fig. 1. During the test period, average DFI increased from 0.07 kg to 0.35 kg (Fig. 1a), and the effect of RFI level was significant from the 3rd day to the end of test ($P < 0.01$). The average number of daily visits and meals increased constantly in the early days, then gradually diminished, and finally stabilized (Fig. 1b, c). Obvious differences in daily visits were found among different RFI level from 4 wk. To the end of test, and these differences magnified with age; however, there was no difference in daily meal number.

The daily feeding rate increased over the entire test period (Fig. 1d, e); in contrast, meal duration gradually decreased with age. Moreover, between-meal pauses was observed to increase until the 11th day, then was maintained at a relative stable value, and finally, in the between-meal pauses of low RFI ducks were slightly longer than high RFI animals.

Feeding tendency

Fig. 2a summarizes the records of all the feeders. The machine with the highest number of visits is the eighth feeder which has 15,102 visits, and the feeder with the lowest records was the first one, with 4422 visit records. Ducks in the top 10% RFI level recorded 14% of the total records, and visit records of the lowest 10% RFI animals were 11% of the total. To further understand the feeding

Table 3 Comparison of measured traits among different RFI levels

Items		Low RFI($n = 33$)		Middle RFI($n = 33$)		High RFI($n = 33$)	
		Mean	S.D.	Mean	S.D.	Mean	S.D.
General Traits	BW 19, kg	1.1	0.1	1.1	0.1	1.2	0.1
	BW 42, kg	3.9	0.3	3.8	0.2	3.9	0.2
	Feed intake, kg	6.5^c	0.7	7.2^b	0.4	8.4^a	1.1
	RFI	−0.9^c	0.59	−0.024^b	0.027	0.9^a	1.1
	FCR	2.3^b	0.2	2.7^a	0.1	2.7^a	0.19
Feeding behavior traits	Total feed time, min	297.8^B	65.1	321.1^{AB}	61.0	346.1^A	98.5
	Daily feed intake, kg/d	0.297^C	0.024	0.324^B	0.025	0.372^A	0.061
	Visit feed intake, g	23	5	23	5	25	5
	Meal feed intake, g	33^b	7	37^b	6	40^a	8
	Visit duration, s	76.1	20.7	75.4	21.3	77.5	22.3
	Meal duration, s	147.5^c	38.8	171.8^a	40.4	170.1^a	41.7
	Daily interval between meals, min	165	26	167	23	161	28
	Daily feeding rate, g/min	19.6	3.0	20.3	4.3	21.7	4.0
	Number of meals per day	8.5	3.3	8.4	3	8.7	3.2
	Number of visits per day	12.9	5.9	13.8	5.8	14.5	6.6

BW 19: average body eight at d 19; BW 42: average body eight at d 42; RFI, residual feed intake; FCR, feed conversion ratio. S.D., standard deviation. Different letters in the same row means significant differences between different groups. Capital: $P < 0.01$; Lowercase: $P < 0.05$

Table 4 Phenotypic correlation between feeding behavior and other traits in different RFI groups

RFI Group	Total Feed Time H	M	L	Daily Feed Intake H	M	L	Daily Feeding Rate H	M	L	Between-Meal Pause H	M	L	Meal Feed Intake H	M	L	Meal Duration H	M	L
BW 19	0	−0.11	0.31	0.19	0.52**	0.33	−0.03	0.24	−0.12	0.14	0.18	0.43*	0.14	0.41*	0.48**	−0.01	0.13	0.01
BW 42	0.21	−0.01	0.32	0.48**	0.79**	0.79**	−0.11	0.32	0.1	0.12	−0.06	0.04	0.25	0.33	0.46**	0.18	0.06	0.08
BWG	0.22	0.05	0.23	0.41*	0.65**	0.73**	−0.10	0.24	0.16	0.03	−0.18	−0.12	0.20	0.15	0.32	0.25	−0.01	0.08
FI	0.41*	−0.02	0.55**	0.77**	0.8**	0.54**	0.16	0.31	0.03	−0.36*	−0.03	0.02	0.18	0.36*	0.26	0.5**	0.08	0.12
RFI	0.36*	0.13	0.38*	0.67**	0.08	−0.03	0.39*	−0.1	−0.04	−0.4*	0.1	−0.05	0.1	0.11	−0.11	0.45**	0.16	0.08
FCR	0.32	−0.09	0.39*	0.35*	0.09	−0.1	0.30	0.04	−0.13	0.10	0.29	0.15	0.08	0.3	0	0.41*	0.15	0.05

BW 19: average body eight at d 19; BW 42: average body eight at d 42; "**"$P < 0.01$;"*" $P < 0.05$. H, M, L mean High RFI group, Middle RFI group and Low RFI group, respectively

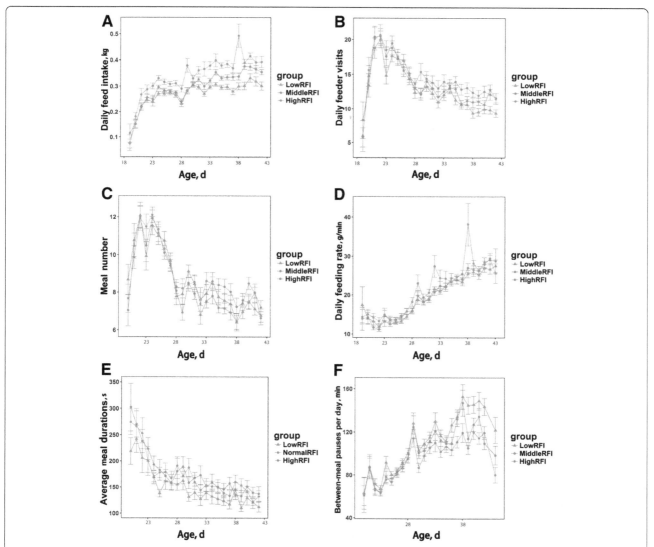

Fig. 1 Evolution of feeding behavior traits during the feeding test period. X axis denotes the age (days), and Y axis denotes daily mean of each feeding behavior trait. Error bars are illustrated on the plots. The data from the last day was removed, since ducks are considered to be under stress during this fasting period. Ducks were 19 d of age at the beginning of the experiment. **a** Daily feed intake (**b**) Number of daily visits (**c**) Number of daily meals (**d**) Daily feeding rate (**e**) Daily meal duration (**f**) Between-meal pause

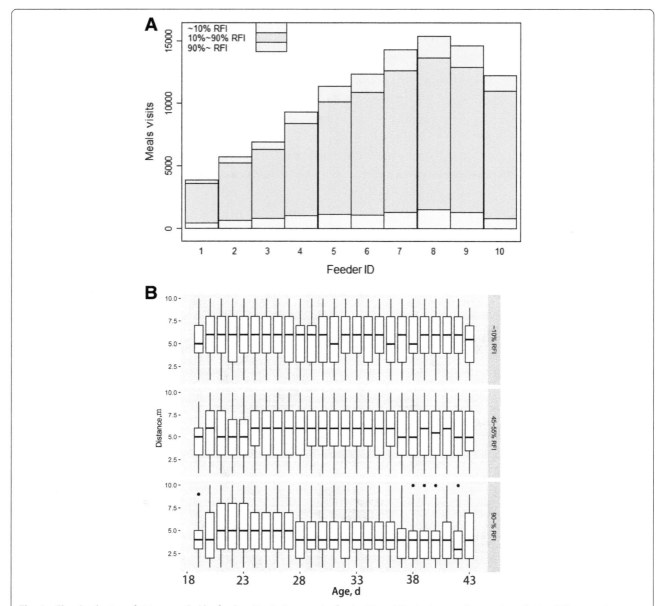

Fig. 2 a The distribution of visits recorded by feeders. X axis denotes the feeder ID, and Y axis denotes the number of visits. Different colors represent three RFI levels of the ducks, respectively. Feeding machine is linearly layout from number one to ten. The first machine was closest to the gate, and the last one was far away from gate. All of invalid visits were removed. **b** Comparison of duck feeding tendencies in different RFI level groups. X axis represents the age of birds, and Y axis represents the distance between feeder and door. The box plot represents the range of feeding activities at different RFI levels. The data of the last day was removed, because at this fasting time the ducks were considered to be under stress. Variance analysis showed that the machine position that high RFI ducks tended to eat in was significantly lower (i.e. closer to the gate) than other ducks, over the entire time period ($P < 0.01$). The initial bird age was 19 d

tendencies of ducks, the distance to the gate was used to define meal positions (Fig. 2b). The results show that animals were concentrated in middle positions (4 to 5 m from gate) for the first two days. After an adaption time, more ducks preferred to eat at 7 ~ 8 m from the gate; however, high RFI ducks frequently visited feeders at positions of 3 to 4 m from the gate, in which frequency was 1.70 per day, higher than 1.57 per day for low RFI animals. The feeder machines visited by high RFI ducks was closer to gate than for low and middle RFI ducks.

Distribution of daily feeding time

To investigate the daily feeding time of ducks, the feeding records at different time points every day are summarized in Fig. 3. The results showed that duck feeding was mainly concentrated at noon and in the afternoon during the

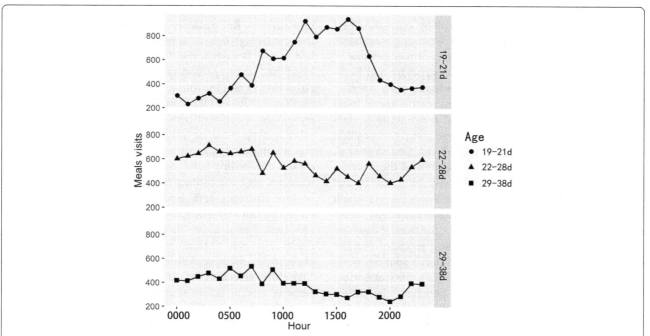

Fig. 3 The feeding visit distribution during the day. X axis denotes the time points (hr), Y axis denotes the number of visits. The results for three different periods are illustrated: 19–21 day-old; 22–28 day-old; 29–38 day-old; the initial bird age was 19 d. The number of each RFI group is 33

adaptation period, and reached the highest peak between 1200 h at noon and 1600 h in the afternoon. In the pregrowth period, feeding activity was observed during the entire 24 h period, and the number of visits increased significantly from 2300 h at night to 0500 h in the early morning. Compared to the first 10 d, the number of feeding visits was reduced during the growth period, but the tendency (frequent activities from 2300 h at night to 0700 h in the early morning) was similar to the pregrowth period, and was concentrated from midnight to the following morning.

Discussion

Meal determination

Feeding behavior consists of different feeding activities, which are referred to feeder visits. Bley and Bessei [12] found that Pekin ducks exhibited an average of 29 visits per day at 4 wk of age, decreasing to fewer than 15 visits per day at 7 wk of age. Basso et al. [11] obtained a similar trend with 27 visits per day at 3 wk, rapidly decreasing to 15 visits, and then stabilizing. In this study, we found that the number of visits was gradually reduced from an average 25 visits per day at 3 wk of age to 14 visits at 4 wk, and then was stable at about 13 visits until the end of the study at 7 wk (Fig. 1b).

Our results are consistent with the two studies above. Actually a meal is not always equal to one visit, but rather contained a few visits and short relaxation pauses called within-meal pauses for animal relaxing or drinking;

frequently, meals are separated by non-feeding intervals called between-meal pauses which are much longer than within-meal pause [11]. Therefore, a meal criterion is needed for grouping different feeding events into defined meals [15, 16]. Howie et al. [17] obtained a meal criterion of 1725 s for Pekin ducks by fitting a truncated log-normal model, and Basso et al. [11] found the meal criterion of 1790 s for Mule ducks using same method; however, Drouilhet et al. [18] estimated the meal criterion as 2208 s in their study in which the feeder entrance had a door. In our study, the log distribution of visit intervals is similar to the trends reported by Howie et al. [15] and Basso et al. [11] (Additional file 2: Figure S2); hence, we used Howie's method for our data set and calculated a meal criterion of 1,132 s for Pekin ducks. This value is, lower than the meal criterion reported by others, which is likely due to the differences in feeder construction and in duck population. We found that meals consisted on average of 1.7 visits and that the number of meal per day had a similar trend in three RFI-level. Basso et al. [11] and Drouilhet et al. [18] found that the number of meals tended to decrease with age; however, between-meal pauses tended to increase. In boilers, Howie et al. [15] also found same trend. Our results showed that the number of meals tended to decrease with age (from 12 meal/d to 7 meal/d) after feeder adaption time and the between-meal pauses tended to increase (from 60 min/d to 2 h/d more). The results are in agreement with previous studies.

Feeding tendency at different RFI levels

Feeding tendency is possibly affected by environmental factors, especially under a wide range of farming conditions. These changes may be due to management mechanisms or system structures. For example, sward structure and herbage quality affected cattle's grazing behavior [19]. Another study showed that the average time cattle spent lying increased with increasing number of concentrate feeders [20]. Funston et al. [21] found different group composition affected cattle feeding tendencies. In ducks, Drouilhet et al. [18] used a feeder with a door closing behind the animals, and reported that this affected the visit intervals. In our study, most ducks tended to eat at the inside feeders which are far away from the gate. However, the innermost feeder was not the best feeding position, since the feeding machine slightly near the center have the highest recorded number of visits (Fig. 2a). This suggests that ducks spontaneously move away from the gate where animals are susceptible to environmental interference. In contrast, the innermost machine is limited by proximity to the fence and this could greatly reduce activity space for ducks. Our results also show that high RFI ducks tend to eat near the gate, where high RFI ducks were more susceptible to outside interface (Fig. 2b), and this always occurs during the entire period. Some studies indicate that selection for feed efficiency might reduce behavioral reactivity [22], and evidence reveals that animals with better feed efficiency have lower fearfulness. Genetic research with mammals and poultry showed that low RFI animals were less active [23, 24]. Drouilhet, et al. [18] reported that selection for low RFI produces mule ducks that are slightly better adapted to human activity, in agreement with a previous study [25]. Compare to their results, our study suggests that high RFI individuals have lower adaptive ability to the new feeding environment. However, there is not enough evidence to determine the clear reason for this effect. Moreover, we found that feeding time points for adaptation during the first 3 d were different than at other time periods (Fig. 3). Ducks mainly were eating at noon and early afternoon during first 3 d while adapting to the feeders. After adaptation, ducks returned to middle of the night eating habits. These behaviors may be improved by feeder structure modification and selection for adaption response.

Feeding behavior at different RFI levels

The relationship between feeding behavior and feeding efficiency depends on population and species, and there are obvious differences in feeding behavior between mammals and poultry selected for feed efficiency. Studies with cattle reported that high RFI cattle consume more feed, but with similar feeding rates, compared to low RFI animals [26, 27]. For pigs, Young et al. [28] found that low RFI animals had shorter feeding times with a lower DFI compared to the high RFI pigs, and that the feeding rates of low RFI animals were significantly faster than the control group. A similar result was obtained in a hen study; Braastad and Katle [23] found that low RFI birds had shorter feeding times than high RFI birds. Drouilhet et al. [18] found that low RFI mule ducks had significantly lower DFI than high RFI ducks, but no differences were significant between high and low RFI lines for other feeding behavior traits. In this study, we found there was a significant phenotypic correlation between feed efficiency and feeding behaviors except for BMP (Table 2). It indicated that the feeding frequency has a limited impact on feed efficiency, total feeding time and daily feed intake are more critical. The Low RFI ducks spent less feeding time and had a lower daily feed intake than High RFI ducks. However these correlative relationships were not observed in different RFI-level sub-populations (Table 4). Therefore, selection for RFI may change the association between feeding behavior and other traits.

Our results were different with those of Drouilhet, et al. [18]; this discrepancy could be caused by feeder structure and selection of meal criterion. Additionally, when compared day by day, ducks in the three different RFI levels had a similar number of daily visits at 4 wk of age; however, after 10 d of the test period, low RFI ducks had a significantly lower visit number than high RFI ducks, and these differences were magnified with age (Fig. 1b). Over the entire test period, these changes were too microscopic to achieve statistical significance. Our results gave a possibility that selection for RFI might change the unknown mechanisms that underlie control of feeding behavior, but more genetic evidences were needed to support the hypothesis. This showed that selection for feeding behavior is a valuable approach for trait improvement.

Conclusions

In this study, we selected 358 male Pekin ducks and measured their feeding behavior, feed efficiency traits in order to analyze relationships between feeding behavior and feed efficiency. The results illustrate that the feeding tendency of Pekin ducks is obviously different at different RFI levels, and that high RFI ducks prefer the feeders closed to side barriers. The results also suggests that certain feeding behaviors have the potential to improve other economic traits with an improvement in feed efficiency. Further research will focus on genetic factors to gain insight into the relationship between feeding behavior and feed efficiency.

Abbreviations

BMP: Between-meal pauses; BW: Body weight; BWG: Body weight gain; DFI: Daily feed intake; DFR: Daily feeding rate; FCR: Feed conversion rate; FI: Feed intake; MD: Meal duration; MFI: Feed intake per meal; RFI: Residual feed intake; RFID: Radio Frequency Identification; TFT: Total feeding time

Acknowledgments
We are grateful to Maxwell Hincke (University of Ottawa, Canada) for a careful review of the manuscript.

Funding
The work was supported by the National Scientific Supporting Projects of China (2015BAD03B06), National Waterfowl-industry Technology Research System (CARS-42) and the Program for Changjiang Scholars and Innovative Research Team in University (IRT_15R62).

Authors' contributions
FZ carried out the experiment and drafted the manuscript. ZCH conceived the study, participated in its design and coordination, and helped draft the manuscript. FBL, JPH, FXY helped sample and experiment. All authors read and approved the final manuscript.

Competing interests
The authors declare that they have no competing interests.

Author details
[1]National Engineering Laboratory for Animal Breeding and MOA Key Laboratory of Animal Genetics and Breeding, Department of Animal Genetics and Breeding, China Agricultural University, Beijing 100193, China. [2]Beijing Golden Star Duck Inc., Beijing 100076, China.

References
1. Bezerra LR, Sarmento JL, Neto SG, de Paula NR, Oliveira RL. Do Rego WM. Residual feed intake: a nutritional tool for genetic improvement. Trop Anim Health Prod. 2013;45(8):1649–61.
2. Koch RM, Gregory KE, Chambers D, Swiger LA. Efficiency of feed use in beef cattle. J Animal Sci. 1963;22(2):486–94.
3. Basso B, Bordas A, Dubos F, Morganx P, Marie-Etancelin C. Feed efficiency in the laying duck: appropriate measurements and genetic parameters. Poult Sci. 2012;91(5):1065–73.
4. Kennedy BW, van der Werf JH, Meuwissen TH. Genetic and statistical properties of residual feed intake. J Anim Sci. 1993;71(12):3239–50.
5. Pingel H. Influence of breeding and management on the efficiency of duck production. Lohmann information. 1999;22:7–13.
6. Larzul C, Guy G, Bernadet MD. Feed efficiency, growth and carcass traits in female mule ducks. Archiv Fur Geflugelkunde. 2004;68(6):265–8.
7. Howie JA, Tolkamp BJ, Avendano S, Kyriazakis I. The structure of feeding behavior in commercial broiler lines selected for different growth rates. Poult Sci. 2009;88(6):1143–50.
8. Houpt KA. Animal behavior and animal welfare. J Am Vet Med Assoc. 1991;198(8):1355–60.
9. Pennisi E. Genetics. A genomic view of animal behavior. Science. 2005; 307(5706):30–2.
10. Schwartzkopf-Genswein KS, Atwood S, Mcallister TA. Relationships between bunk attendance, intake and performance of steers and heifers on varying feeding regimes. Appl Anim Behav Sci. 2002;76(3):179–88.
11. Basso B, Lague M, Guy G, Ricard E, Marie-Etancelin C. Detailed analysis of the individual feeding behavior of male and female mule ducks. J Anim Sci. 2014;92(4):1639–46.
12. Bley TA, Bessei W. Recording of individual feed intake and feeding behavior of pekin ducks kept in groups. Poult Sci. 2008;87(2):215–21.
13. Howie JA, Avendano S, Tolkamp BJ, Kyriazakis I. Genetic parameters of feeding behavior traits and their relationship with live performance traits in modern broiler lines. Poult Sci. 2011;90(6):1197–206.
14. Bian B, Hao J, Yang F, Hu S, Hou Z. Study on feeding behavior and the relationship between residual feed intake and some economic traits in Peking ducks. China Poultry. 2014;36(16):8–11.
15. Howie JA, Tolkamp BJ, Avendaño S, Kyriazakis I. A novel flexible method to split feeding behaviour into bouts. Appl Anim Behav Sci. 2009;116(116):101–9.
16. Yeates MP, Tolkamp BJ, Allcroft DJ, Kyriazakis I. The use of mixed distribution models to determine bout criteria for analysis of animal behaviour. J Theor Biol. 2001;213(3):413–25.
17. Howie JA, Tolkamp BJ, Bley T, Kyriazakis I. Short-term feeding behaviour has a similar structure in broilers, turkeys and ducks. Br Poult Sci. 2010;51(6):714–24.
18. Drouilhet L, Monteville R, Molette C, Lague M, Cornuez A, Canario L, et al. Impact of selection for residual feed intake on production traits and behavior of mule ducks. Poult Sci. 2016;95(9):1999–2010.
19. Stejskalova M, Hejcmanova P, Pavlu V, Hejcman M. Grazing behavior and performance of beef cattle as a function of sward structure and herbage quality under rotational and continuous stocking on species-rich upland pasture. Anim Sci J. 2013;84(8):622–9.
20. Gonzalez LA, Ferret A, Manteca X, Ruiz-de-la-Torre JL, Calsamiglia S, Devant M, et al. Effect of the number of concentrate feeding places per pen on performance, behavior, and welfare indicators of Friesian calves during the first month after arrival at the feedlot. J Anim Sci. 2008;86(2):419–31.
21. Funston RN, Kress DD, Havstad KM, Doornbos DE. Grazing behavior of rangeland beef cattle differing in biological type. J Anim Sci. 1991;69(4): 1435–42.
22. Canario L, Mignon-Grasteau S, Dupont-Nivet M, Phocas F. Genetics of behavioural adaptation of livestock to farming conditions. Animal. 2013;7(3):357–77.
23. Braastad BO, Katle J. Behavioural differences between laying hen populations selected for high and low efficiency of food utilisation. Br Poult Sci. 1989;30(3):533–44.
24. Luiting P, Urff EM. Optimization of a model to estimate residual feed consumption in the laying hen. Livest Prod Sci. 1991;27(4):321–38.
25. Arnaud I, Mignon-Grasteau S, Larzul C, Guy G, Faure JM, Guemene D. Behavioural and physiological fear responses in ducks: genetic cross effects. Animal. 2008;2(10):1518–25.
26. Lancaster PA, Carstens GE, Ribeiro FR, Tedeschi LO, Jr CD. Characterization of feed efficiency traits and relationships with feeding behavior and ultrasound carcass traits in growing bulls. J Anim Sci. 2009;87(4):1528–39.
27. Golden JW, Kerley MS, Kolath WH. The relationship of feeding behavior to residual feed intake in crossbred Angus steers fed traditional and no-roughage diets. J Anim Sci. 2008;86(1):180–6.
28. Young JM, Cai W, Dekkers JC. Effect of selection for residual feed intake on feeding behavior and daily feed intake patterns in Yorkshire swine. J Anim Sci. 2011;89(3):639–47.

17

SIRT1-dependent modulation of methylation and acetylation of histone H3 on lysine 9 (H3K9) in the zygotic pronuclei improves porcine embryo development

Katerina Adamkova[1], Young-Joo Yi[2], Jaroslav Petr[3], Tereza Zalmanova[1,3], Kristyna Hoskova[1,3], Pavla Jelinkova[1], Jiri Moravec[4], Milena Kralickova[5,6], Miriam Sutovsky[7], Peter Sutovsky[7,8] and Jan Nevoral[1,5,6*] (iD)

Abstract

Background: The histone code is an established epigenetic regulator of early embryonic development in mammals. The lysine residue K9 of histone H3 (H3K9) is a prime target of SIRT1, a member of NAD^+-dependent histone deacetylase family of enzymes targeting both histone and non-histone substrates. At present, little is known about SIRT1-modulation of H3K9 in zygotic pronuclei and its association with the success of preimplantation embryo development. Therefore, we evaluated the effect of SIRT1 activity on H3K9 methylation and acetylation in porcine zygotes and the significance of H3K9 modifications for early embryonic development.

Results: Our results show that SIRT1 activators resveratrol and BML-278 increased H3K9 methylation and suppressed H3K9 acetylation in both the paternal and maternal pronucleus. Inversely, SIRT1 inhibitors nicotinamide and sirtinol suppressed methylation and increased acetylation of pronuclear H3K9. Evaluation of early embryonic development confirmed positive effect of selective SIRT1 activation on blastocyst formation rate ($5.2 \pm 2.9\%$ versus $32.9 \pm 8.1\%$ in vehicle control and BML-278 group, respectively; $P \leq 0.05$). Stimulation of SIRT1 activity coincided with fluorometric signal intensity of ooplasmic ubiquitin ligase MDM2, a known substrate of SIRT1 and known limiting factor of epigenome remodeling.

Conclusions: We conclude that SIRT1 modulates zygotic histone code, obviously through direct deacetylation and via non-histone targets resulting in increased H3K9me3. These changes in zygotes lead to more successful pre-implantation embryonic development and, indeed, the specific SIRT1 activation due to BML-278 is beneficial for in vitro embryo production and blastocyst achievement.

Keywords: Embryonic development, Epigenetics, H3K9 methylation, SIRT1, Sirtuin

Background

Correct formation of maternal and paternal pronuclei in the fertilized mammalian oocyte, the zygote, is required for the first mitotic cell cycle, subsequent zygotic gen-

ome activation and successful development of early embryo [1, 2]. Many events, such as protamine-histone replacement [3, 4], protein recycling through ubiquitin-proteasome system (UPS) [5, 6] and correct establishment of euchromatin and heterochromatin [7, 8], lead to genome-wide alterations required for the biogenesis of pronuclei. In addition to these essential genomic and cellular events, pronuclei undergo epigenetic changes, i.e. DNA methylation as well as histone methylation and acetylation, collectively termed the histone code establishment [9–13]. Epigenetic changes in the early zygote include DNA demethylation in both the maternal and paternal pronucleus [14] as well as parent-of-origin specific modifications of pronuclear histone code [9].

* Correspondence: jan.nevoral@lfp.cuni.cz
Katerina Adamkova and Young-Joo Yi contributed equally to this work.
Peter Sutovsky and Jan Nevoral are co-senior authors and project co-directors.
[1]Department of Veterinary Sciences, Faculty of Agriculture, Food and Natural Resources, Czech University of Life Sciences Prague, 6-Suchdol, Prague, Czech Republic
[5]Laboratory of Reproductive Medicine of Biomedical Center, Charles University, Pilsen, Czech Republic
Full list of author information is available at the end of the article

However, up-stream factors of histone code in zygote and their influence on embryo development and blastocyst quality are poorly understood.

Sirtuins (SIRTs) are a family of NADP⁺-dependent histone-deacetylases including 7 isoforms with specific subcellular localization patterns [15]. Among them, SIRT1 is the most potent regulator of histone code, present notably in the nucleus and it enhances cell viability by regulating epigenome remodeling [16, 17]. The expression of SIRTs in mammalian oocytes and embryos have been observed [18–22], and the essential role of SIRT1 in oocyte maturation and early embryonic development has been established [19, 23]. Accordingly, beneficial effect of red grape flavonoid resveratrol, a cell protectant/antioxidant substance and a strong activator of SIRT1, on oocyte quality and success of embryonic development is well-known [24–27]; however, we lack the understanding of mechanisms by which SIRT1 enhances oocyte maturation, fertilization and early embryonic development.

Based on somatic cell studies, SIRT1 is able to remove the acetyl group from lysine residues of several histones, resulting in deacetylation of histone H1 on lysine K26 [28, 29], H3 on K9, K14 and K56 [28, 30], and H4 on K8, K12 and K16 [28, 31]. Acetylation of H3K9 is an established marker of translational activity, but it is also frequently associated with DNA damage [32]. Deacetylation of H3K9 makes it available for methyl group addition by histone methyltransferases [33–36]. The involvement of UPS, through the participation of Mouse double minute 2 homolog (MDM2), an E3-type ubiquitin ligase, in SIRT1-mediated H3K9 methylation is indicated [37] and remains the lone consideration of SIRT1 mechanism in the nucleus.

Based on the above knowledge, we hypothesized that SIRT1 affects acetylation-methylation pattern of H3K9 in formatting porcine zygote pronuclei. We also predicted that the SIRT1-modulated H3K9 zygotic histone code establishment will enhance early embryonic development measured by development to blastocyst and blastocyst quality.

Methods
Collection and in vitro maturation (IVM) of porcine oocytes
Porcine ovaries were obtained from 6- to 8-month-old non-cycling gilts (a crossbreed of Landrace x Large White) at the local slaughterhouse (Jatky Plzen a.s., Plzen, Czech Republic) and transported to laboratory at 39 °C. Cumulus-oocyte complexes (COCs) were collected from ovarian follicles with a diameter of 2–5 mm by aspiration with a 20-gauge needle and handled in HEPES-buffered Tyrode lactate medium containing 0.01% (w/v) polyvinyl alcohol (TL-HEPES-PVA). Only fully grown oocytes with evenly dense cytoplasm, surrounded by compact cumuli, were selected for IVM and

washed in maturation medium. The medium used for IVM was modified tissue culture medium (mTCM) 199 (Gibco, Life Technologies, UK) supplemented with 0.1% PVA, 3.05 mmol/L D-glucose, 0.91 mmol/L sodium pyruvate, 0.57 L-cysteine, 0.5 μg/mL LH (Sigma-Aldrich, USA), 0.5 μg/mL FSH (Sigma), 10 ng/mL epidermal growth factor (EGF; Sigma), 10% porcine follicular fluid, 75 μg/mL penicillin G and 50 μg/mL streptomycin. After 22 h of culture, the COCs were cultured in TCM199 without LH and FSH for an additional 22 h. The COCs were cultured in 500 μL of the medium covered by mineral oil in a four-well Petri dish (Nunc, Denmark), at 39 °C and 5% CO_2 in air [38].

In vitro fertilization (IVF) and culture (IVC) of porcine oocytes and zygotes
After 44 h of IVM, cumulus cells were removed with 0.1% hyaluronidase in TL-HEPES-PVA and the metaphase II (MII) oocytes with extruded first polar body were selected for IVF. The oocytes were washed three-times with modified Tris-buffered medium (mTBM) [38] with 0.2% bovine serum albumin (BSA; A7888; Sigma) and placed into 100 μL drops of mTBM, covered with mineral oil in a 35 mm Petri dish. The dishes were allowed to equilibrate at 38.5 °C and 5% CO_2 for 30 min before spermatozoa were added for fertilization. Spermatozoa were prepared as follows: 1 mL liquid semen preserved in BTS-based extender was washed twice in phosphate buffered saline (PBS) with 0.1% PVA (PBS-PVA) at 1,500 rpm for 5 min. The last wash was supplemented with MitoTracker CMTRos (400 nmol/L; M7510, Invitrogen) for 10 min at 39 °C, used to tag sperm mitochondria that associate with the paternal pronucleus inside the fertilized oocytes. Labeled spermatozoa were resuspended in mTBM ($2.5–5 \times 10^7$ spermatozoa per mL) and 1 μL of this sperm suspension was added to the medium containing the oocytes to give a final sperm concentration of 2.5 or 5×10^5 spermatozoa per mL. Oocytes were co-incubated with spermatozoa for 5 to 6 h at 38.5 °C and 5% CO_2 in air. For zygote acquisition, oocytes were thereafter washed and transferred into 100 μL PZM3 medium [39] containing 0.4% BSA (A6003; Sigma), for further culture for 22 h. Simultaneously, presumed zygotes were cultured in 500 μL of PZM3 medium for 144 h to reach blastocyst stage. PZM3 medium contained different concentrations of SIRT1 activators or inhibitors dissolved in DMSO (in its final concentration 0.1%) as described below. The IVF and IVC studies were repeated three to five times for each treatment regimen.

Sirtuin activation and inhibition
Activation and inhibition of SIRT1 was performed in PZM3 during IVC of early zygote development.

Activators of SIRT1 included resveratrol (3.0, 6.25, 12.5 μmol/L; non-selective sirtuin activator, Abcam, ab120726) and BML-278 (3.0 μmol/L; selective SIRT1 activator, Abcam, ab144536). Inhibitors of SIRT1 included nicotinamide (2.5, 5.0, 7.5 mmol/L; non-selective sirtuin inhibitor, Abcam, ab120864) and sirtinol (10 μmol/L; selective SIRT1 and SIRT2 deacetylase inhibitor, Abcam, ab141263). The effective concentrations of BML-278 (N-Benzyl-3,5-dicarbethoxy-4-phenyl-1,4-dihydropyridine) and sirtinol were chosen based on preliminary experiments conducted to optimize the concentrations of resveratrol and nicotinamide (data not shown). All compounds were dissolved in DMSO, the concentration of which never bypassed 0.5%, as also used for vehicle controls.

Immunofluorescence and imaging of zygotes

After 22 h of IVC, presumed zygotes were treated with 0.5% pronase for zona pellucida removal and processed as described by Yi et al. [40], with modifications. Briefly, zygotes were fixed in 2% formaldehyde and permeabilized in 0.1% Triton-X-100 in PBS (PBS-TX) for 40 min. Thereafter, zygotes were blocked in 5% normal goat serum (NGS) in PBS-TX for 25 min. A mixture of mouse monoclonal anti-histone H3 tri-methylated K9 antibody (H3K9me3; ab71604, Abcam, UK; 1:200) and rabbit monoclonal anti-histone H3 acetylated K9 antibody (H3K9ac; ab32129, Abcam, Cambridge, UK; 1:200) was applied overnight at 4 °C. Subsequently, the oocytes were washed twice in 1% NGS in BPS-TX before being incubated with fluorescein isothiocyanate (FITC)-conjugated goat anti mouse (GAM) IgG (cat. #62–6511; Invitrogen, Thermo Fisher Scientific Inc., Waltham, MA, USA; 1:200) and cyanine dye (Cy5)-conjugated goat anti-rabbit (GAR) IgG (111–175-144; Jackson ImmunoResearch Laboratories Inc., West Grove, PA, USA), for 40 min at room temperature. Thereafter, the oocytes were washed twice and mounted into slides in a Vectashield medium with 4′6′-diamidino-2-phenylindole (DAPI; Vector Laboratories Inc., Burlingame, CA, USA). SIRT1 was detected by adapting the above protocol for mouse monoclonal antibody clone 1F3 (ab104833, Abcam). Images were acquired using the Ti-U microscope (Nikon Co., Tokyo, Japan) with Clara Interline CCD camera (Andor Technology PLC, Belfast, Northern Ireland) operated by NIS Elements Ar software (Nikon Co., Tokyo, Japan). Negative controls were performed by omitting specific antibodies and these slides were processed at comparable settings. The image analysis was performed by NIS Elements. Signal intensities of pronuclear H3K9ac and H3K9me3 were scaled by a basal signal intensity of corresponding zygote cytoplasm and expressed as a relative acetylation/methylation ratio.

Evaluation of fertilization and embryonic development

Within the imaging of 22 h zygotes, an assessment of penetration rate, creation of paternal pronucleus/pronuclei and monospermic fertilization was performed. Embryo cleavage was assessed by microscopy at 48 h of IVC; after 144 h of IVC, embryos were fixed in 2% formaldehyde for 40 min at room temperature (RT), washed three times with PBS, permeabilized with PBS-Triton X-100 for 30 min, and stained with 2.5 μg/mL DAPI (DNA staining; Molecular Probes, Eugene, OR, USA) for 40 min. Embryo cleavage, blastocyst formation, and cell number per blastocyst were assessed under an Eclipse Ci fluorescence microscope (Nikon Co., Tokyo, Japan).

Quantification of fluorescent immunolabeling intensity of the ooplasmic ubiquitin ligase MDM2

Presumed zygotes were subjected to immunofluorescence protocol described above. A mixture of rabbit polyclonal anti-MDM2 antibody (ab137413, Abcam, UK; 1:200) and mouse monoclonal anti-Glyceraldehyde 3-phosphate dehydrogenase antibody (GAPDH; ab8245, Abcam, UK; 1:200) was used. Only zygotes with visible pronuclei were used for further analysis. Images were acquired and analysed as was described. The MDM2 signal intensity was normalized to signal intensity of GAPDH, considered a housekeeping factor with constant expression in pig oocytes [41, 42].

Western blotting

In vitro matured MII oocytes, cells undergoing to in vitro fertilization assay, and their cumulus cells were used for western blot analysis. Samples were prepared and processed using the method set out by Nevoral et al. [42], with slight modifications. In brief, denuded oocytes and their cumulus cells were precipitated in 80% aceton for at least 60 min and then lysed in 20 μL of Laemmli buffer containing Triton-X-100 (0.003%, v/v) and SDS (0.001%, v/v), enriched with Complete Mini Protease Inhibitor Cocktail (Roche, Switzerland). Samples were boiled and subjected to SDS-PAGE electrophoresis in 12.5% separating gels and blotted using Trans-Blot TurboTM Transfer System (Biorad Laboratories, Steenvoorde, France) onto a PVDF membrane (GE Healthcare Life Sciences, Amersham, UK). After blocking in 5% non-fat milk in TBS with 0.5% Tween-20 (TBS-T) for 60 min at room temperature, the membrane was incubated with mouse monoclonal anti-SIRT1 (ab110304, Abcam, 1:1,000) or rabbit polyclonal anti-MDM2 (ab137413, Abcam, UK; 1:1,000) diluted in TBS-T for 60 min at room temperature. Mouse monoclonal anti-GAPDH loading control antibody (ab8245, Abcam, UK; 1:2,000) was used under same conditions. Subsequently, the membrane was incubated with horseradish peroxidase (HRP)-conjugated goat anti-mouse or anti-rabbit IgG in TBS (1:3,000; Invitrogen, USA) for 60 min at room temperature. Proteins with adequate molecular weight were detected using the ECL Select Western Blotting Detection Reagent (GE Healthcare Life Sciences, Amersham, UK) and visualised by ChemiDocTM MP System (Biorad, France).

Statistics

Immunofluorescence data are presented as the mean ± S.E.M. of at least 20 zygotes per experimental group. All analysis were performed in at least three independent repetitions. The general linear models (GLM) procedure in SAS package 9.3 (SAS Institute Inc., Cary, NC, USA) was used, followed by Sheffe's test and Duncan's multiple range test for zygote analysis and evaluation of early embryonic development, respectively. $P \leq 0.05$ was considered to be statistically significant.

Results

SIRT1 is accumulated in zygotic pronuclei

The aim of this experiment was to demonstrate the presence of SIRT1 and verify its subcellular localization in porcine zygotes. Immunocytochemical analysis demonstrated the presence of SIRT1 in both paternal and maternal pronuclei (Fig. 1a) while the intensity of cytoplasmic SIRT1 fluorescence was weak. Such localization was not observed in negative control samples where the primary antibody was omitted. A band of anticipated mass (80.9 kDa; UniProtKB (A7LKB1)) was detected with the same anti-SIRT1 antibody by Western blotting (see Additional file 1).

Resveratrol increases H3K9 methylation in zygotic pronuclei

To evaluate the effect of SIRT1 activation on pronuclear H3K9 acetylation and methylation, the presumed zygotes were treated by resveratrol, a non-selective activator of SIRTs, and examined by epifluorescence microscopy and relative fluorescence intensity measurement (Fig. 2a, b). Specific antibodies against modified H3K9 modification, validated for zygote imaging by a number of previous studies [43–45], and epifluorescence microscopy followed by image analysis were used. Fertilization rate, paternal pronucleus (PPN) formation and monospermic fertilization rate were simultaneously examined.

A significantly increased intensity of H3K9me3 in both pronuclei was observed in zygotes treated with 3 μmol/L and 12.5 μmol/L resveratrol (1.39 and 1.32 fold, respectively, compared to the vehicle control). Interestingly, the 6.25 μmol/L resveratrol did not show a significant difference from control. Contrary to H3K9 methylation, H3K9 acetylation decreased after resveratrol treatment. Although the H3K9 methylation was affected, resveratrol treatments had no effect on IVF indicators (see Additional file 2).

Nicotinamide protects pronuclear H3K9 from deacetylation

The effect of the inhibition of SIRTs on H3K9 acetylation and methylation in the presumed zygotes treated with nicotinamide (non-selective SIRTs' inhibitor) was assessed by epifluorescence microscopy and pixel intensity measurement (Fig. 3a, b). In contrast to resveratrol, nicotinamide protected acetylation of H3K9 (1.42 and 2.28-fold increase with 5.0 and 7.5 mmol/L nicotinamide, respectively, compared to vehicle control) and decreased H3K9me3 at 7.5 mmol/L concentration (relative H3K9me3 pixel intensity of 1.00 ± 0.00 versus 0.51 ± 0.05), differences were statistically significant.

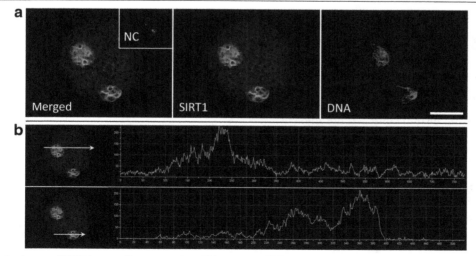

Fig. 1 Representative image of SIRT1 immunofluorescence in a 22-h zygote. **a** SIRT1 is exclusively localized in zygote pronuclei. **b** Yellow arrows indicate the level of signal intensity profiles in the respective maternal and paternal pronuclei. NC: negative control where the primary antibody was omitted. Scale bar represents 50 μm

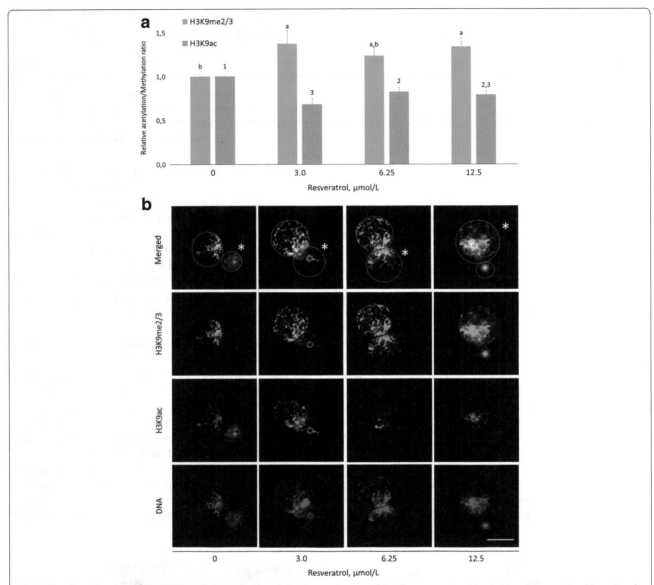

Fig. 2 The effect of resveratrol on H3K9 methylation and acetylation in zygotic pronuclei. **a** Fluorescent signal intensities relative to signal intensity of vehicle control (= 1) and **b** representative images of H3K9me3 and H3K9ac in both pronuclei. [a,b,c,1,2,3]Significant differences in H3K9me3 and H3K9ac, respectively, among experimental groups ($P \leq 0.05$). Asterisks indicate paternal pronucleus. Scalebar represents 25 μm

Similar to resveratrol, nicotinamide did not influence the outcomes of IVF (see Additional file 2).

Specific modulation of SIRT1 activity amplifies the modification of H3K9

Selective SIRT1 activator and inhibitor, BML-278 and sirtinol, respectively, were used for modulation of H3K9 methylation and acetylation in the treated and control zygotes (Figs. 4 and 5). Treatment with BML-278 increased the H3K9me3 and reciprocally decreased H3K9ac (change of 1.87 ± 0.18-fold and 0.58 ± 0.05-fold in H3K9me3 and H3K9ac, respectively, compared with control; $P \leq 0.05$),

(Fig. 4). In reverse, sirtinol significantly decreased H3K9me3 and increased H3K9ac to 0.70 ± 0.08 and 1.28 ± 0.12-fold change over control, respectively (Fig. 5). Similar to resveratrol and nicotinamide, the fertilization and monospermy/polyspermy rates were not affected (see Additional file 2).

SIRT1-modulation of H3K9 at the zygote stage improves subsequent embryonic development

To assess the effect of SIRT1-modulating treatments on preimplantation embryo development and blastocyst quality, fertilized oocytes were cultured in PZM3

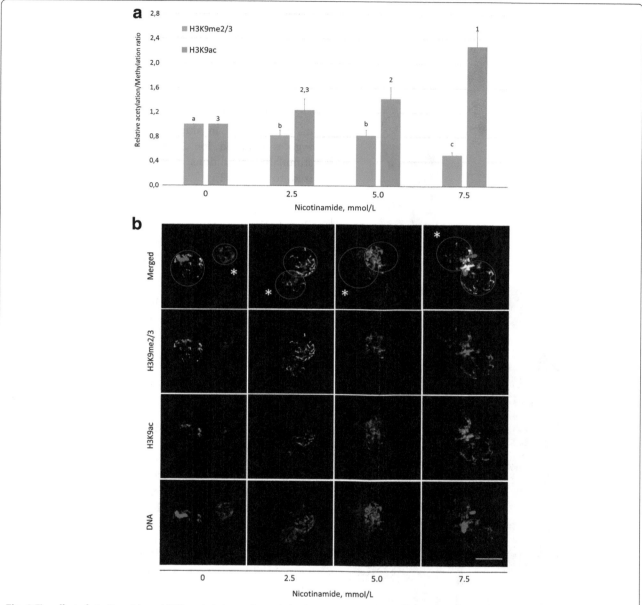

Fig. 3 The effect of nicotinamide on H3K9 methylation and acetylation in zygotic pronuclei. **a** Fluorescent signal intensities relative to vehicle control (= 1) and **b** representative images of H3K9me3 and H3K9ac in both pronuclei. [a,b,c,1,2,3]Significant differences in H3K9me3 and H3K9ac, respectively, among experimental groups ($P \leq 0.05$). Asterisk indicates paternal pronucleus. Scalebar represents 25 μm

medium with varying concentrations of SIRT1 activator or inhibitor for 144 h.

A significantly lower percentage (39.2 ± 5.2%) of cleaved zygotes was observed in embryos cultured with 12.5 μmol/L resveratrol compared to vehicle control treatment (65.9 ± 8.3%; $P \leq 0.05$) or to other activator/inhibitor treatments (Table 1). The inhibition of SIRT1 by sirtinol did not impact blastocyst formation rate. However, the rate of blastocyst formation was significantly increased when embryos were cultured in the presence of 3 μmol/L BML-278 (32.9 ± 8.1; $P \leq 0.05$). There

were no significant differences in average cell number per blastocyst among treatment groups (Additional file 3). Representative blastocyst of each treated group is shown in Additional file 3.

Specific activation of SIRT1 reduces the fluorescent immunolabeling intensity of the ooplasmic ubiquitin ligase MDM2

Based on published evidence of SIRT1 mediating the regulation of MDM2 and vice versa [37, 46], we detected MDM2 in matured MII oocytes due to western blotting

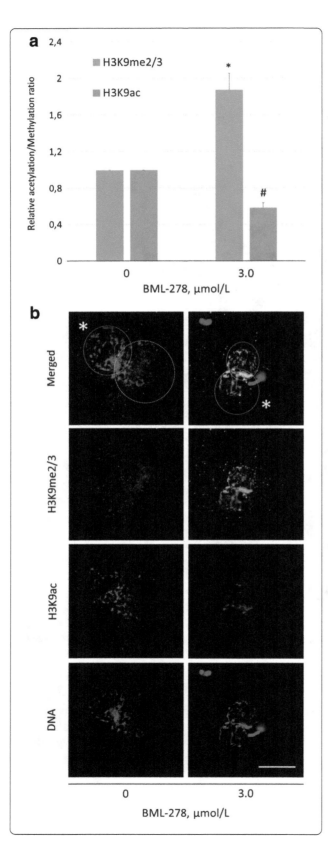

Fig. 4 The effect of BML-278 on H3K9 methylation and acetylation in the zygotic pronuclei. **a** Fluorescent signal intensities relative to vehicle control (= 1) and **b** representative images of H3K9me3 and H3K9ac in both pronuclei. *#Significant difference between control and treated group ($P \leq 0.05$). Asterisk indicates paternal pronucleus. Scalebar represents 25 μm

(see Additional file 1) and measured the fluorescent signal intensity of in situ immunolabeled MDM2 in the zygotes cultured for 22 h under control and SIRT1 stimulating conditions. The signal intensity was normalized to GAPDH. A significant reduction of GAPDH-normalized MDM2 signal intensity (1.13 ± 7.4 versus 0.62 ± 7.7 in vehicle control and treated zygotes; $P \leq 0.05$) was observed after SIRT1 activation with BML-278 (3 μmol/L) (Fig. 6), indicating significant decrease of MDM2 protein amount.

Discussion

Histone deacetylase SIRT1 has been shown to deacetylate a number of histone protein Lys-residues including but not limited to H1K26 [28, 29], H3 K9, H3K14 and H3K56 [28, 30], and H4K8, H4K12 and H4K16 [28, 31]. Moreover, SIRT1 activity leads to methylation of H3K9, and this lysine residuum represents a dual target of SIRT1 through both direct histone targeting and indirect non-histone targeting of other enzymes involved in the modifications of H3K9 (reviewed in [47]). Our study is the first to show that the SIRT1 accumulates in the pronuclei of a mammalian zygote and favors methylation of pronuclear H3K9. Our observations are in accordance with predominant SIRT1 presence in cell nucleus and an overall positive effect of SIRT1 activity on cell lifespan, accompanied with histone methylation, as described [28, 31]. Furthermore, we increased the percentage of blastocyst formation by embryo treatment with a specific, SIRT1 activating synthetic compound, BML-278 [48]. This observation agrees with previously documented beneficial effect of resveratrol, a naturally occurring general sirtuin activator, on preimplantation embryo development [24–26]. Our study thus solidifies previously anecdotal evidence of SIRT1 involvement in the establishment and remodeling of zygotic epigenome. In particular, we show that pharmacological modulation of SIRT1 alters pronuclear histone code in a manner affecting the developmental potential of the preimplantation embryo.

Treatment of porcine zygotes with resveratrol, a non-selective, SIRT-family wide activator increased H3K9me3 methylation after 22 h of IVC. Inversely, such treatment decreased H3K9 acetylation supporting the assumption that histone methylation and acetylation replace each other [49, 50]. Indeed, nicotinamide, a non-selective sirtuin inhibitor, had an opposite effect on pronuclear H3K9

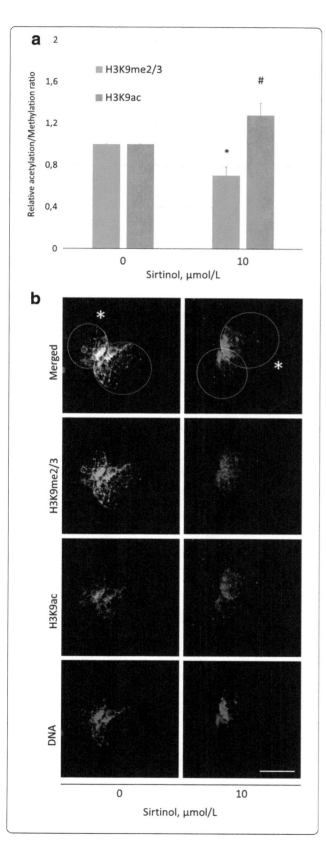

a

b

Sirtinol, μmol/L

Fig. 5 The effect of sirtinol on H3K9 methylation and acetylation in zygotic pronuclei. **a** Fluorescent signal intensities relative to vehicle control (= 1) and **b** representative images of H3K9me3 and H3K9ac in both pronuclei. *#Significant differences between control and treated group ($P \leq 0.05$). Asterisk indicates paternal pronucleus. Scalebar represents 25 μm

modification, wherein H3K9me3 was decreased and H3K9ac was protected in the treated zygotes. While the involvement of SIRT1 in aforementioned modifications of pronuclear H3K9 is plausible, such findings are based on wide-spectrum activation or inhibition of SIRT family, while other effects should be considered as well, such as the antioxidant/reactive oxygen-scavenger action of resveratrol [51]. Hence, BML-278 and sirtinol, the selective SIRT1 activator and inhibitor, respectively, were used. Favoring SIRT1 over SIRT2 or SIRT3 as a substrate [48], BML-278 treatment lead to higher H3K9 methylation and reduced H3K9 acetylation; sirtinol had an opposite, complementary effect. These observations agree with previous studies on sirtinol [19]. Compared to naturally occurring compounds, synthetic small molecules such as BML-278 and sirtinol have a more specific and more prominent effect on SIRT1 activity without secondary targets. Notably, BML-278 can be used for pharmacological modulation of SIRT1 signaling. On the other hand, molecular mechanism of BML-278 remains unknown and there are few alternative SIRT1 activators with suggested resveratrol-like allosteric mechanism mediating interactions between SIRT1 and its substrates [52, 53]. Activating effect of BML-278 is substrate-selective at a level similar to resveratrol [54, 55].

Early zygote features partially demethylated chromatin, resulting from pronuclear histone asymmetry [56, 57] and ongoing, active DNA demethylation [58–60, 14]. Based on the differential tagging of pronuclei with MitoTracker, we found the hyperacetylation of paternal pronucleus on H3K9 and higher H3K9me3 in maternal pronucleus, but no-detectable SIRT1 effect on the asymmetry of pronuclear histone code. The pattern of post-translationally modified histones in pronuclei mirrors DNA methylation status and points out the positive correlation of DNA- and histone H3K9 methylation, indicated earlier [61, 62]. Accordingly, DNA methyltransferase DNMT1 is regulated by H3K9me3-heterochromatin protein 1α complex and crosstalk between DNMTs and histone methyltransferases is obvious [63–65]. Although histone acetylation in general accompanies gene expression and in some cases causes DNA fragmentation [66], H3K9me3 is specifically associated with gene silencing and heterochromatin establishment [67, 68]. These changes are beneficial for chromatin stability and cell longevity; however, they suppress gene expression [67, 69, 70]. With respect to major zygotic genome activation

Table 1 Embryonic development and blastocyst formation after 144 h IVC with SIRT1 activators or inhibitors

		No. of fertilized oocytes	No. of cleaved oocytes (mean % ± SEM)	No. of blastocysts (mean % ± SEM)	Mean cell No. per blastocyst (mean ± SEM)
DMSO, % (v/v)	0.5	38	28 (65.9 ± 8.3)[a]	2 (5.2 ± 2.9)[bc]	36.0 ± 5.0[a]
Resveratrol, µmol/L	3	67	47 (67.7 ± 10.6)[a]	4 (6.2 ± 2.8)[bc]	39.3 ± 11.3[a]
	6.25	51	36 (69.9 ± 6.4)[a]	0 (0.0 ± 0.0)[c]	–
	12.5	50	20 (39.2 ± 5.2)[b]	0 (0.0 ± 0.0)[c]	–
Nicotinamide, mmol/L	2.5	50	39 (76.7 ± 6.9)[a]	7 (16.7 ± 8.8)[abc]	30.9 ± 4.2[a]
	5	70	45 (64.1 ± 1.4)[a]	7 (9.7 ± 6.2)[bc]	36.7 ± 4.5[a]
	7.5	69	45 (66.7 ± 4.6)[a]	13 (21.4 ± 8.4)[ab]	34.9 ± 3.1[a]
BML-278, µmol/L	3	68	44 (62.7 ± 7.1)[a]	18 (32.9 ± 8.1)[a]	38.4 ± 4.2[a]
Sirtinol, µmol/L	10	69	44 (63.5 ± 4.6)[a]	4 (6.5 ± 3.7)[bc]	23.5 ± 1.3[a]

[a,b,c]Different superscripts within the same column were significantly different at $P \leq 0.05$

(MZGA) when gene transcriptional activity is desired, SIRT1-derived histone methylation and gene silencing seems to be strictly selective in not affecting promoters and genes of which the expression is essential for early embryonic development.

In addition to histone targets of SIRT1, non-histone substrates may be involved in SIRT1 action in the zygote [71, 72]. Suppressor of variegation 3–9 homolog 1 (SUV39H1) is one of non-histone targets of SIRT1, responsible for SIRT1-mediated increase of H3K9me3 [33, 35]. SIRT1-regulated modification of SUV39H1 activity occurs by two mechanisms: SUV39H1 deacetylation on K266 [35] and indirect protection of SUV39H1 protein via suppression of its polyubiquitination by Mouse double minute 2 homolog (MDM2), which is an E3-type ubiquitin ligase responsible for proteasomal degradation of SUV39H1 [37]. The second mechanism points to a crosstalk of SIRT1 and UPS [73, 74] during SIRT1-modulation of the histone code. However, current knowledge of SIRT1 involvement in ubiquitin ligation and UPS-mediated proteolysis remains incomplete. Therefore, we performed the MDM2 fluorescent signal intensity measurement after BML-278 treatment. Our findings of decreased MDM2 signal intensity support the role of SIRT1 in MDM2-SUV39H1-H3K9me3 signaling, indicated earlier [37]. Based on such evidence, we can consider SIRT1 as a negative regulator of MDM2, presumably affecting MDM2 via lysine deacetylation and thus exposure of lysine residues for ubiquitination-mediated autocatalytic loop [37, 75, 76]. Such a scenario agrees with observation of SIRT-promoted protein degradation of β-TrPC E3 ligase [77].

Our study thus provides the partial understanding of how resveratrol improves embryonic development in vitro via SIRT1 activity modulation. The effect of resveratrol was compared with sirtuin inhibitors and BML-278, a specific SIRT1 activator. Surprisingly, resveratrol showed no positive effect on development to blastocyst, while nicotinamide, a non-selective

sirtuin family inhibitor actually increased the blastocyst formation rate. These results indicate non-specific effects of nicotinamide and are in agreement with the observations of improved embryonic development after treatment with trichostatine, an inhibitor of histone deacetylases [78–80]. Compared to resveratrol, BML-278, a specific SIRT1 activator, significantly improved early embryonic development. The wide scale modification of histone code favoring methylation of specific loci is a plausible epigenetic mechanism of SIRT1 action on early embryonic development. Moreover, SIRT1 factors in the sperm-derived transgenerational inheritance [81] and further experiments focused on specific SIRT1-affected loci during embryogenesis and SIRT1-modulated epigenetic memory are needed.

Improving blastocyst formation rate through SIRT1 stimulation could be applied to optimization of assisted reproductive technologies such as somatic cell nuclear transfer (SCNT) in animals and assisted fertilization in human infertility patients, wherein the maximizing of embryo developmental competence is paramount to the success of subsequent embryo transfer. Based on our observation, the adequate modulation of SIRT1 in the in vitro fertilized oocytes and embryos could also benefit the production of genetically modified pigs for biomedical research and production agriculture [82, 83]. This work thus offers a new approach to increasing the efficiency of reproductive biotechnologies for creation of biomedical animal models as well as for human assisted reproduction.

Further experiments will be needed to more precisely understand molecular mechanisms, including SIRT1 signal pathways, leading to successful pre-implanted embryonic development.

Conclusions

Specific activation of SIRT1 via BML-278 modulated zygotic histone code, expressed by methylated and acetylated

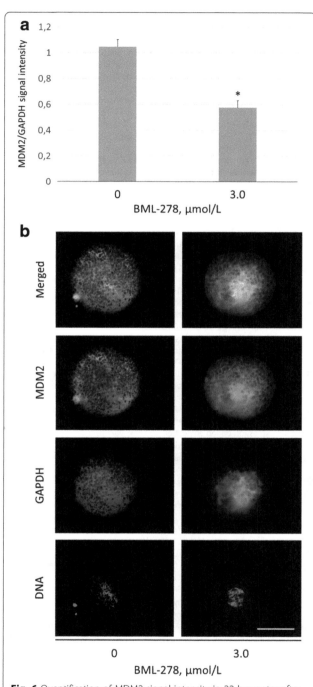

Fig. 6 Quantification of MDM2 signal intensity in 22 h zygotes after treatment with BML-278. **a** Fluorescent signal intensity of MDM2 was normalized against housekeeping protein GAPDH. **b** Representative images show MDM2 and GADPH in control and treated zygotes. *Significant difference between control and treated group (*P* ≤ 0.05). Scalebar represents 50 μm

described changes on histone code and non-histone targets in one-cell zygote. BML-278 as a novel compound and its molecular action through SIRT1 signalization in porcine embryo represents a promising tool utilizable in biotechnology and assisted reproduction of animals and human.

Additional file

Additional file 1: Detection of SIRT1 (A) and MDM2 (B) in matured MII oocytes and their cumulus cells.

Additional file 2: Results of IVF after 22 h of IVC with SIRT1 activators and inhibitors.

Additional file 3: Embryonic development and blastocyst formation after treatment with SIRT1 activators and inhibitors.

Abbreviations

DNMT: DNA methyltransferase; EGF: Epidermal growth factor; FSH: Follicle stimulating hormone; GAPDH: Glyceraldehyde-3-phosphate dehydrogenase; H3K9ac: Acetylation of histone H3 on lysine K9; H3K9me3: Trimethylation of histone H3 on lysine K9; IVC: In vitro culture of embryos; IVF: In vitro fertilization; LH: Luteinizing hormone; MDM2: Mouse double minute 2 homolog, E3-type ubiquitin ligase; MII: Metaphase II; mTBM: Modified Tris-buffered medium; mTCM: Modified tissue culture medium; MZGA: Major zygotic genome activation; PPN: Paternal pronucleus; PZM3: Porcine zygote medium; SCNT: Somatic cell nuclear transfer; SIRT1: Silent mating type information regulator 2 homolog 1, NADP + −dependent histone-deacetylase; SUV39H1: Suppressor of variegation 3–9 homolog 1, histone –lysine N-methyltransferase; TL-HEPES-PVA: HEPES-buffered Tyrode lactate medium with polyvinyl alcohol; UPS: Ubiquitin-proteasome system; β-TrPC E3 ligase: β-Transducin repeat-containing protein

Acknowledgements

We would like to thank Ms. Hee-Jung Lee for help with experiments and Ms. Kathryn Craighead for manuscript editing.

Funding

This work was supported by Agriculture and Food Research Initiative Competitive (Grant no. 2015–67,015-23,231) from the USDA National Institute of Food and Agriculture and by seed funding from the Food for the twenty-first Century program of the University of Missouri to P.S. Work in Y-J.Y. laboratory was supported by Basic Science Research Program through the National Research Foundation of Korea (NRF), funded by the Ministry of Education (NRF-2013R1A6A3A04063769), M.K. and J.N. was supported by Charles University (PROGRES Q-39) and the National Sustainability Program I (NPU I) Nr. LO1503 provided by the Ministry of Education, Youth and Sports of the Czech Republic, K.A., T.Z., K.H., J.P. and J.N. were supported by the National Agency of Agriculture Sciences (NAZV QJ1510138), the Czech Ministry of Agriculture (MZeRO 0714).

Authors' contributions

KZ, PJ and JN carried out the experiments and analysis of zygotes, TZ and KH participated in zygote production, YJY performed blastocyst production and following analysis, JM did proteomic analysis, MS participated the introduction of zygote analysis, KZ, YJY and JN did manuscript writing, JP, JN and PS designed experiments and edited the manuscript. MK, JP and PS provided wherewithal for all experiments. All authors read and approved the final manuscript.

Competing interests

The authors declare that they have no competing interests.

H3K9 in zygote pronuclei. In addition to increased pronucleic H3K9me3, activity of MDM2, a non-histone target of SIRT1, has been described after BML-278 treatment. The positive effect of 3.0 μmol/L BML-278 on blastocyst achievement is obvious and, assumably, is a result of

Author details

[1]Department of Veterinary Sciences, Faculty of Agriculture, Food and Natural Resources, Czech University of Life Sciences Prague, 6-Suchdol, Prague, Czech Republic. [2]Division of Biotechnology, Safety, Environment and Life Science Institute, College of Environmental and Bioresource Sciences, Chonbuk National University, Iksan 54596, South Korea. [3]Institute of Animal Science, 10-Uhrineves, Prague, Czech Republic. [4]Proteomic Laboratory, Biomedical Center of Faculty of Medicine in Pilsen, Charles University, Pilsen, Czech Republic. [5]Laboratory of Reproductive Medicine of Biomedical Center, Charles University, Pilsen, Czech Republic. [6]Department of Histology and Embryology, Faculty of Medicine in Pilsen, Charles University, Pilsen, Czech Republic. [7]Division of Animal Science, University of Missouri, Columbia, MO, USA. [8]Departments of Obstetrics, Gynecology and Women's Health, University of Missouri, Columbia, MO, USA.

References

1. Li L, Lu X, Dean J. The maternal to zygotic transition in mammals. Mol Asp Med. 2013;34:919–38.
2. Langley AR, Smith JC, Stemple DL, Harvey SA. New insights into the maternal to zygotic transition. Development. 2014;141:3834–41.
3. Nakazawa Y, Shimada A, Noguchi J, Domeki I, Kaneko H, Kikuchi K. Replacement of nuclear protein by histone in pig sperm nuclei during in vitro fertilization. Reproduction. 2002;124:565–72.
4. Ajduk A, Yamauchi Y, Ward MA. Sperm chromatin remodeling after intracytoplasmic sperm injection differs from that of in vitro fertilization. Biol Reprod. 2006;75:442–51.
5. Huo LJ, Fan HY, Liang CG, LZ Y, Zhong ZS, Chen DY, et al. Regulation of ubiquitin-proteasome pathway on pig oocyte meiotic maturation and fertilization. Biol Reprod. 2004;71:853–62.
6. Sutovsky P, Manandhar G, McCauley TC, Caamaño JN, Sutovsky M, Thompson WE, et al. Proteasomal interference prevents zona pellucida penetration and fertilization in mammals. Biol Reprod. 2004;71:1625–37.
7. Winking H, Gerdes J, Traut W. Expression of the proliferation marker Ki-67 during early mouse development. Cytogenet Genome Res. 2004;105:251–6.
8. van der Heijden GW, Derijck AA, Ramos L, Giele M, van der Vlag J, de Boer P. Transmission of modified nucleosomes from the mouse male germline to the zygote and subsequent remodeling of paternal chromatin. Dev Biol. 2006;298:458–69.
9. Adenot PG, Mercier Y, Renard JP, Thompson EM. Differential H4 acetylation of paternal and maternal chromatin precedes DNA replication and differential transcriptional activity in pronuclei of 1-cell mouse embryos. Development. 1997;124:4615–25.
10. Rouquier S, Taviaux S, Trask BJ, Brand-Arpon V, van den Engh G, Demaille J, et al. Distribution of olfactory receptor genes in the human genome. Nat Genet. 1998;18:243–50.
11. Torres-Padilla ME, Bannister AJ, Hurd PJ, Kouzarides T, Zernicka-Goetz M. Dynamic distribution of the replacement histone variant H3.3 in the mouse oocyte and preimplantation embryos. Int J Dev Biol. 2006;50:455–61.
12. Lindeman LC, Andersen IS, Reiner AH, Li N, Aanes H, Østrup O, et al. Prepatterning of developmental gene expression by modified histones before zygotic genome activation. Dev Cell. 2011;21:993–1004.
13. Van De Werken C, Van Der Heijden GW, Eleveld C, Teeuwssen M, Albert M, Baarends WM, et al. Paternal heterochromatin formation in human embryos is H3K9/HP1 directed and primed by sperm-derived histone modifications. Nat Commun. 2014;5:5868.
14. Guo F, Li X, Liang D, Li T, Zhu P, Guo H, et al. Active and passive demethylation of male and female pronuclear DNA in the mammalian zygote. Cell Stem Cell. 2014;15:447–58.
15. Tatone C, Di Emidio G, Vitti M, Di Carlo M, Santini S Jr, D'Alessandro AM, et al. Sirtuin Functions in Female Fertility: Possible Role in Oxidative Stress and Aging. Oxidative Med Cell Longev. 2015;2015:659687.
16. Cohen HY, Miller C, Bitterman KJ, Wall NR, Hekking B, Kessler B, et al. Calorie restriction promotes mammalian cell survival by inducing the SIRT1 deacetylase. Science. 2004;305:390–2.
17. Hayakawa T, Iwai M, Aoki S, Takimoto K, Maruyama M, Maruyama W, et al. SIRT1 suppresses the senescence-associated secretory phenotype through epigenetic gene regulation. PLoS One. 2015;10:e0116480.
18. Kawamura Y, Uchijima Y, Horike N, Tonami K, Nishiyama K, Amano T, et al. Sirt3 protects in vitro-fertilized mouse preimplantation embryos against

19. Kwak SS, Cheong SA, Yoon JD, Jeon Y, Hyun SH. Expression patterns of sirtuin genes in porcine preimplantation embryos and effects of sirtuin inhibitors on in vitro embryonic development after parthenogenetic activation and in vitro fertilization. Theriogenology. 2012a;78:1597–610.
20. Di Emidio G, Falone S, Vitti M, D'Alessandro AM, Vento M, Di Pietro C, et al. SIRT1 signaling protects mouse oocytes against oxidative stress and is deregulated during aging. Hum Reprod. 2014;29:2006–17.
21. Sato D, Itami N, Tasaki H, Takeo S, Kuwayama T, Iwata H. Relationship between mitochondrial DNA copy number and SIRT1 expression in porcine oocytes. PLoS One. 2014;9:e94488.
22. Zhao HC, Ding T, Ren Y, Li TJ, Li R, Fan Y, et al. Role of Sirt3 in mitochondrial biogenesis and developmental competence of human in vitro matured oocytes. Hum Reprod. 2016;31:607–22.
23. Riepsamen A, Wu L, Lau L, Listijono D, Ledger W, Sinclair DA, et al. Nicotinamide impairs entry into and exit from meiosis I in mouse oocytes. PLoS One. 2015;10: e0126194.
24. Lee K, Wang C, Chaille JM, Machaty Z. Effect of resveratrol on the development of porcine embryos produced in vitro. J Reprod Dev. 2010;56:330–5.
25. Kwak SS, Cheong SA, Jeon Y, Lee E, Choi KC, Jeung EB, et al. The effects of resveratrol on porcine oocyte in vitro maturation and subsequent embryonic development after parthenogenetic activation and in vitro fertilization. Theriogenology. 2012b;78:86–101.
26. Takeo S, Sato D, Kimura K, Monji Y, Kuwayama T, Kawahara-Miki R, et al. Resveratrol improves the mitochondrial function and fertilization outcome of bovine oocytes. J Reprod Dev. 2014;60:92–9.
27. Itami N, Shirasuna K, Kuwayama T, Iwata H. Resveratrol improves the quality of pig oocytes derived from early antral follicles through sirtuin 1 activation. Theriogenology. 2015;83:1360–7.
28. Vaquero A, Scher M, Lee D, Erdjument-Bromage H, Tempst P, Reinberg D. Human SirT1 interacts with histone H1 and promotes formation of facultative heterochromatin. Mol Cell. 2004;16:93–105.
29. Oberdoerffer P, Michan S, McVay M, Mostoslavsky R, Vann J, Park SK, et al. SIRT1 redistribution on chromatin promotes genomic stability but alters gene expression during aging. Cell. 2008;135:907–18.
30. Das C, Lucia MS, Hansen KC, Tyler JK, et al. Nature. 2009;459:113–7.
31. Vaquero A, Sternglanz R, Reinberg D. NAD+-dependent deacetylation of H4 lysine 16 by class III HDACs. Oncogene. 2007b;26:5505–20.
32. Khobta A, Anderhub S, Kitsera N, Epe B. Gene silencing induced by oxidative DNA base damage: association with local decrease of histone H4 acetylation in the promoter region. Nucleic Acids Res. 2010;38:4285–95.
33. Peters AH, Kubicek S, Mechtler K, O'Sullivan RJ, Derijck AA, Perez-Burgos L, et al. Partitioning and plasticity of repressive histone methylation states in mammalian chromatin. Mol Cell. 2003;12:1577–89.
34. Ait-Si-Ali S, Guasconi V, Fritsch L, Yahi H, Sekhri R, Naguibneva I, et al. A Suv39h-dependent mechanism for silencing S-phase genes in differentiating but not in cycling cells. EMBO J. 2004;23:605–15.
35. Vaquero A, Scher M, Erdjument-Bromage H, Tempst P, Serrano L, Reinberg D. SIRT1 regulates the histone methyl-transferase SUV39H1 during heterochromatin formation. Nature. 2007a;450:440–4.
36. Park KE, Johnson CM, Wang X, Cabot RA. Differential developmental requirements for individual histone H3K9 methyltransferases in cleavage-stage porcine embryos. Reprod Fertil Dev. 2011;23:551–60.
37. Bosch-Presegué L, Raurell-Vila H, Marazuela-Duque A, Kane-Goldsmith N, Valle A, Oliver J, et al. Stabilization of Suv39H1 by SirT1 is part of oxidative stress response and ensures genome protection. Mol Cell. 2011;42:210–23.
38. Abeydeera LR, Wang WH, Cantley TC, Prather RS, Day BN. Presence of beta-mercaptoethanol can increase the glutathione content of pig oocytes matured in vitro and the rate of blastocyst development after in vitro fertilization. Theriogenology. 1998;50:747–56.
39. Yoshioka K, Suzuki C, Tanaka A, Anas IM, Iwamura S. Birth of piglets derived from porcine zygotes cultured in a chemically defined medium. Biol Reprod. 2002;66:112–9.
40. Yi YJ, Sutovsky M, Song WH, Sutovsky P. Protein deubiquitination during oocyte maturation influences sperm function during fertilization, antipolyspermy defense and embryo development. Reprod Fertil Dev. 2015;27:1154–67.
41. Kuijk EW, du Puy L, van Tol HT, Haagsman HP, Colenbrander B, Roelen BA. Validation of reference genes for quantitative RT-PCR studies in porcine oocytes and preimplantation embryos. BMC Dev Biol. 2007;7:58.

42. Nevoral J, Žalmanová T, Zámostná K, Kott T, Kučerová-Chrpová V, Bodart JF, et al. Endogenously produced hydrogen sulfide is involved in porcine oocyte maturation in vitro. Nitric Oxide. 2015;51:24–35.

43. Hou J, Liu L, Zhang J, Cui XH, Yan FX, Guan H, et al. Epigenetic modification of histone 3 at lysine 9 in sheep zygotes and its relationship with DNA methylation. BMC Dev Biol. 2008;8:60.

44. Kan R, Jin M, Subramanian V, Causey CP, Thompson PR, Coonrod SA, et al. Potential role for PADI-mediated histone citrullination in preimplantation development. BMC Dev Biol. 2012;12:19.

45. Oliveira CS, Saraiva NZ, de Souza MM, Tetzner TA, de Lima MR, Garcia JM, et al. Effects of histone hyperacetylation on the preimplantation development of male and female bovine embryos. Reprod Fertil Dev. 2010;22:1041–8.

46. Peng L, Yuan Z, Li Y, Ling H, Izumi V, Fang B, et al. Ubiquitinated sirtuin 1 (SIRT1) function is modulated during DNA damage-induced cell death and survival. J Biol Chem. 2015;290:8904–12.

47. Martínez-Redondo P, Vaquero A. The diversity of histone versus nonhistone sirtuin substrates. Genes Cancer. 2013;4(3–4):148–63.

48. Mai A, Valente S, Meade S, Carafa V, Tardugno M, Nebbioso A, et al. Study of 1,4-dihydropyridine structural scaffold: discovery of novel sirtuin activators and inhibitors. J Med Chem. 2009;52:5496–504.

49. Nicolas E, Roumillac C, Trouche D. Balance between acetylation and methylation of histone H3 lysine 9 on the E2F-responsive dihydrofolate reductase promoter. Mol Cell Biol. 2003;23:1614–22.

50. Stewart MD, Li J, Wong J. Relationship between histone H3 lysine 9 methylation, transcription repression, and heterochromatin protein 1 recruitment. Mol Cell Biol. 2005;25:2525–38.

51. Kong Q, Ren X, Hu R, Yin X, Jiang G, Pan Y. Isolation and purification of two antioxidant isomers of resveratrol dimer from the wine grape by counter-current chromatography. J Sep Sci. 2016;39:2374–9.

52. Hubbard BP, Sinclair DA. Small molecule SIRT1 activators for the treatment of aging and age-related diseases. Trends Pharmacol Sci. 2014;35:146–54.

53. Cao D, Wang M, Qiu X, Liu D, Jiang H, Yang N, et al. Structural basis for allosteric, substrate-dependent stimulation of SIRT1 activity by resveratrol. Genes Dev. 2015;29:1316–25.

54. Baur JA. Biochemical effects of SIRT1 activators. Biochim Biophys Acta. 2010;1804:1626–34.

55. Lakshminarasimhan M, Rauh D, Schutkowski M, Steegborn C. Sirt1 activation by resveratrol is substrate sequence-selective. Aging (Albany NY). 2013;5:151–4.

56. Lepikhov K, Walter J. Differential dynamics of histone H3 methylation at positions K4 and K9 in the mouse zygote. BMC Dev Biol. 2004;4:12.

57. Ma XS, Chao SB, Huang XJ, Lin F, Qin L, Wang XG, et al. The Dynamics and Regulatory Mechanism of Pronuclear H3k9me2 Asymmetry in Mouse Zygotes. Sci Rep. 2015;5:17924.

58. Oswald J, Engemann S, Lane N, Mayer W, Olek A, Fundele R, et al. Active demethylation of the paternal genome in the mouse zygote. Curr Biol. 2000;10:475–8.

59. Park JS, Jeong YS, Shin ST, Lee KK, Kang YK, Dynamic DNA. methylation reprogramming: active demethylation and immediate remethylation in the male pronucleus of bovine zygotes. Dev Dyn. 2007;236:2523–33.

60. Reis Silva AR, Adenot P, Daniel N, Archilla C, Peynot N, Lucci CM, et al. Dynamics of DNA methylation levels in maternal and paternal rabbit genomes after fertilization. Epigenetics. 2011;6:987–93.

61. Liu H, Kim JM, Aoki F. Regulation of histone H3 lysine 9 methylation in oocytes and early pre-implantation embryos. Development. 2004;131:2269–80.

62. Timoshevskiy VA, Herdy JR, Keinath MC, Smith JJ. Cellular and Molecular Features of Developmentally Programmed Genome Rearrangement in a Vertebrate (Sea Lamprey: Petromyzon marinus). PLOS Genet. 2016;12:e1006103.

63. Lehnertz B, Ueda Y, Derijck AA, Braunschweig U, Perez-Burgos L, Kubicek S, et al. Suv39h-mediated histone H3 lysine 9 methylation directs DNA methylation to major satellite repeats at pericentric heterochromatin. Curr Biol. 2003; 13:1192–200.

64. Sun L, Huang L, Nguyen P, Bisht KS, Bar-Sela G, Ho AS, et al. DNA methyltransferase 1 and 3B activate BAG-1 expression via recruitment of CTCFL/BORIS and modulation of promoter histone methylation. Cancer Res. 2008;68:2726–35.

65. Rai K, Jafri IF, Chidester S, James SR, Karpf AR, Cairns BR, et al. Dnmt3 and G9a cooperate for tissue-specific development in zebrafish. J Biol Chem. 2010;285:4110–21.

66. Bikond Nkoma G, Leduc F, Jaouad L, Boissonneault G. Electron microscopy analysis of histone acetylation and DNA strand breaks in mouse elongating spermatids using a dual labelling approach. Andrologia. 2010;42:322–5.

67. Wang F, Kou Z, Zhang Y, Gao S. Dynamic reprogramming of histone acetylation and methylation in the first cell cycle of cloned mouse embryos. Biol Reprod. 2007;77:1007–16.

68. Keniry A, Gearing LJ, Jansz N, Liu J, Holik AZ, Hickey PF, et al. Setdb1-mediated H3K9 methylation is enriched on the inactive X and plays a role in its epigenetic silencing. Epigenetics Chromatin. 2016;9:16.

69. Ryu HY, Rhie BH, Ahn SH. Loss of the Set2 histone methyltransferase increases cellular lifespan in yeast cells. Biochem Biophys Res Commun. 2014;446:113–8.

70. Li S, Liu L, Li S, Gao L, Zhao Y, Kim YJ, et al. SUVH1, a Su(var)3-9 family member, promotes the expression of genes targeted by DNA methylation. Nucleic Acids Res. 2016;44:608–20.

71. Yu X, Zhang L, Wen G, Zhao H, Luong LA, Chen Q, et al. Upregulated sirtuin 1 by miRNA-34a is required for smooth muscle cell differentiation from pluripotent stem cells. Cell Death Differ. 2015;22:1170–80.

72. Zhao H, Yang L, Cui H, et al. Biochem Biophys Res Commun. 2015;464:1163–70.

73. Chen IY, Lypowy J, Pain J, Sayed D, Grinberg S, Alcendor RR, et al. Histone H2A.z is essential for cardiac myocyte hypertrophy but opposed by silent information regulator 2alpha. J Biol Chem. 2006;281:19369–77.

74. Han L, Zhao G, Wang H, Tong T, Chen J. Calorie restriction upregulated sirtuin 1 by attenuating its ubiquitin degradation in cancer cells. Clin Exp Pharmacol Physiol. 2014;41:165–8.

75. Roxburgh P, Hock AK, Dickens MP, Mezna M, Fischer PM, Vousden KH. Small molecules that bind the Mdm2 RING stabilize and activate p53. Carcinogenesis. 2012;33:791–8.

76. Nihira NT, Ogura K, Shimizu K, North BJ, Zhang J, Gao D, et al. Acetylation-dependent regulation of MDM2 E3 ligase activity dictates its oncogenic function. Sci Signal. 2017;10:eaai8026.

77. Woo SR, Byun JG, Kim YH, Park ER, Joo HY, Yun M, et al. SIRT1 suppresses cellular accumulation of β-TrCP E3 ligase via protein degradation. Biochem Biophys Res Commun. 2013;441:831–7.

78. Inoue K, Oikawa M, Kamimura S, Ogonuki N, Nakamura T, Nakano T, et al. Trichostatin A specifically improves the aberrant expression of transcription factor genes in embryos produced by somatic cell nuclear transfer. Sci Rep. 2015;5:10127.

79. Jee BC, Jo JW, Lee JR, Suh CS, Kim SH, Moon SY. Effect of trichostatin A on fertilization and embryo development during extended culture of mouse oocyte. Zygote. 2012;20:27–32.

80. Jeseta M, Petr J, Krejcová T, Chmelíková E, Jílek F. In vitro ageing of pig oocytes: effects of the histone deacetylase inhibitor trichostatin A. Zygote. 2008;16:145–52.

81. Rodgers AB, Morgan CP, Leu NA, Bale TL. Transgenerational epigenetic programming via sperm microRNA recapitulates effects of paternal stress. Proc Natl Acad Sci U S A. 2015;112:13699–704.

82. Mao J, Zhao MT, Whitworth KM, Spate LD, Walters EM, O'Gorman C, et al. Oxamflatin treatment enhances cloned porcine embryo development and nuclear reprogramming. Cell Reprogram. 2015;17:28–40.

83. Whitworth KM, Mao J, Lee K, Spollen WG, Samuel MS, Walters EM, et al. Transcriptome analysis of pig in vivo, in vitro-fertilized, and nuclear transfer blastocyst-stage embryos treated with histone deacetylase inhibitors postfusion and activation reveals changes in the lysosomal pathway. Cell Reprogram. 2015;17:243–58.

18

Stress and immunological response of heifers divergently ranked for residual feed intake following an adrenocorticotropic hormone challenge

A. K. Kelly[1*], P. Lawrence[3], B. Earley[2], D. A. Kenny[2] and M. McGee[3]

Abstract

Background: When an animal is exposed to a stressor, metabolic rate, energy consumption and utilisation increase primarily through activation of the hypothalamic-pituitary-adrenal (HPA) axis. Changes to partitioning of energy by an animal are likely to influence the efficiency with which it is utilised. Therefore, this study aimed to determine the physiological stress response to an exogenous adrenocorticotropic hormone (ACTH) challenge in beef heifers divergently ranked on phenotypic residual feed intake (RFI).

Results: Data were collected on 34 Simmental weaning beef heifers the progeny of a well characterized and divergently bred RFI suckler beef herd. Residual feed intake was determined on each animal during the post-weaning stage over a 91-day feed intake measurement period during which they were individually offered adlibitum grass silage and 2 kg of concentrate per head once daily. The 12 highest [0.34 kg DM/d] and 12 lowest [−0.48 kg DM/d] ranking animals on RFI were selected for use in this study. For the physiological stress challenge heifers (mean age 605 ± 13 d; mean BW 518 ± 31.4 kg) were fitted aseptically with indwelling jugular catheters to facilitate intensive blood collection. The response of the adrenal cortex to a standardised dose of ACTH (1.98 IU/kg metabolic $BW^{0.75}$) was examined. Serial blood samples were analysed for plasma cortisol, ACTH and haematology variables. Heifers differing in RFI did not differ ($P = 0.59$) in ACTH concentrations. Concentration of ACTH peaked ($P < 0.001$) in both RFI groups at 20 min post-ACTH administration, following which concentration declined to baseline levels by 150 min. Similarly, cortisol systemic profile peaked at 60 min and concentrations remained continuously elevated for 150 min. A RFI × time interaction was detected for cortisol concentrations ($P = 0.06$) with high RFI heifers had a greater cortisol response than Low RFI from 40 min to 150 min relative to ACTH administration. Cortisol response was positively associated with RFI status ($r = 0.32$; $P < 0.01$). No effect of RFI was evident for neutrophil, lymphocytes, monocyte, eosinophils and basophil count. Plasma red blood cell number (6.07 vs. 6.23; $P = 0.02$) and hematocrit percentage (23.2 vs. 24.5; $P = 0.02$) were greater for low than high RFI animals.

Conclusions: Evidence is provided that feed efficiency is associated with HPA axis function and susceptibility to stress, and responsiveness of the HPA axis is likely to contribute to appreciable variation in the efficiency feed utilisation of cattle.

Keywords: Beef cattle, Cortisol, Feed efficiency, Residual feed intake, Stress response

* Correspondence: alan.kelly@ucd.ie
[1]School of Agriculture and Food Science, College of Health and Agricultural Sciences, University College Dublin, Belfield, Dublin 4, Ireland
Full list of author information is available at the end of the article

Background

Selecting and propagating cattle genetics that are more efficient at converting feed into carcass gain is an important element of a profitable and sustainable livestock industry [1]. As a moderately heritable trait, independent of growth and body size, RFI is now a favoured measure of feed efficiency for livestock [2]. Independence of RFI from production also affords the opportunity to unravel the inherent variation in biological processes underpinning differences in inter-animal efficiency. Various biological processes have been identified as possible contributors to variation in feed efficiency [2–5], but no one process predominates.

Susceptibility to stress and biological differences in an animal's stress response is proposed [6, 7] as probable drivers of variation in feed efficiency that warrant further investigation. Indeed, an association between stress susceptibility and energetic inefficiency is conceivable, given that one of the noted biological responses to a stressful stimuli is to increase metabolic rate, energy consumption and decrease immunity, through alterations in the functioning of neuro-endocrine-immune pathways [8, 9].

Stress response of animals is generally determined by measuring glucocorticoids and HPA axis activity [10]. The activity of the HPA axis can be stimulated by way of a physiological stress challenge (following exogenous administration of ACTH and/or corticotrophin-releasing hormone; CRH) and the stress response of the HPA can be assessed at the pituitary and adrenal levels. Knott et al. [6] reported that RFI and cortisol response were positively associated, in rams selected based on their response to an ACTH physiological stress challenge. In laying hens originating from a multi-generational divergent RFI selection line, Luiting et al. [11] reported that the inefficient hen lines had a greater cortisol response following an adrenal ACTH challenge compared to their efficient contemporaries. However, from a cattle perspective there is a dearth of published information on this topic. Therefore, the objective of this study was to determine whether endocrine and hematologic responses following an exogenous ACTH challenge differ between beef heifers divergent for phenotypic RFI.

Methods

All procedures involving animals in this study were conducted under an experimental licence from the Irish Department of Health and Children in accordance with the cruelty to Animals Act 1876 and the European Communities (Amendment of Cruelty to Animals Act 1876) Regulation 2002 and 2005 (http://www.dohc.ie/other_-health_issues/uaeosp).

Animal management and determination of RFI

Performance and RFI data were collected on 34 Simmental weaning beef heifers which were the progeny of a well characterized and divergently bred RFI suckler beef herd established in Teagasc Grange, as described by [12–14]. During the post-weaning stage individual daily feed intake was recorded on heifers (mean initial BW = 299 ± 4 kg, mean initial age = 258 ± 27 d) over a 91-day period, following which RFI co-efficients were calculated for each animal. Heifers were housed in a slatted floor building with a Calan gate feeding system (American Calan Inc., Northwood, NH) and were given an adaptation period of 14 d to acclimatise to their diet and environment before recording individual intake began (RFI measurement period). Heifers were individually offered grass silage and 2 kg of concentrate once per day (at 0800 h). The concentrate offered contained 430 kg rolled barley, 430 kg beet pulp, 80 kg soya bean meal, 45 kg molasses and 15 kg minerals and vitamins per tonne. Grass silage allocation was based on approximately 1.1 times the previous day's intake. The chemical composition and in vitro dry matter digestibility of the grass silage and concentrate offered is outlined in Table 1.

Heifers were weighed (prior to feeding) on two consecutive days at the beginning and end of the period of RFI measurement and every 21 d during the experimental period. Average daily gain was computed as the coefficient of the linear regression of weight (kg) on time using the REG procedure of the Statistical Analysis Systems (SAS Institute Inc., Cary, NC). Mid-test metabolic body weight (MBW) was estimated from the intercept and slope of the regression line after fitting a linear regression through all metabolic body weight (BW$^{0.75}$) observations. Feed conversion ratio (F:G) of each animal was computed as the ratio of daily DMI to ADG. Residual feed intake was calculated for each animal as the difference between actual DMI and expected DMI. Expected DMI was computed for each animal using a multiple regression model, regressing DMI on MBW and ADG. The model used was

$$Y_j = \beta_0 + \beta_1 MLW_j + \beta_2 ADG_j + e_j,$$

Where Y_j is the average DMI of the jth animal, β_0 is the partial regression intercept, β_1 is the partial regression coefficient on MBW$^{0.75}$, β_2 is the partial regression coefficient on ADG, and e_j is the uncontrolled error of the jth animal. The coefficient of determination (R^2) from this model was equal to 0.66 ($P < 0.001$) and the model was subsequently used to predict DMI for each animal.

Heifers were turned out to pasture after the RFI experimental period ended. For the duration of the grazing season (7 mo), heifers were rotationally grazed as one

Table 1 Chemical composition of the dietary ingredients offered during the residual feed intake experimental period

Variable	RFI experimental period	
	Silage	Concentrate
Dry matter, g/kg[a]	315 ± 33.9	822 ± 0.7
Composition of DM, g/kg DM unless otherwise stated		
pH	3.9 ± 0.15	ND[b]
In vitro DMD	693 ± 47.3	726 ± 44.0
In vitro DOMD[c]	625 ± 37.1	676 ± 44.6
OMD[d]	677 ± 47.5	726 ± 46.9
Ash	75 ± 11.7	73 ± 1.6
Crude protein	121 ± 10.3	115 ± 11.6
NDF	532 ± 54.7	244 ± 31.4
Starch	ND	230 ± 56.3
ME, MJ/kg DM[e]	10.41 ± 0.45	11.56 ± 0.36
Fermentation characteristics[f], g/kg		
Lactic acid	102 ± 49.8	ND
Acetic acid	33 ± 7.7	ND
Propionic acid	2.9 ± 0.66	ND
Butyric acid	4.5 ± 2.39	ND
Ethanol	13.4 ± 2.82	ND
Ammonia N, g/kg total N	76 ± 10.9	ND

[a]Corrected for loss of volatiles during oven drying
[b]ND not determined
[c]Digestible OM in the total DM, measured in vitro
[d]OM digestibility, measured in vitro
[e]Estimated based on in vitro digestible OM in total DM (AFRC, 1993)
[f]Expressed as g/kg volatile corrected DM

group under a moderate stocking rate on predominately perennial ryegrass (*Lolium perenne* L.) pasture until housing at the end of October, when the grazing season ended [mean ADG of 0.80 kg/d (SEM = 0.03)]. For the physiological stress challenge experiment the 12 highest [mean (0.48 ± 0.08) kg/d]; High RFI and 12 lowest [mean – (0.50 ± 0.08) kg/d]; Low RFI ranking animals on post weaning RFI were selected.

Jugular vein catheterisation

During the ACTH stress challenge the 24 heifers (mean age 605 ± 13 d; mean BW 516 ± 31.4 kg) were housed in a slatted-floor facility and were offered grass silage ad libitum. Start weight ($P = 0.52$) and age ($P = 0.88$) were not different between the high and low RFI groups. As part of standard husbandry and research management practices animals were accustomed to stockpersons and routine handling, approximately 14-day prior to the ACTH challenge. To facilitate intensive blood collection and minimize handling stress during blood sampling, heifers were fitted aseptically with indwelling jugular catheters on d –1. The procedure was using 12-gauge Anes spinal needles (Popper and Sons, Inc., New Hyde

Park, NY) and polyvinyl tubing (approximately 1.47 mm i.d.; Ico Rally Corp., Palo Alto, CA; catalog No. SVL 105–18 CLR) attached to an 18-gauge needle at the blood collection end. After catheterization, catheter patency was maintained by flushing with 3.5% sodium citrate after each blood collection.

ACTH challenge

The response of the adrenal cortex to a standardised dose of ACTH (1.98 IU/kg metabolic $BW^{0.75}$) was examined. The dose chosen of exogenous ACTH and was previously established by our group [15] to be effective in adequately stimulating HPA responses, with resultant elevations in cortisol not sufficient to cause negative effects on animal health. Dexamethasone (20 μg/kg BW; Faulding Pharmaceuticals Plc, UK) was administered intramuscularly (i.m.) on d –1, 12.40 h prior to heifers undergoing ACTH challenge (Synacthen Ampoules, Novartis Pharmaceutical Ltd., UK) on the following morning (d 0). In cattle, DEX is rapidly absorbed and the elimination half-life of DEX has been reported by Toutain et al. to ranges from 291 to 335 min [16, 17]. Dexamethasone was administered to equalize systemic concentrations of cortisol [9] in animals across the RFI groups, in an effort to facilitate a more equitable examination of ACTH on cortisol release [18, 19].

On d 0, serial heparinised blood samples were collected at –40, –20, 0, 20, 40, 60, 80, 100, 120, 150, 180, 210, 240, 270, 300, 330, and 390 min relative to the time of ACTH administration to each heifer. Plasma samples were separated by centrifugation at $1600 \times g$ at 8 °C for 15 min and subsequently stored at –20 °C until assayed.

Cortisol concentration was determined for all heifers at all blood sampling time points, whereas ACTH concentration was determined only at blood sampling time points –40, –20, 20, 80, 120, 180, and 390 min relative to ACTH administration. Plasma cortisol concentrations were determined using the Correlate-EIA kits from Assay Designs (Ann Arbor, MI, USA). The inter- and intra- assay coefficients of variation (CV) were 6.25% and 4.89%, respectively. Plasma ACTH concentrations were measured using a double antibody human ACTH radioimmunoassay (RIA; DiaSorin Inc., Stillwater, MN, USA) previously validated for bovine ACTH [15]. The inter- assay and intra- assay CVs were 8.85% and 7.67%, respectively.

Hematology

An additional blood sample was collected into a 6-mL evacuated tube containing K_3EDTA at –40, –20, 0, 20, 80, 150, 270, 330, and 390 min relative to CRH administration for determination of blood hematology variables; white blood cell (WBC) number, red blood cell (RBC) number, hemoglobin (HGB) concentration, hematocrit percentage (HCT %), platelet count (PLT) and total circulating neutrophil, lymphocyte, monocyte, eosinophil, and basophil numbers were determined within 1 h of collection according to the procedures described by Lynch et al. [19].

Statistical analysis

Data were checked for normality and homogeneity of variance by histograms, QQ-plots, and formal statistical tests as part of the UNIVARIATE procedure of SAS (version 9.1.3; SAS Institute, 2006). Data that were not normally distributed were transformed by raising the variable to the power of lambda. The appropriate lambda value was obtained by conducting a Box-Cox transformation analysis using the TRANSREG procedure of SAS. The transformed data were used to calculate P-values. The corresponding least squares means and SE of the non-transformed data are presented in the results for clarity. To evaluate the response over time for plasma concentrations of cortisol, ACTH and haematological variables, these were analysed using repeated measures ANOVA (MIXED procedure), with terms for RFI group, bleed time, pen and their interactions included in the model. Heifer baseline plasma concentrations (mean of −40, −20 and 0 min prior to ACTH administration for cortisol and haematology variables and −40 and −20 for plasma ACTH concentration) were included as covariates. The type of variance-covariance structure used was chosen depending on the magnitude of the Akaike information criterion (AIC) for models run under compound symmetry, unstructured, autoregressive, heterogeneous 1st order autoregressive, or Toeplitz variance-covariance structures. The model with the least AIC value was selected. The differences between mean values for the two RFI groups were determined by F-tests using Type III sums of squares. The PDIFF option and the Tukey test were applied as appropriate to evaluate pairwise comparisons between RFI group means. A probability of $P < 0.05$ was selected as the level of significance and statistically tendencies were reported when $P < 0.10$. Spearman correlation coefficients between traits were determined using PROC CORR of SAS.

Results

Feed efficiency

Based upon the results of the post-weaning determination of RFI, the low RFI heifers selected for the ACTH challenge had a 19% reduced DMI compared to the high selected RFI animals for the same level of performance. Residual feed intake averaged 0.00 kg DM/d (SD = 0.43) and ranged from −0.87 to 1.02 kg DM/d, equating to a difference of 1.89 kg DM/d between the most and least efficient ranking heifers used in the ACTH challenge . Least squares means for RFI ($P < 0.001$) and F:G

Table 2 Effect of residual feed intake (RFI) ranking on feed intake, efficiency and animal performance in beef heifers

	High RFI	Low RFI	SEM	P-value
Number of Animals	12	12	–	–
Residual feed intake, kg/d	0.34	−0.48	0.037	0.001
Feed conversion ratio, kg DM/kg gain	8.13	6.19	0.834	0.03
DMI, kg/d	5.92	4.98	0.127	0.001
Mid-test metabolic BW, kg$^{0.75}$	75	77	1.6	0.27
ADG, kg/d	0.71	0.76	0.028	0.92
Final BW, kg	340	355	9.8	0.21

($P = 0.03$; Table 2) were greater for high RFI than for low RFI animals. Heifers in the high RFI and low RFI groups did not differ in final BW ($P = 0.21$) and ADG ($P = 0.92$). Within the group of weanling heifers RFI was positively correlated with DMI ($r = 0.59$; $P < 0.001$). Dry matter intake was positively correlated ($P < 0.001$) with ADG ($r = 0.55$) and negatively correlated ($P < 0.05$) with F:G ($r = −0.27$). Feed conversion ratio was negatively correlated ($P < 0.001$) with ADG during the RFI measurement period ($r = −0.85$).

Exogenous ACTH challenge

Plasma ACTH and cortisol responses to administration of exogenous ACTH for both RFI groupings are presented Figs. 1 and 2, respectively. Heifers differing in RFI did not differ ($P = 0.59$) in ACTH concentration. Concentration of ACTH peaked ($P < 0.001$) in both RFI groups at 20 min post-ACTH administration, following which concentration declined to baseline levels by 150 min. Similarly, cortisol systemic profile peaked at 60 min and concentrations remained continuously elevated for 150 min. A RFI × time interaction was detected for cortisol concentrations ($P = 0.06$). High RFI heifers had a greater cortisol response than Low RFI from 40 min to 150 min relative to ACTH administration (Fig. 3), whereas there was no difference between RFI grouping subsequently. Cortisol response, from ACTH adrenal stimulation was positively associated with RFI status ($r = 0.32$; $P < 0.01$).

Haematology

No effect of RFI grouping or a RFI × time interaction ($P > 0.10$; Table 3) was evident for neutrophil, lymphocytes, monocyte, eosinophils and basophil count. Mean plasma RBC number ($P = 0.02$) and HCT percentage ($P = 0.02$) were greater for low RFI than high RFI animals (RBC = 6.07 vs. 6.23; HCT = 23.2 vs. 24.5 for high, and low RFI groups, respectively). An effect of time ($P < 0.0001$; Table 3) was found neutrophil, monocyte,

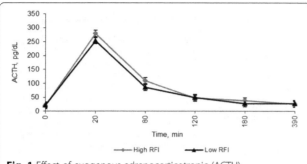

Fig. 1 Effect of exogenous adrenocorticotropic (ACTH) administration (1.98 IU/kg metabolic BW$^{0.75}$) on plasma ACTH concentration (pg/dL) in beef heifers differing in phenotypic RFI. The values are expressed as Lsmeans with SEM. RFI, $P = 0.59$; Time, $P < 0.001$; RFI × Time interaction, $P = 0.13$. SEM: High RFI = 4.98; Low RFI = 6.55

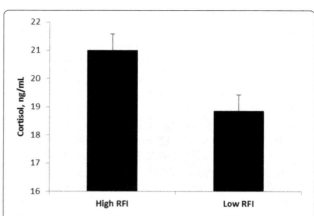

Fig. 3 Mean plasma cortisol stress response from 40 min to 150 min relative to exogenous adrenocorticotropic (ACTH) administration (1.98 IU/kg metabolic BW$^{0.75}$) in beef heifers differing in phenotypic residual feed intake (RFI). The values are expressed as Lsmeans with SEM. RFI, $P = 0.006$. SEM: High RFI = 0.57; Low RFI = 0.52

eosinophils and basophil count and for PLT number and HGB with values generally increasing for leucocytes and monocytes, and decreasing for PLT and HGB as time progressed.

Discussion

Understanding the biological mechanisms that control feed efficiency is fundamental to the future of profitable livestock production systems [1, 20]. Susceptibility to stress and the cross regulatory responses between the neuro-endocrine and immune systems have long been postulated as potential processes that contributes to inter- animal variation in feed efficiency in cattle [18]. However, despite the credible link between these traits there is a dearth of published work in this area, which is surprising given the importance of feed efficiency and susceptibility to stress (the implications on health and behaviour) to overall animal production efficiency [2]. Therefore, this study aimed to further enhance our knowledge of the biological control of feed efficiency in

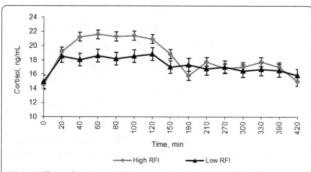

Fig. 2 Effect of exogenous adrenocorticotropic (ACTH) administration (1.98 IU/kg metabolic BW$^{0.75}$) on plasma cortisol concentration (ng/mL) in beef heifers differing in phenotypic residual feed intake (RFI). The values are expressed as Lsmeans with SEM. RFI, $P = 0.45$; Time, $P < 0.001$; RFI × Time interaction, $P = 0.06$. SEM: High RFI = 0.712; Low RFI = 0.688

cattle by investigating the sensitivity of the HPA axis (a putative regulator of energetic efficiency) to a physiological ACTH stress challenge.

HPA axis and feed efficiency

Activation of the HPA axis is the main defining feature of the stress response [21] and one of the main biological outcomes in response to a stressor is a rise in metabolic rate [8, 22]. Elevated metabolic rate can influence appetite and energy partitioning and utilisation mainly through altering catabolic process such as increasing the rate of lipolysis and protein degradation [8, 23]. These biological pathways have all been postulated as inherent regulators of an animal's energetic efficiency [2–5].

To monitor animal's stress response measuring glucocorticoid hormones and HPA axis activity is the standard approach. Through dynamic testing methodologies the HPA axis can be pharmacologically stimulated with the use of exogenous CRH or ACTH, and response at both the pituitary and adrenal levels can be evaluated. Indeed, HPA axis challenges by way of exogenous CRH or ACTH stimulation have been shown by our group previously [15, 24] and others [25, 26] to be appropriate for investigation of the bovine stress response.

The adrenal cortex was successfully stimulated by ACTH administration in this study and the cortisol response and maximum peak time (60 min) was consistent with previous ACTH challenge experiments [9, 15, 27]. For the ACTH response no difference in plasma concentrations were detected between the divergent RFI phenotypes throughout the physiological challenge, concurring with the results that Adago [28] reported in Braham Cattle. Cortisol is the principal stress biomarker and the

Table 3 Effect of exogenous adrenocorticotropic (ACTH) administration[a] on haematology variables in beef heifers differing in phenotypic residual feed intake (RFI)

Cell type[b]	RFI Group	Time (min) relative to ACTH administration							SEM	RFI	Time	RFI × Time
		0	20	80	150	270	330	390				
Neutrophils, × 10³ cells/µL	High	9.1	11.7	10.9	11.4	11.9	11.9	12.2	0.36	0.90	0.001	0.92
	Low	8.5	11.3	10.8	12.2	12.9	11.9	11.7	0.41			
Lymphocytes, × 10³ cells/µL	High	3.36	3.51	3.3	2.92	3.21	3.41	4.02	0.159	0.99	0.71	0.78
	Low	3.88	3.38	3.16	3.07	3.55	3.32	3.27	0.175			
Monocytes, × 10³ cells/µL	High	0.69	0.39	0.53	0.39	0.45	0.47	0.58	0.04	0.99	0.01	0.95
	Low	0.7	0.5	0.68	0.39	0.49	0.38	0.42	0.045			
Eosinophils, × 10³ cells/µL	High	0.18	0.08	0.21	0.12	0.12	0.12	0.36	0.027	0.85	0.001	0.06
	Low	0.19	0.21	0.06	0.13	0.19	0.31	0.24	0.031			
Basophils, × 10³ cells/µL	High	0.05	0.06	0.05	0.06	0.06	0.06	0.07	0.003	0.61	0.001	0.3
	Low	0.05	0.05	0.05	0.06	0.07	0.06	0.06	0.003			
WBC, × 10³ cells/µL	High	16.6	14.7	15.3	14	14.8	14.7	16.3	0.3	0.35	0.53	0.7
	Low	15.4	16.2	16.1	15.3	16.6	15.4	15.3	0.34			
RBC, × 10⁶ cells/µL	High	6.19	6.33	6.04	5.66	6.08	6.03	6.17	0.093	0.02	0.1	0.81
	Low	5.89	6.66	6.42	6.41	6.53	6.43	6.57	0.107			
HGB, g/dL	High	9.46	9.6	9.14	8.63	9.25	9.25	9.54	0.124	0.08	0.04	0.8
	Low	8.77	9.94	9.6	9.51	9.78	9.62	9.81	0.142			
HCT, %	High	23.6	24.3	23.1	21.8	23.3	22.9	23.6	0.36	0.02	0.06	0.87
	Low	22.4	24.5	24.5	24.4	24.8	24.5	25.1	0.42			
MCHC, g/dL	High	40.3	39.5	40	44.8	40.2	40.7	40.9	0.63	0.23	0.71	0.92
	Low	44.3	38.7	38.7	38.8	39.2	39.1	38.8	0.74			
PLT, × 10³ cells/µL	High	366.6	312.7	458.9	327	420.5	300.7	346.4	22.82	0.69	0.001	0.76
	Low	309.3	329.4	405.2	269.1	436.3	274.4	310.4	26.11			

[a]Exogenous ACTH (1.98 IU/kg metabolic BW⁰⁷⁵)

[b]*WBCP* White blood cell number, *RBC* red blood cell number, *HGB* hemoglobin concentration, *HCTH* hematocrit percentage, *MCHC* mean corpuscular hemoglobin concentration, *PLT* platelet number

biological endpoint for the investigation of HPA function. Knott et al. [29] found that rams selected for greater cortisol concentration following an ACTH challenge also had a higher RFI than their lower cortisol response counterparts. Those authors also attributed up to 35% of the variation in RFI to the changes in cortisol concentrations, indicative of a stress response. Similarly, in a multi-generational selection experiment for RFI in chickens, Luiting and Verstegen [11] reported that inefficient hens had a greater cortisol response following an adrenal ACTH challenge. In cattle, Richardson et al. [30] observed a trend for a positive relationship between blood cortisol concentration of steers and their sire's estimated breeding value for RFI. Although, admittedly when phenotypic RFI and plasma cortisol concentrations were compared in the Richardson study a negative or inverse relationship was detected between the traits. However, it could also be argued that the ability to truly quantify stress responsiveness or HPA sensitivity in that study may have been somewhat limited, as a physiological stressor/challenge was not imposed on the

divergent RFI animals. Results from the current experiment agree with the findings in sheep [6, 29] and chicken [11] indicating that energetic inefficiency is associated with greater susceptibility to stress and is a likely underlying driver along with other biological processes [2] of variation in the trait. Indeed, our work clearly shows a greater cortisol stress response in High RFI heifers following activation of the HPA axis with exogenous ACTH (adrenal stimulation). In contrast to these findings Kelly et al. [18], failed to show an association between feed efficiency and stress biomarkers when focusing on neuroendocrine responses to a physiological CRH stress challenge. In that study HPA axis stimulation was via exogenous CRH and a more diminished adrenal cortex response was observed relative to that measured in the present study, which may explain the disagreement in results between our study and that of Kelly et al. [18]. Indeed, in all other published feed efficiency studies that investigated physiological stressors [6, 11], ACTH which was the sectretagogue used to examine adrenal stimulation of HPA and subsequent stress response. Foote et al. [31] also showed in cattle that efficient or low RFI heifers have lower concentrations of fecal corticosterone than the high RFI heifers and fecal corticosterone was identified as a useful physiological indicator for feed efficiency in finishing beef cattle. This finding complements our results, in that corticosterone is an informative representative of adrenal activity and is a glucocorticoid produced in a pattern similar to cortisol in response to ACTH release.

Immune response and feed efficiency

The central nervous system regulation of the immune system acts principally via the HPA [32, 33]. Neutrophils, as well as many other immune cells, are well known targets of stress hormones, possessing receptors for catecholamines and glucocorticoids secreted during an acute stress response. A profound leukocytosis, with marked neutrophilia, has often been observed in association with elevated circulating glucocorticoids [8, 23] indicating a disruption of neutrophil homeostasis in response to stress in cattle. This cross regulation between neuro-endocrine-immune systems is critical for homeostasis and has profound effects on the health, behaviour and performance of animals [33]. In general, the data from the current trial failed to show a relationship between feed efficiency, and the principal indicators of immune-competence in response to a physiological stressor. When Kelly et al. [18] investigated neuroendocrine responses to an exogenous physiological CRH physiological challenge in animals of differing RFI status, they did not find an association between haematological subpopulations and RFI. Additionally, Lawrence et al. [13] reported that high and low RFI heifers did not differ in any of the haematology variables measured, except at parturition when lymphocyte and monocyte counts were positively associated with RFI. Theis et al. [34] did not detect any relationship between phenotypic RFI with a similar set of haematological variables as was reported on in this study. However, recent work from our group [35] has shown that feed efficient pigs had lower gene expression profiles for intestinal innate immune response genes following a lipopolysaccharide (LPS) challenge compared to their inefficient contemporaries, suggesting that feed efficient animals have an altered immune response to infection. Indeed, mounting effective innate immune responses is metabolically costly to the animal which demands a nutrient tradeoff at the expense of other biological demanding processes and as such could be an underlying contributor to variation in energetic efficiency [36]. Richardson et al. [37] reported a genetic association between sire estimated breeding value (EBV) for RFI with white blood cell count, lymphocyte count and haemoglobin level, in crossbred beef steers. Gomes et al. [38] also reported that white blood cell count was lower and haemoglobin concentration was higher in high RFI heifers than low RFI heifers. In our study, we did detect that red blood cell count and hematocrit percentage was greater in Low RFI compared to High RFI animals, potentially indicating some deviations in oxygen requirement or carbon dioxide transport within their haem structure due to stress sensitivity [39].

Conclusions

The data presented provide evidence that animals differing in feed efficiency have an altered HPA axis function in response to a physiological stressor (ACTH adrenal stimuli) and susceptibility to stress and responsiveness of the HPA axis is likely to contribute to appreciable variation in the efficiency feed utilisation of cattle. As such, these findings add greater clarity and focus to the body of published work examining the biological control of feed efficiency in cattle.

Abbreviations
ACTH: Adrenocorticotropic hormone; ADG: Average daily gain; BW: Bodyweight; CRH: Corticotrophin-releasing hormone; CV: Co-efficient of variation; DMI: Dry matter intake; EBV: Estimated breeding value; F:G: Feed conversion ratio; HCT %: Hematocrit percentage; HGB: Hemoglobin; HPA: Hypothalamic-pituitary-adrenal; LPS: Lipopolysaccharide; MBW: Metabolic body weight; RBC: Red blood cell; RFI: Residual feed intake; WBC: white blood cell

Acknowledgements
The authors of this study gratefully acknowledge skilled technical assistance from Mr. Eddie Mulligan (Teagasc Animal Bioscience Research Centre, Grange) and the farm manager and his staff at Teagasc, Grange.

Authors' contributions
MM and AK conceived the study, conducted the statistical analysis and prepared the manuscript. BE conducted the laboratory analyses and prepared the manuscript. PL managed the animal model and assisted with manuscript preparation. DK assisted with diet formulation, sample recovery, statistical analysis and manuscript preparation. All authors read and approved the final manuscript.

Competing interests
The authors declare that they have no competing interests.

Author details
[1]School of Agriculture and Food Science, College of Health and Agricultural Sciences, University College Dublin, Belfield, Dublin 4, Ireland. [2]Animal and Bioscience Research Department, Animal & Grassland Research and Innovation Centre, Teagasc, Dunsany, Co. Meath, Ireland. [3]Livestock Systems Research Department Teagasc, Animal & Grassland Research and Innovation Centre, Grange, Dunsany, Co. Meath, Ireland.

References

1. Moore S, Mujibi F, Sherman E. Molecular basis for residual feed intake in beef cattle. J Anim Sci. 2009;87:E41–7.

2. Herd R, Arthur P. Physiological basis for residual feed intake. J Anim Sci. 2009;87:E64–71.

3. Kelly AK, Waters SM, McGee M, Fonseca RG, Carberry C, Kenny DA. mRNA expression of genes regulating oxidative phosphorylation in the muscle of beef cattle divergently ranked on residual feed intake. Physiol Genomics. 2011;43:12–23.

4. Kelly A, Waters S, McGee M, Browne J, Magee D, Kenny D. Expression of key genes of the somatotropic axis in longissimus dorsi muscle of beef heifers phenotypically divergent for residual feed intake. J Anim Sci. 2013;91:159–67.

5. Kelly A, McGee M, Crews D, Fahey A, Wylie A, Kenny D. Effect of divergence in residual feed intake on feeding behavior, blood metabolic variables, and body composition traits in growing beef heifers. J Anim Sci. 2010;88:109–23.

6. Knott S, Cummins L, Dunshea F, Leury B. Rams with poor feed efficiency are highly responsive to an exogenous adrenocorticotropin hormone (ACTH) challenge. Domest Anim Endocrinol. 2008;34:261–8.

7. Richardson E, Herd R. Biological basis for variation in residual feed intake in beef cattle. 2. Synthesis of results following divergent selection. Anim Prod Sci. 2004;44:431–40.

8. Burton JL, Madsen SA, Chang LC, Weber PS, Buckham KR, van Dorp R, et al. Gene expression signatures in neutrophils exposed to glucocorticoids: A new paradigm to help explain "neutrophil dysfunction" in parturient dairy cows. Vet Immunol Immunopathol. 2005;105:197–219.

9. Gupta S, Earley B, Crowe M. Pituitary, adrenal, immune and performance responses of mature Holstein× Friesian bulls housed on slatted floors at various space allowances. Vet J. 2007;173:594–604.

10. Ralph C, Tilbrook A. The hypothalamo-pituitary-adrenal (HPA) axis in sheep is attenuated during lactation in response to psychosocial and predator stress. Domest Anim Endocrinol. 2016;55:66–73.

11. Luiting P, Urff E, Verstegen M. Between-animal variation in biological efficiency as related to residual feed consumption. NJAS Wagen J Life Sci. 1994;42:59–67.

12. Lawrence P, Kenny D, Earley B, McGee M. Intake of conserved and grazed grass and performance traits in beef suckler cows differing in phenotypic residual feed intake. Livest Sci. 2013;152:154–66.

13. Lawrence P, Kenny DA, Earley B, Crews DH, McGee M. Grass silage intake, rumen and blood variables, ultrasonic and body measurements, feeding behavior, and activity in pregnant beef heifers differing in phenotypic residual feed intake1. J Anim Sci. 2011;89:3248–261.

14. Fitzsimons C, Kenny D, Deighton M, Fahey A, McGee M. Methane emissions, body composition, and rumen fermentation traits of beef heifers differing in residual feed intake. J Anim Sci. 2013;91:5789–800.

15. Gupta S, Earley B, Ting S, Leonard N, Crowe M. Technical Note: Effect of corticotropin-releasing hormone on adrenocorticotropic hormone and cortisol in steers. J Anim Sci. 2004;82:1952–6.

16. Gaignage P, Lognay G, Bosson D, Vertongen D, Dreze P, Marlier M, et al. Dexamethasone bovine pharmacokinetics. Eur J Drug Metab Pharmacokinet. 1991;16:219–21.

17. Toutain P, Brandon R, Alvinerie M, Garcia-Villar R, Ruckebusch Y. Dexamethasone in cattle: pharmacokinetics and action on the adrenal gland. J Vet Pharmacol Ther. 1982;5:33–43.

18. Kelly A, Earley B, McGee M, Fahey A, Kenny D. Endocrine and hematological responses of beef heifers divergently ranked for residual feed intake following a bovine corticotropin-releasing hormone challenge. J Anim Sci. 2016;94:1703–11.

19. Lynch EM, Earley B, McGee M, Doyle S. Effect of abrupt weaning at housing on leukocyte distribution, functional activity of neutrophils, and acute phase protein response of beef calves. BMC Vet Res. 2010;6:39.

20. Kelly A, McGee M, Crews D, Sweeney T, Boland T, Kenny D. Repeatability of feed efficiency, carcass ultrasound, feeding behavior, and blood metabolic variables in finishing heifers divergently selected for residual feed intake. J Anim Sci. 2010;88:3214–25.

21. Smagin GN, Heinrichs SC, Dunn AJ. The role of CRH in behavioral responses to stress. Peptides. 2001;22:713–24.

22. Buckham Sporer K, Weber P, Burton J, Earley B, Crowe M. Transportation of young beef bulls alters circulating physiological parameters that may be effective biomarkers of stress. J Anim Sci. 2008;86:1325–34.

23. Gomes RDC, Sainz RD, Leme PR. Protein metabolism, feed energy partitioning, behavior patterns and plasma cortisol in Nellore steers with high and low residual feed intake. Rev Bras Zootec. 2013;42:44–50.

24. Fisher A, Verkerk G, Morrow C, Matthews L. The effects of feed restriction and lying deprivation on pituitary–adrenal axis regulation in lactating cows. Livest Prod Sci. 2002;73:255–63.

25. Verkerk G, Macmillan K. Adrenocortical responses to an adrenocorticotropic hormone in bulls and steers. J Anim Sci. 1997;75:2520–5.

26. Fisher A, Crowe M, O'Kiely P, Enright W. Growth, behaviour, adrenal and immune responses of finishing beef heifers housed on slatted floors at 1.5, 2.0, 2.5 or 3.0 m² space allowance. Livest Prod Sci. 1997;51:245–54.

27. Gupta S, Earley B, Crowe M. Effect of 12-hour road transportation on physiological, immunological and haematological parameters in bulls housed at different space allowances. Vet J. 2007;173:605–16.

28. Agado BJ. Serum concentrations of cortisol induced by exogenous adrenocorticotropic hormone (ACTH) are not predictive of residual feed intake (RFI) in Brahman cattle. PhD Diss. 2011.

29. Knott S, Cummins L, Dunshea F, Leury B. Feed efficiency and body composition are related to cortisol response to adrenocorticotropin hormone and insulin-induced hypoglycemia in rams. Domest Anim Endocrinol. 2010;39:137–46.

30. Richardson E, Herd R, Archer J, Arthur P. Metabolic differences in Angus steers divergently selected for residual feed intake. Anim Prod Sci. 2004;44:441–52.

31. Foote A, Hales K, Tait R, Berry E, Lents C, Wells J, et al. Relationship of glucocorticoids and hematological measures with feed intake, growth, and efficiency of finishing beef cattle. J Anim Sci. 2016;94:275–83.

32. Buckingham JC. Glucocorticoids: exemplars of multi-tasking. Br J Pharmacol. 2006;147:S258–68.

33. Carroll J, Burdick Sanchez N, Bill E. Kunkle interdisciplinary beef symposium: Overlapping physiological responses and endocrine biomarkers that are indicative of stress responsiveness and immune function in beef cattle. J Anim Sci. 2014;92:5311–8.

34. Theis C, Carstens G, Hollenbeck R, Kurz M, Bryan T, Slay LJ, et al. Residual feed intake in beef steers: II. Correlations with hematological parameters and serum cortisol. In: Proceed-American Soc Anim Sci Western Sec; 2002. p. 483–6.

35. Vigors S, O'Doherty JV, Kelly AK, O'Shea CJ, Sweeney T. The Effect of Divergence in Feed Efficiency on the Intestinal Microbiota and the Intestinal Immune Response in Both Unchallenged and Lipopolysaccharide Challenged Ileal and Colonic Explants. PLoS One. 2016;11:e0148145.

36. van der Most PJ, de Jong B, Parmentier HK, Verhulst S. Trade-off between growth and immune function: a meta-analysis of selection experiments. Funct Ecol. 2011;25:74–80.

37. Richardson E, Herd R, Colditz I, Archer J, Arthur P. Blood cell profiles of steer progeny from parents selected for and against residual feed intake. Anim Prod Sci. 2002;42:901–8.

38. Gomes RDC, Siqueira RFD, Ballou MA, Stella TR, Leme PR. Hematological profile of beef cattle with divergent residual feed intake, following feed deprivation. Pesq Agrop Brasileira. 2011;46:1105–11.

39. Jones ML, Allison RW. Evaluation of the ruminant complete blood cell count. Vet Clin North Am Food Anim Pract. 2007;23:377–402.

19

Impacts of prenatal nutrition on animal production and performance: a focus on growth and metabolic and endocrine function in sheep

Prabhat Khanal[1,2]* and Mette Olaf Nielsen[1]

Abstract

The concept of foetal programming (FP) originated from human epidemiological studies, where foetal life nutrition was linked to health and disease status later in life. Since the proposal of this phenomenon, it has been evaluated in various animal models to gain further insights into the mechanisms underlying the foetal origins of health and disease in humans. In FP research, the sheep has been quite extensively used as a model for humans. In this paper we will review findings mainly from our Copenhagen sheep model, on the implications of late gestation malnutrition for growth, development, and metabolic and endocrine functions later in life, and discuss how these implications may depend on the diet fed to the animal in early postnatal life. Our results have indicated that negative implications of foetal malnutrition, both as a result of overnutrition and, particularly, late gestation undernutrition, can impair a wide range of endocrine functions regulating growth and presumably also reproductive traits. These implications are not readily observable early in postnatal life, but are increasingly manifested as the animal approaches adulthood. No intervention or cure is known that can reverse this programming in postnatal life. Our findings suggest that close to normal growth and slaughter results can be obtained at least until puberty in animals which have undergone adverse programming in foetal life, but manifestation of programming effects becomes increasingly evident in adult animals. Due to the risk of transfer of the adverse programming effects to future generations, it is therefore recommended that animals that are suspected to have undergone adverse FP are not used for reproduction. Unfortunately, no reliable biomarkers have as yet been identified that allow accurate identification of adversely programmed offspring at birth, except for very low or high birth weights, and, in pigs, characteristic changes in head shape (dolphin head). Future efforts should be therefore dedicated to identify reliable biomarkers and evaluate their effectiveness for alleviation/reversal of the adverse programming in postnatal life. Our sheep studies have shown that the adverse impacts of an extreme, high-fat diet in early postnatal life, but not prenatal undernutrition, can be largely reversed by dietary correction later in life. Thus, birth (at term) appears to be a critical set point for permanent programming in animals born precocial, such as sheep. Appropriate attention to the nutrition of the late pregnant dam should therefore be a priority in animal production systems.

Keywords: Adipose tissue, Endocrine function, Foetal programming, Metabolic function, Sheep

* Correspondence: prabhat.khanal@medisin.uio.no
[1]Department of Veterinary and Animal Sciences, Faculty of Health and Medical Sciences, University of Copenhagen, Grønnegårdsvej 3, 1st floor, DK-1870 Frederiksberg C, Denmark
[2]Current address: Department of Nutrition, Faculty of Medicine, Transgenic Animal and Lipid Storage, Norwegian Transgenic Centre (NTS), University of Oslo, Oslo, Norway

Background

The term 'foetal metabolic programming' was defined in the early 1990s as a phenomenon linking long-term adverse health consequences in animal species with adverse nutritional exposures in utero [1, 2]. In the past, foetal programming (FP) and its long-term impacts have been evaluated particularly from a human health and disease perspective [3, 4], and such studies have revealed that FP has implications for a wide range of body functions, which are also key determinants of animal productivity. However, knowledge about the potential long-term implications of FP for animal productivity is still scare. Such knowledge is needed in order to assign the best management strategies (postnatal feeding, culling, etc.) to minimize implications of adverse FP for animal productivity and avoid possible trans-generational transfer of undesirable FP outcomes. In this review we will primarily focus on what has been found in sheep, where the long-term implications of foetal life malnutrition for development and metabolic and endocrine functions later in life have been extensively studied. Furthermore, we will also evaluate to what extent the diet fed in postnatal life can influence the phenotypic manifestation of the prenatal FP. In this regards, observations from other species will only be included where appropriate.

Animal experimental approaches to the study of foetal programming

In the past, early nutritional programming has mostly been investigated in rodent models with a focus on the long-term implications for health and disease risk in humans. However, FP is one of the rare areas of research where sheep has also been used quite extensively as a model for humans [5–10] due to the similarities in the foetal developmental trajectory and physiological maturity at birth. Pig is another farm animal commonly used as a model for humans studies regarding FP [11], but the pig is born less physiologically mature than humans and ruminant offspring. Less frequently, non-human primates [12] have been used.

We developed the Copenhagen sheep model [9, 10] to be able to study the long-term impacts of foetal over- and undernutrition in late gestation, and further to study how postnatal manifestations of FP are affected by the diet received in early postnatal life. In the studies based on this model, twin-pregnant ewes were exposed to adequate nutrition (NORM; 100% of Danish daily energy and protein requirements), undernutrition (LOW; 50% of NORM for energy and protein requirements) or overnutrition (HIGH; 150% of energy and 110% of protein requirements) during the last 6 wk of pregnancy (term ~147 d). When the twin lambs were born, they received colostrum within 3 h of birth, and suckled their dams at will for the first 3 days after parturition.

Thereafter, the dam was removed and the lambs artificially reared until 6 mo of age (after puberty) on two different diets: one lamb from each twin pair received a moderate, conventional diet (CONV, consisting of good quality hay sufficient to achieve moderate growth rates of 225 g/d with a milk replacer until 8 wk of age), whereas the other lamb received an energy dense, high-starch-high-fat diet (HCHF, consisting of rolled maize and a dairy cream-milk replacer mix (1:1) fed ad libitum up to a maximum daily intake of 1 kg and 2.5 kg, respectively) (Fig. 1). From 6 mo of age until adulthood at 2–2.5 yr of age, all sheep were fed the same moderate grass/hay based diet. To minimize the potential paternal impacts in regards to foetal programming of maternal nutrition, rams used for mating the ewes prior to an exposure to prenatal nutritional treatments were of the same breed and similar ages and body weights and they were reared under similar management conditions. With this experimental design, it was possible to evaluate whether long-term adverse outcomes of foetal malnutrition and excessive fat deposition in early postnatal life can be reversed by nutritional intervention later in life. Details of the type of feeds used and the nutritional composition of the experimental diets are shown in Table 1.

It should be noted that we characterized the long-term consequences of pre- and postnatal nutrition mismatch scenarios in a ruminant animal species without disrupting rumen fermentation or compromising animal health. Nutritional manipulations in our experiments were done during the late gestation period (third trimester), which is the period of extensive quantitative foetal growth [13] where many endocrine organs and tissues are matured, including adipose tissues [14]. Although nutritional insults during all stages of gestation can influence body functions of the offspring later in life [5, 15, 16], late gestation is the time window, where FP is most likely to occur in precocial farm animal species given birth to multiple offspring, such as sheep, due to the dramatic rise in nutrient requirements for the foetuses in late gestation [13, 17]. Although our studies were designed from a human health perspective, the results obtained allow us to evaluate how programming outcomes may affect both animal growth and metabolic and endocrine functions of importance for animal productivity, including the timing of manifestations. Such knowledge can help to refine nutritional strategies applied in livestock production [18].

Impacts of maternal malnutrition on postnatal growth and organ and tissue development and function in growing animals

Growth characteristics

Historically, birth weight has been used as a marker to identify individuals at risk of having undergone adverse FP [19]. Birth weight is in itself a poor indicator of

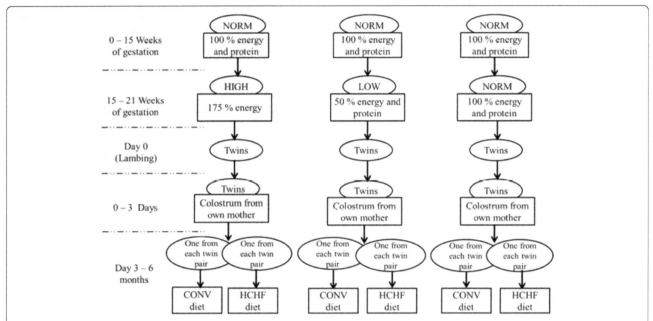

Fig. 1 The experimental design of the Copenhagen sheep model showing different nutritional and dietary interventions during late gestation and early postnatal life in sheep (obtained from Khanal et al., 2014 [10]). Late gestational nutrition groups: HIGH, fulfilling 150% of Danish requirements for energy and 110% of requirements for protein; LOW, fulfilling only 50% of requirements for energy and protein; NORM; fulfilling 100% requirements for energy and protein. Early postnatal (from 3 d after birth until 6 mo of age) nutrition groups: one lamb from each twin pair was allocated to a HCHF diet (high-starch-high fat consisting of a milk replacer-dairy cream mix supplemented with rolled maize), and the other was fed a CONV (conventional/moderate, hay-based diet; growth rate of appr. 225 g/d) diet

nutritional programming, since it provides little information about body composition, adiposity and potentially altered body functions. Moreover, nutritional insults interfering with foetal growth during the earlier stages of gestation may not be reflected in changes in birth weight if catch-up growth occurs during later stages of gestation [6, 20]. However, it has been shown in several studies that undernutrition in late gestation can result in animals being born small for gestational age. In sheep, birth weights have been reduced in different experiments by 10–18% under controlled conditions, when dams were

fed only 50–70% of their daily energy requirements during late gestation (see e.g. [9, 21]), and in goats kid birth weight was reduced by ~10% when dams were exposed to poor grazing conditions during the last 4 wk prepartum [22]. Lower birth weight has in different studies been associated with reduced survival rate [23], poorer growth rate during the suckling period and 24% lower weaning weight at 14 wk of age in sheep [21]. These findings may partly be explained by poorer mammary development of dams malnourished in late gestation [24], leading to a reduction in colostrum and early

Table 1 Different types of experimental feeds and their chemical composition and energy content

Feeds	DM, %	Ash, % of DM	aNDF, % of DM	ADF, % of DM	ADL, % of DM	CP, % of DM	Cfat, % of DM	DE, MJ/ kg DM
Sheep diet during late gestation								
Hay	91.4	5.6	47.7	27	3.1	20.8	4.8	13.7
Barley	89.0	2.3	14	6	1.1	12.5	3.1	17.1
Concentrate	87.7	7.7	25.8	18	2.8	15.3	3.8	12.8
Lamb diet during early postnatal life until puberty								
Hay	93.1	6.8	50.4	32.3	3.5	19.1	3.7	13.5
Maize	89.5	0.6	4.1	<5	0.9	8.5	1.9	16.3
Milk powder	95.6	7.1	-	-	-	22.5	23.6	19.2
Cream	42.9	0.8	-	-	-	4.3	38.0	30.5

These are the types of feeds used in diets for experimental animals in the Copenhagen sheep model; the table was obtained from Khanal et al., 2014 [10] with modifications. DM, dry matter; aNDF, amylase-treated neutral detergent fiber; ADF, acid detergent fiber; ADL, acid detergent lignin; CP, crude protein; Cfat, crude fat; DE, digestible energy lactation milk production [9, 25, 26]. However, in other studies in sheep where the dietary intake of offspring was controlled after birth (artificial rearing), the postnatal growth appeared to be entirely determined by the postnatal and not the prenatal level of nutrition [9, 10, 27, 28]. However, postnatal growth does appear to terminate earlier in individuals with a history of late gestation undernutrition, resulting in smaller adult body size [27]. In cattle, low birth weight (28.6 vs. 38.3 kg) or slow growth till weaning led to lowered body (56 or 46 kg less, respectively) and carcass weights (32 or 40 kg less, respectively) at slaughter at 30 mo of age [29]. Thus, proper attention should be given to ewe nutrition in the late gestational period to ensure not only optimal foetal growth, but also a desired level of colostrum and milk production. It may therefore be beneficial to consider supplementary milk feeding after birth in suckling individuals which were exposed to prenatal undernourishment to improve their postnatal immunization and growth performance.

Skeletal muscle development and function
Proper growth of skeletal muscle and lean carcass mass in slaughter animals are important production traits for the livestock industry. Muscle fibre formation commences during the embryonic stage and, in animal species born precocial, the formation of secondary muscle fibers is concluded during mid-gestation [30]. Thus, in sheep, and other animals born precocial, conclusion of myotube formation and establishment of the final number of muscle fibres occurs prior to the onset of the third trimester [14, 31]. In other farm animal species, myogenesis may occur over a larger part of gestation, for example in pigs, where muscle fibre hyperplasia is not concluded until 95 d of the 114-d gestation period [32]. It must be anticipated, therefore, that foetal myofibre formation in such species may be affected by adverse nutritional insults during a greater part of the gestation. In ruminant animals, foetal undernutrition during the first part of gestation, when myogenesis takes place, has been shown to reduce the formation and number of secondary muscle fibres [15, 31]. In lambs exposed to undernutrition from 28 to 78 d of gestation, the reduced number of total secondary myofibers was reconizable at 8 mo of age [15], and in another study it was shown that undernutrition from 30 to 70 d of gestation altered muscle characteristics (fewer fast fibres and more slow fibres in the *longissimus* and *vastus lateralis* muscles) in new-born lambs [33]. In cows, improving the nutritional status of pregnant cows (improved pasture conditions) during mid to mid-late gestation (120–150 through 180–210 d of gestation; term ~280 d) improved carcass characteristics (tenderness) and also increased live and hot carcass weight in steers [34].

In contrast, exposure to malnutrition after myofibre formation has been completed does not appear to have major implications for muscle development and function postnatally. Thus, nutrient restriction (to 50% of daily requirements) between 85 and 115 d of gestation decreased muscle weight in lambs without affecting muscle fibre number [33]. Similarly, in adult sheep, we have not been able to demonstrate any changes in muscle mass or expression of markers for metabolic function in muscle of adult offspring that could be related to a history of late gestational foetal undernutrition (50% reduction in maternal energy and protein supply relative to recommendations) [35]. In conclusion, whether gestational malnutrition will have implications for myogenesis thus appears to depend on the timing of the nutritional insults relative to the conclusion of myotube formation in utero. The nutritional programming of skeletal muscle development prior to the conclusion of myofibre development appears to be permanent, whereas malnutrition in late gestation in precocial animal species does not have long-term consequences for skeletal muscle development.

Adipose tissue deposition
In precocial animal species such as sheep, the major part of foetal adipogenesis and adipose tissue differentiation takes place during the last part of gestation [36–38]. Thus, if foetal nutrition should have implications for adipose tissue development, it would most likely be during the late gestation period. This has received much less attention in relation to FP in farm animals than muscle development, presumably due to the greater economic importance of the latter.

From studies using our Copenhagen sheep model, we have earlier reported that both prenatal overnutrition (150% of energy and 110% of protein requirements) and undernutrition (50% of energy and protein requirements) during late pregnancy led to changes in fat deposition patterns in adolescent offspring (~6 mo of age) resulting in a greater preference for deposition in the abdominal (mesenteric or perirenal) rather than subcutaneous region when the lambs were fed a high-fat diet in early postnatal life [9, 10]. This could be ascribed to a reduced ability to increase fat deposition in subcutaneous adipose tissue during fatness development (Fig. 2). Moreover, in the 6 months old lambs with a history of late gestation undernourishment, an increased occurrence of collagen and non-collagen extracellular matrix, together with greater numbers of a subpopulation of very small adipocytes (<40 μm diameter) was observed in the subcutaneous fat (Fig. 3) [39]. Our recent data also indicate that altered fat distribution patterns due to late-gestation under- as well as overnutrition, followed by exposure to a high-fat diet in early postnatal life are

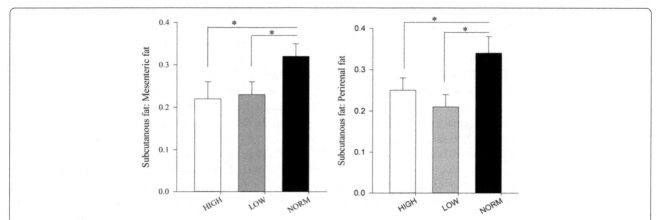

Fig. 2 Impacts of late gestational over- and undernutrition on fat deposition patterns in adolescent (6 months old) offspring (left panel: subcutaneous to mesenteric fat ratio; right panel: subcutaneous to perirenal fat ratio) (obtained from Khanal et al. [10]). For HIGH, NORM and LOW, see legends for Fig. 1

associated with markedly increased perirenal adipocyte hypertrophy (Khanal et al., unpublished data; Fig. 4).

A limited expandability of subcutaneous adipose tissue may give rise to increased intramuscular fat deposition during fattening. This would be consistent with findings in pigs, where low birth weight pigs (1.05 vs. 1.89 kg) had increased lipid deposition (25%) in *semitendinosus muscle* when subcutaneous fat deposition increased, and lean meat content and fibre numbers (19%) were lowered compared to high-birth weight animals [40]. In contrast, other findings in sheep have shown that maternal overnutrition (~50% above maintenance energy requirements) during late pregnancy increased relative subcutaneous fat deposition and leptin expression in subcutaneous and perirenal fat in 1 mo old lambs [7]. The reasons for these apparently conflicting results are not known, but the postnatal diet could have had an influence.

Impacts of maternal malnutrition on metabolic and endocrine function in growing animals

Malnutrition during gestation has been linked to substantial changes in metabolic and endocrine functions postnatally, and in the following sections implications of prenatal nutritional for glucose-insulin homeostasis, hepatic function and other endocrine functions will be addressed.

Glucose-insulin homeostasis

The glucose-insulin regulatory axis has long been known to be an important target of foetal programing in humans. In our Copenhagen sheep model we have shown that this is also the case in sheep, and that both under- and overnutrition in late gestation can change the function of this axis permanently. The digestive system and intermediary metabolism of the ruminants differ from the non-ruminant animals, as ruminants ferment most of the dietary carbohydrates to short-chain volatile fatty acids using microbial activity in the rumen leaving only little glucose available for intestinal absorption [41]. However, like in other animal species, hormonal regulatory mechanisms for the maintenance of stable blood glucose level appear to be quite similar in both ruminant and monogastric animals [42, 43]. In our studies, we have performed in vivo metabolic and endocrine tolerance tests to elucidate glucose-insulin-axis function in sheep.

Late gestation undernutrition (50% of energy and protein requirements) decreased peripheral insulin sensitivity in young lambs, but their ability to clear intravenously administered glucose was maintained due to a compensatory upregulation of insulin secretion [44, 45]. However, upon exposure to a high-fat diet in early postnatal life, the ability to clear glucose was reduced, since the high-fat diet interfered with the compensatory upregulation of glucose secretion from the pancreas [45]. The high-fat diet also gave rise to very high plasma levels of triglyceride (~2.0 vs. 0.5 mmol/L) in the lambs, and development of pancreatic fibrosis [45], which to our knowledge has not been reported in ruminant animals previously.

Late gestation overnutrition also affects glucose-insulin regulation, but in a different way. In our studies, lambs exposed to overnutrition in late gestation had increased postnatal gluconeogenic ability in response to intravenous propionate injection [46] and had higher levels of plasma glucose during 44-h fasting when exposed to an early postnatal high-fat diet [10], which was not observed in lambs that had been exposed to undernutrition in late gestation. The underlying mechanims are not known.

Hepatic function

The liver is an important organ for integration of metabolic pathways, and studies in sheep have shown that

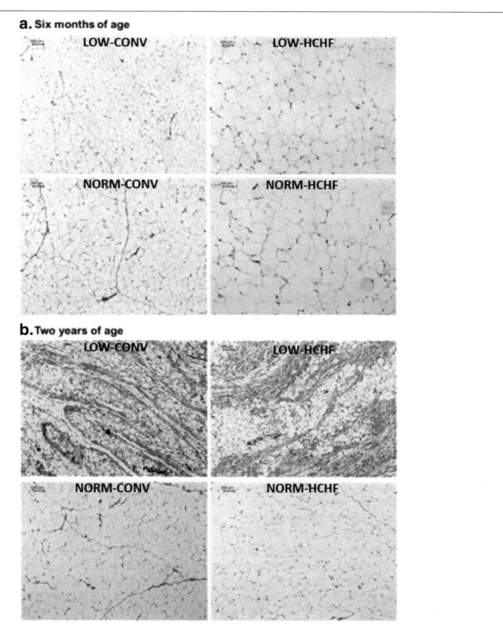

Fig. 3 Morphology of Van Gieson-stained subcutaneous adipose tissue from 6 months old adolescent lambs and 2 years old adult sheep (obtained from Nielsen et al. [39]). Panel A: examples of pictures from the 4 groups of lambs, used to grade cell size (and with negligible collagen infiltration) showing a larger population of very small cells in the LOW/CONV group (bottom left) relative to the other groups, and extensive hypertrophy in adipocytes from HCHF lambs (pictures to the right). Panel B: morphological characteristics observed in slides from adult LOW sheep, which was not restricted to a specific early postnatal diet (pictures at the top) with extensive collagen infiltration (grade 4), which was never observed to the same extent among NORM sheep (max grade assigned = 2). For HIGH, NORM, LOW, CONV, HCHF see legends for Fig. 1

prenatal over- and undernutrition may have long-term and differential impacts on hepatic lipid accumulation, glucose and lactate release, and cholesterol synthesis. Nutrient restriction (50% of requirements) during early pregnancy led to increased hepatic lipid accumulation in obese 1 years old sheep offspring [47], and in 4 months old lambs born to dams that were obese around the periconceptional period, expression of genes encoding

for factors involved in hepatic fatty acid oxidation was increased [48]. The implications for hepatic function in the offspring later in life were, however, not studied. We ourselves observed increased plasma cholesterol levels in prenatally undernourished 6 months old sheep offspring provided a high-starch-high-fat diet during early postnatal life [46], which may be due to upregulated hepatic cholesterol metabolism. As with glucose metabolism, the

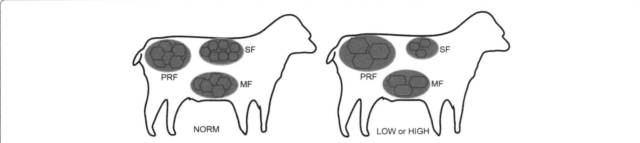

Fig. 4 Impacts of late gestational nutrition on hyperplasia and hypertrophy of different adipose tissue depots in adolescent (6 months old) offspring (obtained from Khanal et al., unpublished data). SF, subcutaneous fat (encircled as green); MF, mesenteric fat (encircled as yellow); PRF, perirenal fat (encircled as red). For HIGH, NORM and LOW, see legends for Fig. 1. Each hexagonal structure represents an individual adipocyte

impacts of an early postnatal high-starch-high-fat diet in our studies were additive to those of prenatal undernutrition with regard to lipid metabolism and hepatic function. This is because the predisposition of higher plasma cholesterol levels due to late gestation undernutrition (4.3 vs. 1.6 mmol/L) was manifested particularly upon additional exposure to a postnatal high-starch-high-fat diet in adolescent lambs [46]. Our studies have thus shown clear indications that late gestation undernutrition impacts on postnatal cholesterol metabolism in sheep offspring but further investigation into the underlying mechanism and its potential long-term implications for farm animals is warranted in future.

Other endocrine functions

Not only glucose-insulin axis function, but also a number of other endocrine systems are subject to programming in foetal life, such as the hypothalamic-pituitary-adrenal (HPA) axis [49], the growth hormone (GH)-insulin-like growth factor 1 (IGF-1) axis [50], leptin regulation [51] and the hypothalamic-pituitary-thyroid hormone (TH) axis [52]. We have, however, found that the impacts of prenatal nutrition on basal plasma levels and the adaptive responses to fasting in these endocrine systems became manifested in adulthood only, and were not evident in lambs at 6 mo of age [53]; this will be addressed in the following section.

Long-term consequences of early life nutrition in adult animals

In animal production systems, animals are kept until adulthood for special production purposes, such as reproduction and lactation [54]. Much less information is available regarding the implications of FP in adulthood in farm animals, probably due to the fact that it is costly to run such studies for several years. Prevailing evidence suggests that many of the implications of FP become increasingly manifested as the animal approaches adulthood. As already mentioned, low birth weight in lambs can be compensated for early postnatal catch-up growth

so that normal slaughter weights can be achieved. However, adult body composition can be altered as demonstrated by increased adult adiposity and body weight [55], and there is also, as previously mentioned, evidence suggesting that linear growth can be terminated earlier, resulting in smaller adult body size [27]. In our sheep studies, we did, however, not observe changes in neither body size nor proportions or weights of major organs or muscle or adipose tissue mass in adult offspring that had been exposed to over- or undernutrition in their late foetal life [56], except for increased adrenal weights [9]. Irrespective of that, abnormal subcutaneous morphology (more extensive fibrosis and the occurrence of a subpopulation of very small adipocytes) was also clearly evident in prenatally nutrient-restricted adult offspring regardless of postnatal diet [39]. These morphological changes were similar to the previously mentioned morphological changes observed in non-obese 6 months old lambs. We were to our knowledge the first to report that prenatal undernutrition has long-term implications for composition of fatty acids in skeletal muscle, liver and adipose tissues [35, 39, 57]. Thus, sheep with a history of prenatal undernutrition reduced the myristic acid content and increased the C16:0 to C18:0 fatty acid ratio in perirenal fat, an effect which was not observed in lambs [39]. The underlying reasons for such a FP of fatty acid composition in tissue lipids are unknown, and this could not be ascribed to differences in the postnatal diet. Tissue levels of myristic acid appear to be rate-limiting for a process termed protein myristoylation, and myristoylation impacts function of appr. 0.5–3% of the human proteome [58], and the consequences of such changes thus merit to be addressed in future studies. Prenatal undernutrition was also associated with increased triglyceride, ceramide and free fatty acid contents in livers of adult sheep, which was not observed in lambs [57].

Although studies focusing on early life nutritional impacts in adult offspring are relatively scarce, it seems reasonable to conclude that (subcutaneous) adipose tissue morphology and expandability (increased extracellular matrix, abundance of very small adipocytes) as well as hepatic and adipose lipid composition appear to be a permanent target of FP induced by late gestational undernutrition.

Glucose-insulin axis function

The functionality of a whole range of endocrine systems is altered in animals subjected to maternal malnutrition, but long-lasting impacts appear to be less pronounced following prenatal overnutrition as compared to undernutrition. Undernutrition during late, but not early, gestation in sheep led to impaired insulin sensitivity of peripheral tissues (reduced glucose tolerance) in adult offspring alongside increased adipose tissue mass [59]. This is consistent with our previous studies, where we found that a depression of insulin sensitivity [45], reduced pancreatic insulin secretory capacity as well as plasticity of down-regulation of insulin secretion [60] persisted into adulthood in sheep with a history of late gestation undernutrition.

In contrast, the adverse impacts of an early postnatal high-fat diet on glucose-insulin axis function, which were clearly observed in lambs (poorer glucose tolerance, reduced insulin secretion and clearance ability), completely disappeared in adult sheep when they were shifted from a high-starch-high-fat to a normal diet and normalization body fat contents [45].

In ruminants as in monogastrics, insulin is the hormone responsible for stimulating transport of glucose into insulin-sensitive tissues, where skeletal muscle and adipose tissues are the most important, but not into insulin-insensitive tissues including tissues important for reproduction (mammary gland and the conceptus). Thus, poor insulin sensitivity and reduced plasticity of pancreatic insulin secretion in sheep exposed to undernutrition in late foetal life, can undoubtedly influence how different tissues are adjusted or prioritized for the glucose utilization during reproductive cycle, pregnancy, etc. [61] and hence the (re) production potential of animals.

Thyroid-hormone axis function

Studies on prenatal nutritional impacts on TH axis function in farm animals are very scarce, although these hormones affect the adaptation and maintenance of a wide range of body functions under different environmental conditions [62] and also play an important role in ending seasonal reproduction in ewes [63]. In one study, adult hyperthyroidism was observed in adult sheep exposed to late gestation undernutrition and this was associated with increased thyroid expression of genes regulating TH synthesis and deiodination. It also increased the number of TH receptors and deiodinase mRNA expression in different target tissues such as liver, cardiac muscle and longissimus dorsi muscle but decreased the number of TH receptors and deiodinase mRNA expression in adipose tissues [64]. This suggests

that long-term TH axis function is a target of FP in response to foetal undernutrition during late gestation, but its potential influences on animal production traits remain to be established.

Other endocrine functions

In our Copenhagen sheep model, alterations in HPA axis function and leptin response were induced by late gestational undernutrition and became manifested in adulthood regardless of the dietary exposures early in postnatal life. Prenatally undernourished male lambs and adult female animals had, as already mentioned, increased adrenal weights [9], and we observed that the adult sheep also had elevated plasma cortisol levels and responded to fasting with a reduction in the cortisol levels [53], in contrast to an expected increase plasma cortisol levels during fasting. This may indicate hyperactivity of the HPA axis, and confirms the previous finding in which increased HPA axis response was observed in adult sheep offspring exposed to a short duration of undernutrition during late foetal life [5]. The GH-IGF-1 axis and adaptations to leptin also appear to be targets of FP. In our prenatally undernourished sheep total plasma IGF-1 concentrations were unexpectedly increased during fasting (presumably due to extended half-life in the blood), whereas plasma leptin concentrations were higher during fasting from much lower levels than in non-programmed sheep [45].

Thus, all the hypothalamic-pituitary axes hitherto studied (TH, GH-IGF-1, HPA) have been shown to be targets of FP. However, the phenotypic manifestation of this programming may not become manifested until the animals approach adulthood, and the consequences for productive functions in adulthood are not well-known. The hypothalamus is a main target for leptin, a hormone produced in white adipose tissues, and hypothalamic binding of leptin can induce changes in all hypothalamic-pituitary endocrine axes in addition to its role in down-regulation of feed intake [65] Considering that FP also induces abnormalities of adipose tissue morphology (fibrosis and very small adipocytes), this has led us to hypothesize that FP may target the entire leptin-hypothalamic-pituitary axis.

It is not known, to what extent overnutrition in late foetal life can predispose for similar long-term impacts on this axis, but it can, as previously reviewed, predispose for increased fat deposition, and the development of leptin resistance, with associated defects in a number of endocrine systems affecting hypothalamic appetite regulators and metabolic function [66].

Reproductive function

Considering that the nutritional history in foetal life has implications for all aspects of later HPA axes previously studied, it is not surprising that reproductive development during foetal and neonatal life is also affected [67, 68] with consequences for subsequent reproductive function in adulthood. There are, however, relatively few studies on these issues, and it is still not clear to what extent the changes in reproductive function are a consequence of FP targeting reproductive organs directly or there may be indirect effects of altered functions of other endocrine systems and changes in energy metabolism.

It has been shown that both over- (ad libitum feeding) and undernutrition (60% of dietary requirements) during a period of 8 wk prior to oocyte collection in ewes led to reduced oocyte competence and fertilization and poor early embryonic development [69]. Additionally, uundernutrition during the early stages of pregnancy, before and during the period of folliculogenesis, delayed foetal ovarian follicular development in sheep [70]. These findings may explain impaired reproductive funtion in sheep offspring observed in other studies. For example, prenatal undernutrition (50% of energy requirements) during the first 95 d of gestation reduced the ovulation rate in female adult sheep [71]. Furthermore, maternal undernutrition during mid- to late gestation led to a reduction in the number of large corpora lutea in female sheep offspring [72]. Nutrition of ewes during late pregnancy or lactation can also influence subsequent lifetime reproductive performance of the female offspring through impact on the ability to sustain pregnancy, i.e. avoidance of embryo or foetal loss [73].

Although it appears evident that maternal undernutrition, both in the preconceptional period and during gestation, can have adverse effects on the overall reproductive function of the offspring, much remains to be understood about the impacts of prenatal overnutrition and gestational stage-specific influences on the development of reproductive function. Studies are also required to ascertain whether lactation performance may be affected by suboptimal nutrition during foetal life.

Epigenetic changes due to maternal nutrition during gestation

Epigenetic regulation of gene expression, i.e. DNA methylation, histone modification etc., could be a potential mechanism linking foetal malnutrition to subsequent phenotypic changes in postnatal life (see reviews [74, 75]). Periconceptional undernutrition in sheep has been shown to induce epigenetic changes, namely histone acetylation and promoter methylation, of foetal hypothalamic genes including glucocorticoid receptors and proopiomelanocortin genes [76], which ultimately affects food intake and energy expenditure after birth. Additionally, periconceptional undernutrition has been associated with epigenetic changes in the adrenal *IGF2/H19* genes coexisting with adrenal overgrowth in offspring [77], which may predispose for postnatal susceptibility to stress. Periconceptional restriction of

maternal vitamin B and methionine supply led to altered methylation at CpG islands in the foetal sheep liver and with increased adult body weight and fatness of the offspring [78]. The detailed molecular biological mechanisms underlying epigenetic modifications in response to foetal life malnutrition are still poorly understood. Future studies are needed to identify the impacts of prenatal malnutrition at different gestational stages on tissue-specific epigenetic changes and long-term implications of such epigenetic modifications induced in foetal for animal production and performance.

Can dietary intervention later in life reverse the adverse programming outcomes of early life nutrition?

An important issue in animal production is to what extent undesirable effects of early life malnutrition can be minimized or completely reversed by dietary or other interventions later in life. Unfortunately, studies investigating the possibility of reversing undesirable FP outcomes with dietary interventions later in life are scarce and not encouraging. We have shown in sheep that it was possible to effectively reverse the adverse outcomes (in terms of increased body fat, higher plasma lipid profiles, poor glucose-insulin homeostasis etc.) induced by an unhealthy fatness-inducing diet fed in early postnatal life if the diet was changed later in life to a normal (for sheep) grass-based diet. Late gestation undernutrition, however, induced permanent programming outcomes [9, 45], particularly on lipid and urea metabolism as described previously, and these implications were more strongly manifested in adult sheep than lambs, irrespective of changes in the postnatal diet.

Extremely few studies have focussed on the long-term impacts of late gestation overnutrition in farm animal species, but it appears from our studies that the possibility of recovery from undesirable nutritional programming outcomes is more likely in individuals exposed to late gestational over rather than undernutrition. The alterations observed in body fat composition and glucose-insulin homeostasis in young lambs with a history of foetal overnutrition did not persist into adulthood [56].

Conclusion and future perspectives

Foetal or developmental programming can have lifelong impacts on the health and disease status of farm animals (Table 2), thus affecting the economy of livestock production (Fig. 5). Indeed, it has earlier been reported that foetal programming can be treated as a management tool to improve the livestock productivity in commercial farming but long-term programing impacts specific to different gestational stages and their interactions with postnatal nutritional environment are known [79]. Here, we highlight that foetal programming may be induced during any time point

Table 2 Major impacts of foetal programming due to abnormal nutrition applied at different stages of gestation and under various experimental conditions in sheep

Experimental conditions (gestational age and nutritional environment)	Primary changes in postnatal life	Reference
Growth characteristics		
Late gestational (105 d to term) overnutrition (150% energy and 110% protein) or undernutrition (50% energy and protein) + Early postnatal high-fat diet (0 d to 6 mo)	Reduced birth weight due to prenatal undernutrition, but no impacts due to prenatal overnutrition; Increased abdominal and perirenal fat deposition relative to subcutaneous fat by prenatal under- and overnutrition	[10]
Late gestational (105 d to term) undernutrition (50% energy and protein) + Early postnatal high-fat diet (0 d to 6 mo)	Reduced birth weight; Increased TG, ceramide and free fatty acids in liver, increased extracellular matrix content and very small adipocytes proportion in subcutaneous fat, hyperthyroidism and increased adrenal weights in prenatally undernourished adult sheep (2 yr)	[9, 39, 57, 64]
Late gestational (100 d to term) undernutrition (70% of energy requirements)	Reduced birthweight (18%) and weaning weight, but no weight differences in adulthood (26 wk)	[21]
Late gestational (115 d to term) overnutrition (133% energy)	Increased relative subcutaneous deposition in 1 months old lamb	[7]
Late gestational (109 d to term) undernutrition (50% of energy and protein)	Lowered colostrum yield	[24]
Late gestational (105 d to term) undernutrition (50% of energy and protein)	Lowered birth weight, colostrum and milk yield (lactation performance)	[25]
Mid-gestational (85 d to 115 d) undernutrition (50% of energy requirements)	Decreased muscle weights in newborn lambs	[33]
Early to mid-gestational (28 d to 78 d) undernutrition (50% of requirements)	Increased intramuscular fat content in skeletal muscle in 8 mo old offspring	[15]
Early to mid-gestational (30 d to 70 d) undernutrition (50% of energy requirements)	Fewer fast and more slow muscle fibres in newborn lambs	[33]
Early to mid-gestational (30 d to 80 d) undernutrition (50% of energy requirements) + Postnatal obesogenic environment (restricted physical activity) from weaning (10 wk) to 1 yr	Increased hepatic TG accumulation in prenatally undernourished, obese adult sheep (1 yr.)	[47]
Metabolic and endocrine function		
Late gestational (105 d to term) over- (150% energy and 110% protein) or undernutrition (50% energy and protein) + Early postnatal high-fat diet (0 d to 6 mo)	Reduced glucose clearance and increased glucogeneogensis in matched prenatally overnourished high-fat fed lambs; Increased cholesterol levels in mismatched prenatally undernourished high-fat diet fed lambs and adult sheep	[10, 46, 56]
Late gestational (from 105 d to term) undernutrition + Early postnatal high-fat diet (0 d to 6 mo)	Reduced insulin sensitivity and increased insulin secretory responses to glucose in prenatally undernourished lambs; Poor glucose tolerance in mismatched prenatally undernourished high-fat fed lambs (mismatch group); Poor insulin clearance in prenatally undernourished high-fat fed adult sheep	[45]
Late gestational undernutrition (from 105 d to term)	Reduced insulin secretory ability with increased compensatory insulin sensitivity in 19 wk. old lambs	[44]
Late gestation undernutrition (from 110 d to term)	Poor glucose tolerance in adult sheep (1 yr)	[59]
Late gestational overnutrition (from 115 d to term)	Increased leptin expression in subcutaneous and perirenal fat from 1 months old lamb	[7]
Reproductive function		
Early gestational (0 d to 95 d) undernutrition (50% energy)	Reduced ovulation rate in prenatally undernourished adult female sheep (20 mo)	[71]
Early to mid (0 d to 30 d) or mid to late (31 d to 100 d) gestational undernutrition (50% requirements)	Increased number of small follicles in the ovary (early to mid-gestation undernutrition); reduced large corpora lutea (mid to late gestation undernutrition) in 10 mo old female lambs.	[72]

from prior to conception until birth, but the exact manifestation of the foetal programming later in life will depend on the timing of the insult relative to the critical time windows during which embryo formation, placental growth and foetal organogenesis take place.

Serious maternal malnutrition during the earlier parts of gestation can influence the development of reproductive functions and muscle fibre numbers and characteristics, but this is probably not very likely to occur under normal production conditions.

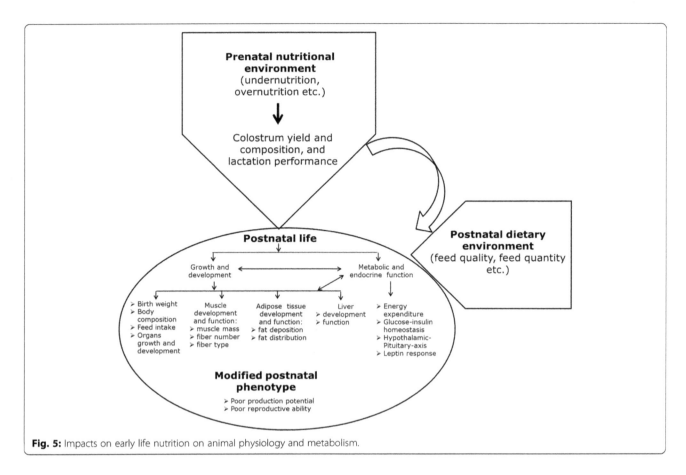

Fig. 5: Impacts on early life nutrition on animal physiology and metabolism.

On the other hand, in animals born precocial, such as sheep, around three quarters of the growth of the foetus [80], and of the mammary gland of the dam [24], occurs during the last 2 mo of gestation. For that reason, adverse FP is much more likely to occur in animal production systems, and with more severe consequences, during late- rather than early- to mid-gestation [8, 81, 82], if adequate attention is not paid to the nutrition of the pregnant dam [83] particularly during multiple pregnancy. Indeed, this may be the case in many parts of the world, where the late gestation period coincides with poor grazing conditions, e.g. during the dry season in tropical countries [84, 85] or the winter in the Nordic and alpine regions [86]. Late gestation malnutrition can have a wide range of both short-term (birth weight, weaning and slaughter weight, and glucose-insulin regulation) and long-term (metabolic and endocrine function, including but not limited to the glucose-pancreatic-hepatic and adipose-hypothalamic-pituitary axis functions, adipose development, fatty acid composition, and reproduction) consequences.

In ruminant production systems, young animals used for meat production are slaughtered within months of birth to obtain the best slaughter result in terms of economic return and meat quality [87]. It can be anticipated that impacts of adverse nutritional programming in utero are of minor quantitative significance at this age, unless the animal has been severely affected during foetal life. However, the timing of abnormal nutrition exposures in utero and the early postnatal nutrition can have implications for lean-to-fat ratios in slaughter animals, since morphogenesis of muscle in precocial species takes place, and may be programmed during the earlier parts of gestation, whilst development of adipose tissue development occurs later in gestation and into early postnatal life [88].

It has earlier been acknowledged that foetal programming in response to severe or prolonged improper nutrition is likely to affect various production traits in commercial sheep farming [79]. Here, we suggest that the best way to manage prenatally programmed animals, particularly undernourished animals (birth weight deviation by 15–20% of the normal range), is to destine them for slaughtering, and not allow them to enter into production processes taking place in adulthood, since the major adverse implications of FP become manifested later in life [45, 89]. Furthermore, there is a risk that undesirable traits may be transferred to future generations due to epigenetic inheritance, and it is therefore advisable to apply proper strategies to avoid the entry of adversely programmed animals into reproduction [90, 91].

Our recent findings suggest that a moderate diet and lower body fat content later in life can prevent or reverse a large part of the impacts induced by fatness development in early postnatal life [56]. FP due to late gestation undernutrition, however, has irreversible lifetime impacts on offspring, which are exacerbated upon transient fatness development in early postnatal life. Further studies are needed to confirm the findings from our studies that late gestational overnutrition has fewer long-term detrimental consequences for animal production than foetal undernutrition.

There are unfortunately no biomarkers other than birth weight, which can be used to reliably identify animals (at an early age) that have undergone FP. Although more studies are needed to assess the long-term quantitative impacts and economic consequences of FP, and to find biomarkers and potential means for reversal of such programming outcomes, commercial animal production should now acknowledge this phenomenon and devise management strategies to ensure its prevention and spread to future generations.

Abbreviations
CONV: Conventional diet; FP: Foetal programming; GH: Growth hormone; HCHF: High-starch-high-fat diet; HIGH: Overnutrition; HPA: Hypothalamic-pituitary-adrenal; IGF-1: Insulin-like growth factor 1; LOW: Undernutrition; NORM: Normal (adequate) nutrition; TG: Triglycerides; TH: Thyroid hormones

Acknowledgements
The authors wish to thank all the members and laboratory technicians involved in the studies based on the Copenhagen sheep model at the Department of Veterinary and Animal Sciences, University of Copenhagen Denmark. The authors also extend their sincere gratitude to Dr. Mark Birtwistle for proofreading the manuscript and providing valuable suggestions.

Funding
The research activities involving the Copenhagen sheep model were supported by the Danish Council for Strategic Research through the research programme of the Centre for Foetal Programming (CFP), Denmark.

Authors' contributions
PK wrote the manuscript under the guidance of MON and both authors read and approved the final version of the manuscript.

Consent for publication
Not applicable.

Competing interests
The authors declare that they have no competing interests.

References
1. Lucas A. Programming by early nutrition in man. CIBA Found Symp. 1991; 156:38–50.
2. Lucas A. Programming by early nutrition: an experimental approach. J Nutr. 1998;128:401S–6S.
3. Hales CN, Barker DJ. Type 2 (non-insulin-dependent) diabetes mellitus: the thrifty phenotype hypothesis. Diabetologia. 1992;35:595–601.
4. Gluckman PD, Hanson MA, Spencer HG. Predictive adaptive responses and human evolution. Trends Ecol Evol. 2005;20:527–33.
5. Bloomfield FH, Oliver MH, Giannoulias CD, Gluckman PD, Harding JE, Challis JRG. Brief undernutrition in late-gestation sheep programs the hypothalamic-pituitary-adrenal axis in adult offspring. Endocrinology. 2003; 144:2933–40.
6. Gopalakrishnan GS, Gardner DS, Rhind SM, Rae MT, Kyle CE, Brooks AN, et al. Programming of adult cardiovascular function after early maternal undernutrition in sheep. Am J Physiol-Endoc M. 2004;287:R12–20.
7. Mühlhäusler BS, Duffield JA, McMillen IC. Increased maternal nutrition increases leptin expression in perirenal and subcutaneous adipose tissue in the postnatal lamb. Endocrinology. 2007;148:6157–63.
8. Ford S, Hess B, Schwope M, Nijland M, Gilbert J, Vonnahme K, et al. Maternal undernutrition during early to mid-gestation in the ewe results in altered growth, adiposity, and glucose tolerance in male offspring. J Anim Sci. 2007;85:1285–4.
9. Nielsen MO, Kongsted AH, Thygesen MP, Strathe AB, Caddy S, Quistorff B, et al. Late gestation undernutrition can predispose for visceral adiposity by altering fat distribution patterns and increasing the preference for a high-fat diet in early postnatal life. Brit J Nutr. 2013;109:2098–110.
10. Khanal P, Husted SV, Axel AMD, Johnsen L, Pedersen KL, Mortensen MS, et al. Late gestation over- and undernutrition predispose for visceral adiposity in response to a post-natal obesogenic diet, but with differential impacts on glucose–insulin adaptations during fasting in lambs. Acta Physiol. 2014; 210:110–26.
11. Nissen PM, Nebel C, Oksbjerg N, Bertram HC. Metabolomics reveals relationship between plasma inositols and birth weight: possible markers for fetal programming of type 2 diabetes. J Biomed Biotechnol. 2011;article ID: 378268, doi: https://doi.org/10.1155/2011/378268.
12. Rutherford JN. Toward a nonhuman primate model of fetal programming: phenotypic plasticity of the common marmoset fetoplacental complex. Placenta. 2012;33:e35–9.
13. Rattray PV, Garrett WN, East NE, Hinman N. Growth, development and composition of the ovine conceptus and mammary gland during pregnancy. J Anim Sci. 1974;38:613–26.
14. Symonds ME, Stephenson T, Gardner DS, Budge H. Long-term effects of nutritional programming of the embryo and fetus: mechanisms and critical windows. Reprod Fert Develop. 2007;19:53–63.
15. Zhu MJ, Ford SP, Means WJ, Hess BW, Nathanielsz PW, Du M. Maternal nutrient restriction affects properties of skeletal muscle in offspring. J Physiol. 2006;575:241–50.
16. Todd SE, Oliver MH, Jaquiery AL, Bloomfield FH, Harding JE. Periconceptional undernutrition of ewes impairs glucose tolerance in their adult offspring. Pediatr Res. 2009;65:409–13.
17. Bell AW. Regulation of organic nutrient metabolism during transition from late pregnancy to early lactation. J Anim Sci. 1995;73:2804–19.
18. Reynolds LP, Borowicz PP, Caton JS, Vonnahme KA, Luther JS, Hammer CJ, et al. Developmental programming: the concept, large animal models, and the key role of uteroplacental vascular development. J Anim Sci. 2010;88:E61–72.
19. Barker DJP. Fetal programming of coronary heart disease. Trends Endocrin Met. 2002;13:364–8.
20. Chadio SE, Kotsampasi B, Papadomichelakis G, Deligeorgis S, Kalogiannis D, Menegatos I, et al. Impact of maternal undernutrition on the hypothalamic-pituitary-adrenal axis responsiveness in sheep at different ages postnatal. J Endocrinol. 2007;192:495–503.
21. Borwick SC, Rae MT, Brooks J, McNeilly AS, Race PA, Rhind SM. Undernutrition of ewe lambs in utero and in early post-natal life does not affect hypothalamic–pituitary function in adulthood. Anim Reprod Sci. 2003;77:61–70.
22. Bajhau HS, Kennedy JP. Influence of pre- and postpartum nutrition on growth of goat kids. Small Rumin Res. 1990;3:227–36.
23. Dalton D, Knight T, Johnson D. Lamb survival in sheep breeds on New Zealand hill country. N Z J Agric Res. 1980;23:167–73.
24. Nørgaard JV, Nielsen MO, Theil PK, Sørensen MT, Safayi S, Sejrsen K. Development of mammary glands of fat sheep submitted to restricted feeding during late pregnancy. Small Rumin Res. 2008;76:155–65.
25. Tygesen MP, Nielsen MO, Norgaard P, Ranvig H, Harrison AP, Tauson AH. Late gestational nutrient restriction: effects on ewes' metabolic and

homeorhetic adaptation, consequences for lamb birth weight and lactation performance. Arch Anim Nutr. 2008;62:44–59.

26. McGovern FM, Campion FP, Sweeney T, Fair S, Lott S, Boland TM. Altering ewe nutrition in late gestation: II. The impact on fetal development and offspring performance. J Anim Sci. 2015;93:4873–82.

27. Schinckel P, Short B. Influence of nutritional level during pre-natal and early post-natal life on adult fleece and body characters. Crop Pasture Sci. 1961;12:176–202.

28. Cleal JK, Poore KR, Boullin JP, Khan O, Chau R, Hambidge O, et al. 2007. Mismatched pre- and postnatal nutrition leads to cardiovascular dysfunction and altered renal function in adulthood. Proc Natl Acad Sci. 2007;104:9529–33.

29. Greenwood P, Cafe L, Hearnshaw H, Hennessy D, Thompson J, Morris S. Long-term consequences of birth weight and growth to weaning on carcass, yield and beef quality characteristics of Piedmontese-and Wagyu-sired cattle. Anim Prod Sci. 2006;46:257–69.

30. Yan X, Zhu MJ, Dodson MV, Du M. Developmental programming of fetal skeletal muscle and adipose tissue development. J Genomics. 2013;1:29–38.

31. Du M, Tong J, Zhao J, Underwood KR, Zhu M, Ford SP, et al. Fetal programming of skeletal muscle development in ruminant animals. J Anim Sci. 2010;88:E51–60.

32. Wigmore PM, Stickland NC. Muscle development in large and small pig fetuses. J Anat. 1983;137:235–45.

33. Fahey AJ, Brameld JM, Parr T, Buttery PJ. The effect of maternal undernutrition before muscle differentiation on the muscle fiber development of the newborn lamb. J Anim Sci. 2005;83:2564–71.

34. Underwood KR, Tong JF, Price PL, Roberts AJ, Grings EE, Hess BW, et al. Nutrition during mid to late gestation affects growth, adipose tissue deposition, and tenderness in cross-bred beef steers. Meat Sci. 2010;86:588–93.

35. Hou L, Kongsted AH, Ghoreishi SM, Takhtsabzy TK, Friedrichsen M, Hellgren LI, et al. Pre- and early-postnatal nutrition modify gene and protein expressions of muscle energy metabolism markers and phospholipid fatty acid composition in a muscle type specific manner in sheep. PLoS One. 2013;8:e65452.

36. Symonds ME, Lomax MA. Maternal and environmental influences on thermoregulation in the neonate. Proc Nutr Soc. 1992;51:165–72.

37. Symonds ME, Stephenson T. Maternal nutrition and endocrine programming of fetal adipose tissue development. Biochem Soc Trans. 1999;27:97–104.

38. Symonds ME, Sebert SP, Hyatt MA, Budge H. Nutritional programming of the metabolic syndrome. Nat Rev Endocrinol. 2009;5:604–10.

39. Nielsen MO, Hou L, Johnsen L, Khanal P, Bechshøft CL, Kongsted AH, et al. Do very small adipocytes in subcutaneous adipose tissue (a proposed risk factor for insulin insensitivity) have a fetal origin? Clin Nutr Exp. 2016;8:9–24.

40. Gondret F, Lefaucheur L, Louveau I, Lebret B. The long-term influences of birth weight on muscle characteristics and eating meat quality in pigs individually reared and fed during fattening. Archiv Tierzucht Dummerstorf. 2005;48:68–73.

41. Annison EF, White RR. Glucose utilization in sheep. Biochem J. 1961;80:162–9.

42. Brockman RP, Laarveld B. Hormonal regulation of metabolism in ruminants; a review. Livest Prod Sci. 1986;14:313–34.

43. Sasaki SI. Mechanism of insulin action on glucose metabolism in ruminants. Anim Sci J. 2002;73:423–33.

44. Husted SM, Nielsen MO, Tygesen MP, Kiani A, Blache D, Ingvartsen KL. Programming of intermediate metabolism in young lambs affected by late gestational maternal undernourishment. Am J Physiol-Endoc M. 2007;293:E548–57.

45. Kongsted AH, Tygesen MP, Husted SV, Oliver MH, Tolver A, Christensen VG, et al. Programming of glucose-insulin homoeostasis: long-term consequences of pre-natal versus early post-natal nutrition insults. Evidence from a sheep model. Acta Physiol. 2014;210:84–98.

46. Khanal P, Axel A, Kongsted A, Husted S, Johnsen L, Pandey D, et al. Late gestation under-and overnutrition have differential impacts when combined with a post-natal obesogenic diet on glucose–lactate–insulin adaptations during metabolic challenges in adolescent sheep. Acta Physiol. 2015;213:519–36.

47. Hyatt MA, Gardner DS, Sebert S, Wilson V, Davidson N, Nigmatullina Y. Suboptimal maternal nutrition, during early fetal liver development, promotes lipid accumulation in the liver of obese offspring. Reproduction. 2011;141:119–26.

48. Nicholas LM, Rattanatray L, Morrison JL, Kleemann DO, Walker SK, Zhang S, et al. Maternal obesity or weight loss around conception impacts hepatic fatty acid metabolism in the offspring. Obesity. 2014;22:1685–93.

49. Phillips DI, Bennett FI, Wilks R, Thame M, Boyne M, Osmond C. Maternal body composition, offspring blood pressure and the hypothalamic-pituitary-adrenal axis. Paediatr Perinat Ep. 2005;19:294–302.

50. Langford K, Blum W, Nicolaides K, Jones J, McGregor A, Miell J. The pathophysiology of the insulin-like growth factor axis in fetal growth failure: a basis for programming by undernutrition? Eur J Clin Investig. 1994;24:851–6.

51. McMillen I, Edwards L, Duffield J, Mühlhäusler B. Regulation of leptin synthesis and secretion before birth: implications for the early programming of adult obesity. Reproduction. 2006;131:415–27.

52. Wilcoxon JS, Redei EE. 2004. Prenatal programming of adult thyroid function by alcohol and thyroid hormones. Am J Physiol-Endoc M. 2004; 287:E318–E26.

53. Kongsted AH, Husted SV, Thygesen MP, Christensen VG, Blache D, Tolver A, et al. Pre- and postnatal nutrition in sheep affects beta-cell secretion and hypothalamic control. J Endocrinol. 2013;219:159–71.

54. Festa-Bianchet M, King WJ. Age—related reproductive effort in bighorn sheep ewes. Ecoscience. 2007;14:318–22.

55. Louey S, Cock ML, Harding R. Long term consequences of low birthweight on postnatal growth, adiposity and brain weight at maturity in sheep. J Reprod Dev. 2005;51:59–68.

56. Khanal P, Johnsen L, Axel AM, Hansen PW, Kongsted AH, Lyckegaard NB, et al. Long-term impacts of foetal malnutrition followed by early postnatal obesity on fat distribution pattern and metabolic adaptability in adult sheep. PLoS One. 2016;11:e0156700.

57. Hou L, Hellgren LI, Kongsted AH, Vaag A, Nielsen MO. Pre-natal undernutrition and post-natal overnutrition are associated with permanent changes in hepatic metabolism markers and fatty acid composition in sheep. Acta Physiol. 2014;210:317–29.

58. Legrand P, Rioux V. The complex and important cellular and metabolic functions of saturated fatty acids. Lipids. 2010;45:941–6.

59. Gardner DS, Tingey K, Van Bon BWM, Ozanne SE, Wilson V, Dandrea J, et al. Programming of glucose-insulin metabolism in adult sheep after maternal undernutrition. Am J Physiol-Endoc M. 2005;289:R947–R54.

60. Husted SM, Nielsen MO, Blache D, Ingvartsen KL. Glucose homeostasis and metabolic adaptation in the pregnant and lactating sheep are affected by the level of nutrition previously provided during her late fetal life. Domest Anim Endocrin. 2008;34:419–31.

61. Bell AW, Bauman DE. Adaptations of glucose metabolism during pregnancy and lactation. J Mammary Gland Biol. 1997;2:265–78.

62. Todini L 2007. Thyroid hormones in small ruminants: effects of endogenous, environmental and nutritional factors. Animal. 2007;1:997-1008.

63. Webster JR, Moenter SM, Woodfill CJI, Karsch FJ. Role of the thyroid gland in seasonal reproduction. II. Thyroxine allows a season-specific suppression of gonadotropin secretion in sheep. Endocrinology. 1991;129:176–83.

64. Johnsen L, Kongsted AH, Nielsen MO. Prenatal undernutrition and postnatal overnutrition alter thyroid hormone axis function in sheep. J Endocrinol. 2013;216:389–402.

65. Ahima RS, Flier JS. Leptin. Annu Rev Physiol. 2000;62:413–37.

66. Mühlhäusler BS. Programming of the appetite-regulating neural network: a link between maternal overnutrition and the programming of obesity? J Neuroendocrinol. 2007;19:67–72.

67. Dupont C, Cordier AG, Junien C, Mandon-Pépin B, Levy R, Chavatte-Palmer P. 2012. Maternal environment and the reproductive function of the offspring. Theriogenology. 2012;78:1405–14.

68. Rhind SM. Effects of maternal nutrition on fetal and neonatal reproductive development and function. Anim Reprod Sci. 2004;82:169–81.

69. Grazul-Bilska A, Borowczyk E, Arndt W, Evoniuk J, O'neil M, Bilski J, et al. Effects of overnutrition and undernutrition on in vitro fertilization (IVF) and early embryonic development in sheep. Sheep Beef Day. 2006;47:56–66.

70. Rae MT, Palassio S, Kyle CE, Brooks AN, Lea RG, Miller DW, et al. Effect of maternal undernutrition during pregnancy on early ovarian development and subsequent follicular development in sheep fetuses. Reproduction. 2001;122:915–22.

71. Rae MT, Kyle CE, Miller DW, Hammond AJ, Brooks AN, Rhind SM. The effects of undernutrition, in utero, on reproductive function in adult male and female sheep. Anim Reprod Sci. 2002;72:63–71.

72. Kotsampasi B, Chadio S, Papadomichelakis G, Deligeorgis S, Kalogiannis D, Menegatos I, et al. Effects of maternal undernutrition on the hypothalamic–pituitary–gonadal axis function in female sheep offspring. Reprod Domest Anim. 2009;44:677–84.

73. Gunn RG, Sim DA, Hunter EA. Effects of nutrition in utero and in early life on the subsequent lifetime reproductive performance of Scottish blackface ewes in two management systems. Anim Sci. 1995;60:223–30.

74. Sookoian S, Gianotti TF, Burgueño AL, Pirola CJ. Fetal metabolic programming and epigenetic modifications: a systems biology approach. Pediatr Res. 2013;73:531–42.

75. Burdge GC, Hanson MA, Slater-Jefferies JL, Lillycrop KA. Epigenetic regulation of transcription: a mechanism for inducing variations in phenotype (fetal programming) by differences in nutrition during early life? Br J Nutr. 2007;97:1036–46.

76. Stevens A, Begum G, Cook A, Connor K, Rumball C, Oliver M, et al. Epigenetic changes in the hypothalamic proopiomelanocortin and glucocorticoid receptor genes in the ovine fetus after periconceptional undernutrition. Endocrinology. 2010;151:3652–64.

77. Zhang S, Rattanatray L, MacLaughlin SM, Cropley JE, Suter CM, Molloy L, et al. Periconceptional undernutrition in normal and overweight ewes leads to increased adrenal growth and epigenetic changes in adrenal IGF2/H19 gene in offspring. FASEB J. 2010;24:2772–82.

78. Sinclair KD, Allegrucci C, Singh R, Gardner DS, Sebastian S, Bispham J, et al. DNA methylation, insulin resistance, and blood pressure in offspring determined by maternal periconceptional B vitamin and methionine status. Proc Natl Acad Sci. 2007;104:19351–6.

79. Kenyon PR, Blair HT. Foetal programming in sheep – effects on production. Small Ruminant Res. 2014;118:16–30.

80. Robinson JJ, McDonald I, Fraser C, Crofts RMJ. Studies on reproduction in prolific ewes. J Agric Sci. 1977;88:539–52.

81. Greenwood PL, Thompson AN, Ford SP. Postnatal consequences of the maternal environment and of growth during prenatal life for productivity of ruminants. In: Greenwood PL, Bell AW, Vercoe PE, Viljoen GJ, editors. In managing the prenatal environment to enhance livestock productivity. Netherlands.: Springer Science; 2010. p. 3–36.

82. Smith NA, McAuliffe FM, Quinn K, Lonergan P, Evans ACO. The negative effects of a short period of maternal undernutrition at conception on the glucose–insulin system of offspring in sheep. Anim Reprod Sci. 2010;121:94–100.

83. Greenwood P, Thompson A. Consequences of maternal nutrition during pregnancy and of foetal growth for productivity of sheep. Rec Adv Anim Nutr. 2007;16:185–96.

84. Degen AA, Benjamin RW, Hoorweg JC. Bedouin households and sheep production in the Negev Desert, Israel. Nomadic Peoples. 2000;4:125–47.

85. McWilliam EL, Barry TN, Lopez-Villalobos N, Cameron PN, Kemp PD. The effect of different levels of poplar (Populus) supplementation on the reproductive performance of ewes grazing low quality drought pasture during mating. Anim Feed Sci Technol. 2004;115:1–18.

86. Dýrmundsson ÓR. Sustainability of sheep and goat production in north European countries—from the Arctic to the alps. Small Ruminant Res. 2006;62:151–7.

87. Cifuni GF, Napolitano F, Pacelli C, Riviezzi AM, Girolami A. Effect of age at slaughter on carcass traits, fatty acid composition and lipid oxidation of Apulian lambs. Small Ruminant Res. 1993;35:65–70.

88. Bonnet M, Cassar-Malek I, Chilliard Y, Picard B. Ontogenesis of muscle and adipose tissues and their interactions in ruminants and other species. Animal. 2010;4:1093–109.

89. Godfrey KM, Barker DJ. Fetal nutrition and adult disease. Am J Clin Nutr. 2000;71:1344s–52s.

90. Bertram C, Khan O, Ohri S, Phillips DI, Matthews SG, Hanson MA. Transgenerational effects of prenatal nutrient restriction on cardiovascular and hypothalamic-pituitary-adrenal function. J Physiol. 2008;586:2217–29.

91. Daxinger L, Whitelaw E. Transgenerational epigenetic inheritance: more questions than answers. Genome Res. 2010;20:1623–8.

20

The effects of starter microbiota and the early life feeding of medium chain triglycerides on the gastric transcriptome profile of 2- or 3-week-old cesarean delivered piglets

Paolo Trevisi[1], Davide Priori[1†], Vincenzo Motta[1], Diana Luise[1], Alfons J. M. Jansman[2], Sietse-Jan Koopmans[2] and Paolo Bosi[1*†] (iD)

Abstract

Background: The stomach is an underestimated key interface between the ingesta and the digestive system, affecting the digestion and playing an important role in several endocrine functions. The quality of starter microbiota and the early life feeding of medium chain triglycerides may affect porcine gastric maturation. Two trials (T1, T2) were carried out on 12 and 24 cesarean-delivered piglets (birth, d0), divided over two microbiota treatments, but slaughtered and sampled at two or three weeks of age, respectively. All piglets were fed orally: sow serum (T1) or pasteurized sow colostrum (T2) on d0; simple starter microbiota (*Lactobacillus amylovorus*, *Clostridium glycolicum* and *Parabacteroides* spp.) (d1-d3); complex microbiota inoculum (sow diluted feces, CA) or a placebo (simple association, SA) (d3-d4) and milk replacer ad libitum (d0-d4). The The T1 piglets and half of the T2 piglets were then fed a moist diet (CTRL); the remaining half of the T2 piglets were fed the CTRL diet fortified with medium chain triglycerides and 7% coconut oil (MCT). Total mRNA from the oxyntic mucosa was analyzed using Affymetrix©Porcine Gene array strips. Exploratory functional analysis of the resulting values was carried out using Gene Set Enrichment Analysis.

Results: Complex microbiota upregulated 11 gene sets in piglets of each age group vs. SA. Of these sets, 6 were upregulated at both ages, including the set of gene markers of oxyntic mucosa. In comparison with the piglets receiving SA, the CA enriched the genes in the sets related to interferon response when the CTRL diet was given while the same sets were impoverished by CA with the MCT diet.

Conclusions: Early colonization with a complex starter microbiota promoted the functional maturation of the oxyntic mucosa in an age-dependent manner. The dietary fatty acid source may have affected the recruitment and the maturation of the immune cells, particularly when the piglets were early associated with a simplified starter microbiota.

Keywords: Development, Microbiota, Pig, Stomach, Transcriptome

Background

The stomach is a key point in the interface between the ingesta and the digestive system, affecting post-gastric efficiency of digestion, playing a role in several endocrine functions and being involved in several diseases of domestic animals and humans. Nevertheless the main molecular-cellular processes that govern its development and homeostasis have not yet been well defined. Current evidence indicates that the stomach can actively interact with the microbiota resident in or in transit to the lower digestive tracts. In pigs, the stomach shows an array of important tools related to immunity and barrier defense, also developmentally controlled: toll-like receptors (TLRs) 2, 3 and 4 [1] among the receptors specialized in detecting microorganisms and activating an innate

---

* Correspondence: paolo.bosi@unibo.it
†Equal contributors
[1]DISTAL, University of Bologna, Viale Fanin 46, 40127 Bologna, Italy
Full list of author information is available at the end of the article

immune response, the polymeric immunoglobulin receptor responsible for the transport of secretory immunoglobulins [1], submucosal and mucosal lymphoid follicles with and without compartmental organization of T and B lymphocytes [2] and plasma cells [3]. This implies the relevance of adequate stimulation of the local microbiota as has been demonstrated in the intestine [4]. This would also have practical relevance for managing the piglets, namely stimulating the diffusion of management solutions favoring the acquisition by neonate piglets of microbiota also out of their litter. Conversely, the acquisition of the dominant importance of specific core microbiota would indicate that certain bacterial pools can be used for the early orientation or acceleration of gut maturation. However, surprisingly, studies regarding the impact of different microbiota on the development of the stomach of mammals are lacking. It has generally been recognized that the availability of a single 'standard' microbiome could circumvent several problems in studies focusing on the interaction of the host and its microbiota [5]. For this purpose, Laycock et al. [6] selected a few bacteria strains as a standardized microbiota owing to their positive influence on a new early microbial colonization in the porcine gut.

Medium chain triglycerides have received research attention due to their potential ability for selecting microbes [7, 8]; they are an interesting nutrient source for very young animals [9], but also potentially interact with the acid-secreting function of the oxyntic mucosa, owing to their acidic nature. It has been observed that constant dietary supplementation with different dietary organic acids can affect (Formate [10]) or not (Butyrate [11]) the morphology of the oxyntic mucosa of the young pig, reducing the relative number of parietal cells responsible for acid secretion, and the expression of the protein marker of this function, H^+/K^+-ATPase. This could be due to the reduced gastrin-dependent growth stimulation of the oxyntic cells after the increase in feed acidity [12].

Two studies were designed to elucidate the molecular networks in the oxyntic mucosa of the young pig which characterize the long-term individual response after various associations with a simplified or complex starter microbiota during their first days of life. Furthermore, an additional aim in the second study was to evidence the possible action of a dietary source of medium chain triglycerides on the transcriptome of the oxyntic mucosa.

Methods

Two different trials (Fig. 1) were carried out on 12 and 24 piglets obtained from from 4 and 6 sows respectively [(Great York × Pietrain) × 'Dalland' cross] by cesarean delivery (CD) (d 0); they were divided over two microbiota association treatments, and were then slaughtered and sampled at wk 2 (Trial 1) or wk 3 (Trial 2) of age, respectively. Piglets from different litters were balanced over the various groups based on litter, body weight and gender and in each pen/room 3 piglets per group were housed.

The piglets were housed in two pens per clean, non-sterile room having an automatic feeding system for supplying a moist diet, and were balanced for birth weight (BW) and litter of origin. At 1 h and 5 h after birth, each piglet received 110 mL sow serum (Trial 1) or 90 mL pasteurized (30 min at 60 °C) sow colostrum by oral gavage [13, 14] (Trial 2). All the piglets received a simple starter microbiota consisting of *Lactobacillus amylovorus* (3.6×10^7 CFU), *Clostridium glycolicum* (5.7×10^7 CFU) and *Parabacteroides* spp. (4.8×10^7 CFU) by oral inoculation (2 mL) on d1, d2 and d3 after birth. These bacteria are the most frequently identified of the phylogenetic groups in the intestines of piglets as have previously been proposed as minimal intestinal colonization microbiota for gnotobiotic pigs [6].

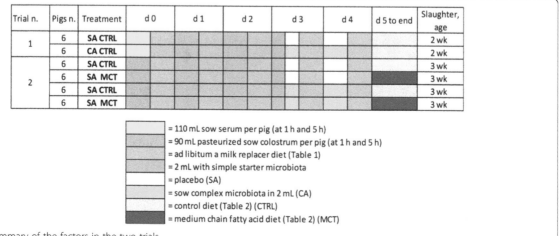

Trial n.	Pigs n.	Treatment	d 0	d 1	d 2	d 3	d 4	d 5 to end	Slaughter, age
1	6	SA CTRL							2 wk
	6	CA CTRL							2 wk
2	6	SA CTRL							3 wk
	6	SA MCT							3 wk
	6	SA CTRL							3 wk
	6	SA MCT							3 wk

  = 110 mL sow serum per pig (at 1 h and 5 h)
  = 90 mL pasteurized sow colostrum per pig (at 1 h and 5 h)
  = ad libitum a milk replacer diet (Table 1)
  = 2 mL with simple starter microbiota
  = placebo (SA)
  = sow complex microbiota in 2 mL (CA)
  = control diet (Table 2) (CTRL)
  = medium chain fatty acid diet (Table 2) (MCT)

Fig. 1 General summary of the factors in the two trials

On d 3 (at 15:30 and 21:00 h) and on d 4 (at 8:00 h), the piglets of treatment CA received a fecal-oral inoculant immediately after feeding in the form of a 2-mL suspension of feces (feces in 0.1 mol/L sterile potassium phosphate buffer (pH 7.2), diluted 1:10). The SA-treated piglets received a placebo treatment (2 mL of a sterile 0.1 mol/L potassium phosphate buffer, pH 7.2).

Average birth weight was 1213 ± 144 g and 1228 ± 164 (Trial 1), and 1273 ± 138 g and 1275 ± 153 g (Trial 2) for the CA and the SA piglets, respectively. The piglets were fed a milk replacer diet ad libitum during a period of 5 d (d0 - d4). It consisted of bovine skimmed milk powder, whey powder, vegetable oil, hydrolyzed wheat protein, wheat starch, sucrose, and a vitamin and mineral premix, and contained 230 g crude protein per kg milk replacer (Table 1). All the piglets in Trial 1 and half of the piglets in Trial 2 were then fed a moist diet (CTRL) based on whey powder, maize, wheat, toasted full fat soybeans, oat flakes, sucrose, soybean meal, vegetable oil, coconut oil, wheat gluten, potato protein, rice protein, and brewer's yeast during the remainder of the study (Table 2). The other half of the piglets in Trial 2 were fed the same diet but containing 7% of coconut oil in place of 4.68% soy oil and of 2.32% palm oil in order to generate a diet fortified in medium chain triglycerides (MCTs) (Table 2). In other words, the 24 piglets of the 2nd trial were allocated by BW to 4 treatments arranged in a 2 × 2 factorial design, with 2 different early associations with microbiota (CA and SA) and two diets (CTRL and MCT). On d 16 (Trial 1), and on d 21 and d 22 (Trial 2), the piglets were euthanized, and the stomach was

Table 1 Analyzed nutrient composition of the artificial milk provided to the piglets during the first 4-day after birth (in g/kg, unless stated otherwise)

Nutrient	Content
Dry matter	957
Ash	78
Crude protein	222
Crude fiber	1
Starch	27
Glucose	29
Sugar	435
Lactose	405
Phosphorus	7.2
Calcium	8.6
Crude fat	227
Total fatty acids (g/kg fat)	913
Saturated fatty acids (g/kg fat)	461
Monounsaturated fatty acids (g/kg fat)	360
Polyunsaturated fatty acids (g/kg fat)	86

Table 2 Ingredient composition of the experimental diets provided to the piglets from day 5 after birth until the end of the study (g/kg)

Ingredients	CTRL	MCT
Maize	200.0	200.0
Whey powder	180.0	180.0
Wheat	150.0	150.0
Oat hulls	80.0	80.0
Barley	70.0	70.0
Wheat gluten meal	50.0	50.0
Soybeans, heat treated	40.0	40.0
Soycomill-P	33.0	33.0
Lactose	30.0	30.0
Protastar	30.0	30.0
Monocalciumphosphate	0.6	0.6
Rice meal	12.0	12.0
Limestone	0.9	0.9
Salt	3.6	3.6
Dicalciumphosphate	19.4	19.4
Sodium bicarbonate	4.5	4.5
Piglet premix 0–5%	5.0	5.0
L-Lysine HCl	5.9	5.9
DL-Methionine	2.0	2.0
L-Threonine	1.1	1.1
L-Tryptophan	0.9	0.9
L-Valine	0.5	0.5
Soy oil	50.6	3.8
Palm oil	30.0	6.8
Coconut oil	0	70.0

sampled always in the center of the area corresponding to the oxyntic mucosa.

RNA isolation and gene array quantification

Total RNA was isolated from the gastric tissue samples using the FastPure RNA kit (TaKaRa Bio Inc., Shiga, Japan). All other procedures were in agreement with the manufacturer's protocol. The RNA purity and integrity were evaluated using Agilent Bioanalyzer 2100 (Agilent Technologies, Palo Alto, CA) just before the transcriptome analysis. The total RNA was hybridized on Affymetrix Porcine Gene 1.1 ST array strips. The hybridized arrays were scanned on a GeneAtlas imaging station (Affymetrix, Santa Clara, CA, USA). Performance quality tests of the arrays including the labelling, hybridization, scanning and background signals using Robust Multichip Analysis were carried out on the CEL files using the Affymetrix Expression Console.

Statistics

The Affymetrix Trascript IDs, in general each characterized by several exonic sequences, were associated with 13,406 human gene names, based on *Sus scrofa* Ensembl (release 79, www.ensembl.org), where data regarding each porcine gene could be directly accessed using the names of genes which were adopted for the present text. Regarding the processed gene expression values, two exploratory functional analyses were applied with Gene Set Enrichment Analysis using the Hallmark (aggregating many gene sets to represent well-defined biological states or processes) and C5.BP (based on Gene Ontology (GO)) catalogues of gene sets from Molecular Signatures Database v5.1 (http://software.broadinstitute.org/gsea/msigdb/). The normalized enrichment score (NES) was calculated for each gene set, and statistical significance was defined when a False Discovery Rate (FDR) % < 25 and *P*-values of NES < 0.05. Enrichment score *P*- values were estimated using a gene set-based permutation test procedure. A new data set was created and added to the Hallmark catalogue which included 19 genes typically characterizing the stomach as compared with the different intestinal sections [15], and were more or equally expressed in the porcine oxyntic mucosa as compared with pyloric mucosa [16]. Reference sequences for all these gene are available on *Sus scrofa* Ensembl except for the following genes that have the Ensembl gene accession number: *ATP4A*, ENSSSCT00000003207; *EGF*, ENSSSCT00000010002, or the NCBI Reference Sequence: *LPL*, NM_214286.1; *TFF1*, XM_003358973.3.

The overlap in enriched GO terms was visualized using the Enrichment Map (http://baderlab.org/Software/EnrichmentMap20) plugin for Cytoscape 2.8.0 (http://www.cytoscape.org/), including the gene sets with a *P*-value <0.005 and an FDR *q*-value <0.10. The nodes were joined if the overlap coefficient was ≥0.5.

Results

As general information, body weight gain was similar among piglet groups both in trial 1 and trial 2. Average birth weight was approx. 1 kg, increased to 2.5 kg at 2 wk of age and to 4 kg at 3 wk of age. Feed intake on a pen base did not differ. The general health score (including diarrhea) was similar among piglets, both for trial 1 and 2, and diarrhea was absent in trial 1 and in trial 2 diarrhea was only incidental and erratic among the 4 groups of piglets as judged on a daily basis from birth till 3 wk of age.

Figure 2a visualizes the nodes of the gene sets of the Hallmark catalogue, enriched (color red) or impoverished (color blue) in the porcine oxyntic mucosa of piglets early associated with complex microbiota (CA) as compared to simplex microbiota (SA) piglets, and sampled at two or three weeks of age (represented by the center and the ring

of each node, respectively). The CA upregulated 9 and 10 gene sets in piglets of two and three weeks of age, respectively as compared to the SA. Of these sets, 4 were upregulated at both ages. The CA downregulated 6 gene sets at both ages, all related to cycling activity, mitosis regulation steps and general protein synthesis. In addition, another 1 and 8 gene sets were downregulated in piglets of two and three weeks of age, respectively as compared to the SA.

In addition to the gene sets of the Hallmark catalogue, represented in Fig. 2a, the new cured gene set oxyntic mucosa was upregulated by the CA at both two and three weeks of age (Table 3). Nevertheless, the upregulation involved more genes in two-week old piglets than in the older piglets. Potassium voltage-gated channel, Isk-related family, member 2 (*KCNE2*), epidermal growth factor (*EGF*), Anion exchanger 2 (*SLC4A2*), Lipoprotein lipase (*LPL*), potassium inwardly rectifying channel, subfamily J, member 15 (*KCNJ15*) and solute carrier family 2 (facilitated glucose transporter), member 4 (*SLC2A4*) genes were enriched at both ages of the piglets.

Figure 2b visualizes the nodes of the gene sets of the C5 catalogue (based on Gene Ontology), differentially enriched in the CA or the SA piglets at 2 or 3 wk of age following the same scheme as Fig. 2a. From the figure, all the gene sets related to cycling activity and mitosis regulation, already shown in Fig. 2a, were depicted. The figure primarily shows that two groups of gene sets, related respectively to response to nutrient stimulus and to transforming the growth factor β (TGFB) pathway were upregulated by the CA in 2-week-old piglets, but not in older pigs; a group of gene sets related to chemokine activity was down- and upregulated with CA in piglets of two and three weeks of age respectively as compared to the SA.

A second gene enrichment study was generated to demonstrate the effect of early association with microbiota of different complexities regarding the enrichment of gene sets expressed in the porcine oxyntic mucosa fed each of the two diets. Figure 3a visualizes the nodes of the gene sets of the Hallmark catalogue, enriched (color red) or impoverished (color blue) in samples from the CA piglets as compared to the SA piglets, and fed the MCT or the CTRL diet (represented by the center and the ring of each node, respectively). As compared to the SA piglets, the CA enriched genes in the sets related to both the interferon α and γ response when the CTRL diet was given while the same sets were impoverished by the CA with the MCT diet. The CA downregulated 5 gene sets in both the diets, all related to controlling the cell cycling, cellular transformations and the related stress-induced homeostatic mechanisms of DNA control and replication. Another 3 gene sets were also downregulated by the CA, including mTORC-related protein synthesis and cellular stress response related to the

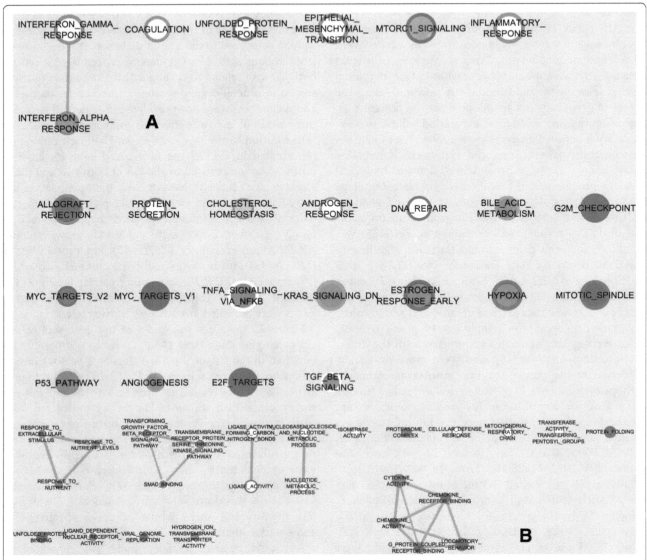

Fig. 2 The nodes of the gene sets enriched in the oxyntic mucosa of piglets early associated with CA as compared to the SA piglets and sampled at two or three weeks of age. **a** The entire picture regarding the Hallmark gene set, summarizing the main biological states or processes; **b** Picture of nodes obtained with the more detailed gene ontology (GO) gene set after the removal of those related to cell mitosis; the nodes represent gene sets. The color on the center of the node represents the results of the trial ending at 2 wk; the color on the ring visualizes the trial ending at 3 wk. The edges represent the link of two or more gene sets sharing the same core group of genes explaining the enrichment of each of the gene sets. Enrichment significance (*P*-value) is conveyed as node color intensity where red stands for upregulation at 2 or 3 wk in the CA piglets, and blue stands for downregulation as compared with the SA piglets. The node size represents the number of genes in the gene set

endoplasmic reticulum. Conversely, 3 gene sets were downgraded by the CA only when the CTRL diet was used, including the set related to protein secretion. The set of genes related to inflammatory response was enriched in the CA piglets fed only the CTRL diet while the TNFα and reactive oxygen species pathway sets were enriched in the SA piglets.

Figure 3b shows the nodes of the gene sets of the C5 catalogue (based on Gene Ontology), differentially enriched in the CA or the SA piglets, supplemented with medium chain fatty acids (MCTs) or not (CTRL),

following the same scheme as Fig. 2a. Compared to the SA, the gene sets related to transmembrane receptor activity, including that of the neurotransmitters, were enriched in the CA piglets, particularly when fed the MCT diet (as demonstrated by the more intense red coloration). With the MCT diet, several gene sets related to anion channel activity were enriched by the CA as compared to the SA.

A third gene enrichment study was generated to demonstrate the effect of the different diets on the enrichment of the gene sets in the oxyntic mucosa from pigs which

Table 3 List of genes enriched or not inside the new cured gene set oxyntic mucosa tested inside the Hallmark catalogue

Gene product	Gene name	Trial 1 (2-week-old piglets)		Trial 2 (3-week-old piglets)	
		Rank enrichment score[a]	Core enrichment	Rank enrichment score[a]	Core enrichment
Potassium voltage-gated channel, Isk-related family, member 2	KCNE2	0.729	Yes	0.330	Yes
Epidermal growth factor	EGF	0.583	Yes	0.185	Yes
Anion exchanger 2	SLC4A2	0.580	Yes	0.176	Yes
Lipoprotein lipase	LPL	0.578	Yes	0.321	Yes
Potassium inwardly rectifying channel, subfamily J, member 15	KCNJ15	0.576	Yes	0.406	Yes
Chloride intracellular channel 6	CLIC6	0.495	Yes	0.070	No
Solute carrier family 2 (facilitated glucose transporter), member 4	SLC2A4	0.479	Yes	0.275	Yes
ATPase, H^+/K^+ exchanging, β polypeptide	ATP4B	0.417	Yes	0.058	No
ATPase, H^+/K^+ exchanging, α polypeptide	ATP4A	0.371	Yes	0.013	No
Insulin-like growth factor binding protein 5	IGFBP5	0.266	Yes	0.055	No
Pheromaxein, subunit C	PHEROC	0.248	Yes	−0.048	No
Acquaporin 4	AQP4	0.236	Yes	0.006	No
Potassium inwardly rectifying channel, subfamily J, member 13	KCNJ13	0.167	No	0.202	Yes
Ghrelin	GHRL	0.094	No	−0.237	No
Trefoil factor 1	TFF1	0.093	No	0.002	No
Pepsinogen C	PGC	0.075	No	−0.072	No
Insulin-like growth factor 1	IGF1	0.067	No	0.401	Yes
Pepsinogen B	PGB	−0.023	No	−0.202	No
Chitinase, acidic	CHIA	−0.089	No	−0.121	No

[a]Reflects the relative degree to which a gene is overrepresented at the top (positive sign) or bottom (negative sign) of the ordered dataset

had an early association with microbiota of different complexities. Figure 4a shows the nodes of the gene sets of the Hallmark catalogue, enriched (color red) or impoverished (color blue) in samples from piglets fed the MCT diet compared to CTRL-fed piglets, and belonging to the CA or the SA group (represented by the ring and the center of each node, respectively). The MCT diet enriched gene sets of INFLAMMATORY_RESPONSE, TNFA_SIGNALING_VIA_NFKB, IL2_STAT5_SIGNALING, COMPLEMENT and also EPITHELIAL_MESENCHYMAL_TRANSITION, whatever the type of early association with the microbiota. As compared to the CTRL diet, feeding with the MCT diet impoverished genes in the sets related to both interferon α and γ response in the CA piglets while the same gene sets were enriched with the MCT diet in the SA piglets. Furthermore, ALLOGRAFT_REJECTION, IL6_JAK_STAT3_SIGNALING, APOPTOSIS, KRAS_SIGNALING_UP were also enriched by MCTs only in the SA piglets.

Several nodes were also observed with the more detailed GO catalogue (Fig. 4b) with the MCT diet in the SA piglets as compared to the CTRL diet: a large network of gene sets related to chemokine and cytokine activity and binding, immune cell activation and, in general, the positive regulation of the immune system process; sparse gene sets including those related to tissue modelling and differentiation.

Discussion

The joint visualization of the results of the two experiments allowed demonstrating that some patterns of stimulation of gene groups are constantly related to the type of early association to microbiota, at least for the age range of our piglets. Conversely, other gene groups were regulated by the early treatment of pigs in a more age dependent manner. In general, early association with a complex starter microbiota tended to anticipate functional maturation while association with a simplified microbiota stimulated the cell cycles in the oxyntic mucosa.

Proliferative response in the oxyntic mucosa
The local action of pathogenic bacteria can overactivate a range of growth signaling pathways within the gastric mucosa, thus altering the physiological balance between growth and apoptosis [17, 18]. On the contrary, in rats, the probiotic bacteria can have a restorative action on an acetic-induced gastric ulcer, reestablishing correct cell proliferation [19]. Bacteria can also produce putrescine

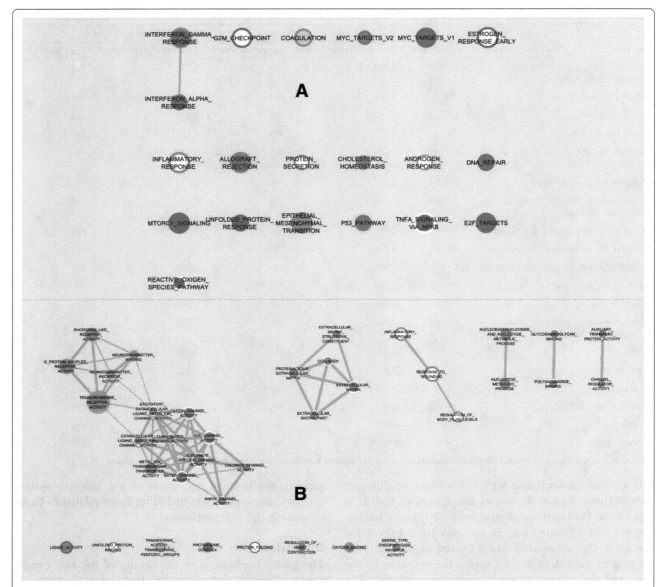

Fig. 3 The nodes of the gene sets enriched in the oxyntic mucosa of piglets early associated with the CA as compared to the SA piglets, and supplemented with medium chain fatty acids (MCTs) or not (CTRL). **a** The entire picture for the Hallmark gene set, summarizing the main biological states or processes; **b** Picture of nodes obtained with the more detailed GO gene set; Nodes represent gene sets. The color on the center of the node represents the MCT diet; the color on the ring represents the CTRL diet. The edges represent the link of two or more gene sets sharing the same core group of genes, thus explaining the enrichment of each of the gene sets. Enrichment significance (P-value) is conveyed as node color intensity where red stands for upregulation in the CA piglets fed the MCT or the CTRL diets, and blue for downregulationas compared with the SA piglets. The node size represents the number of genes in the gene set

and spermidine, the short-chain aliphatic amines involved in the regulation of cell proliferation and differentiation [20]. In our study, the early provision of a complex starter microbiota to the piglets favorably prevented the activation of pathways related to cycling activity and mitosis regulation steps (Fig. 1a) in the gastric mucosa at an age of two and three weeks as compared to the association with a simple microbiota. This may have been related to the observation that a short

encounter with a complex microbiota in early life could be sufficient to influence the intestinal microbiota in the following weeks ([21] which was the same animal experimental design of trial 1. Nevertheless, the reduction of proliferative activity in the CA piglets was not accompanied by an insufficient maturation of the oxyntic mucosa but, on the contrary, several major genes implicated in the activity of the parietal cells or typical of the gastric mucosa were upregulated (Table 1).

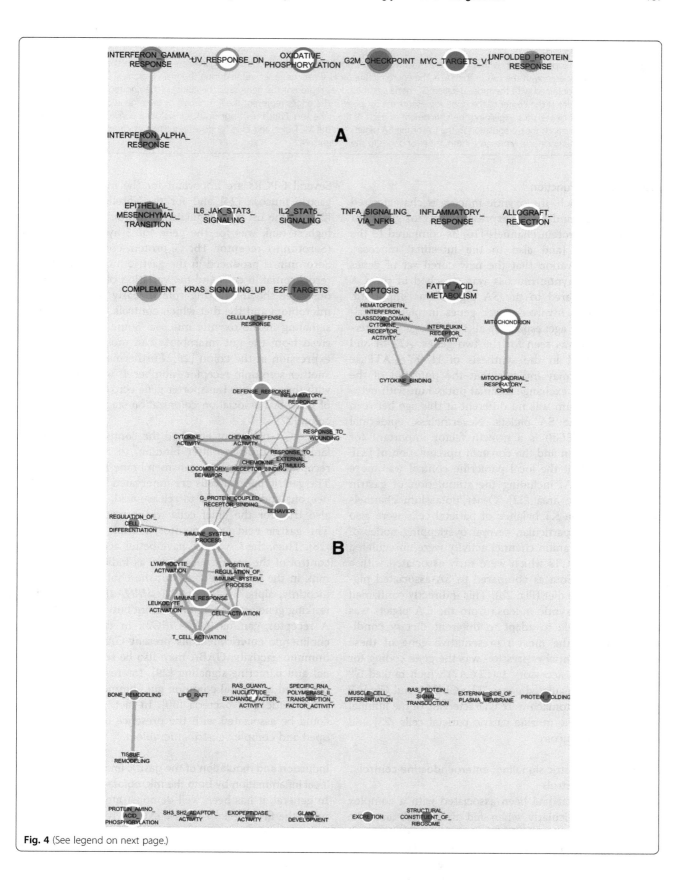

Fig. 4 (See legend on next page.)

(See figure on previous page.)
Fig. 4 The effect of supplemention with a fortified diet with (MCT) or not (CTRL) on the nodes of the gene sets enriched in the oxyntic mucosa of piglets early associated with the CA) or the SA. **a** The entire picture for Hallmark gene set, summarizing the main biological states or processes; **b** Picture of nodes obtained with the more detailed GO gene set; the nodes represent the gene sets. The color on the ring represents the CA piglets while the color at the center of the node represents the SA piglets. The edges represent the link of two or more gene sets sharing the same core group of genes, thus explaining the enrichment of each of the gene sets. Enrichment significance (*P*-value) is conveyed as node color intensity where red stands for upregulation in the CA or the SA piglets fed the MCT diet, and blue for downregulation as compared with the CTRL-fed piglets. The node size represents the number of genes in the genen set

Acidic secretion function

Inside the stomach, the oxyntic mucosa is characterized by a very large and highly expressed array of genes for ion and water protein channels [16] as compared to the pyloric mucosa (and also to the intestinal mucosa). Thus, the observation that the new cured set of genes typical of the oxyntic mucosa was enriched in the CA piglets as compared to the SA piglets was important. The enrichment involved fewer genes in piglets sacrificed at an older age; especially at 3 wk of age no different expression was seen for the two genes, ATP4A and ATP4B, involved in the synthesis of H^+/K^+ –ATPase (Table 1). This may indicate that the potential of the proton pump to exchange luminal potassium with cytoplasmic hydronium was no different at this age between the CA and the SA piglets. Nevertheless, epidermal growth factor (EGF) is a growth factor important for gastric maturation and the constant upregulation of EGF may indicate that the local paracrine control was more active in the CA, including the stimulation of gastrin from the pyloric area [22]. Other potassium channels implicated in the K+ balance of parietal cells were also upregulated. In particular, several overlapping nodes of genes related to anion channel activity were upregulated in piglets fed MCTs which were early associated with a complex microbiota as compared to SA-associated piglets fed the same diet (Fig. 2b). This indirectly confirmed again that the oxyntic mucosa from the CA piglets was already more able to adapt to different dietary conditions. Notably, the most representative gene of these nodes for differential expression was the gene coding for chloride channel accessory 1 (*CLCA1*) which ranked 5th in the list of upregulated genes. The CLCA1 and CLCA2 products in the stomach were characteristic of the luminal membranes of the murine gastric parietal cells [23] and also of surface mucosa.

Maturation of gastric signaling, enteroendocrine control and neuronal controls

In the piglets that had been associated with a complex microbiota, particularly when fed the MCT diet, the connected nodes G_PROTEIN_COUPLED_RECEPTOR (GPCR)_ACTIVITY, RHODOPSIN_LIKE_RECEPTOR_ ACTIVITY and NEUROTRANSMITTER_ _RECEPTOR_ ACTIVITY were enriched of activated genes (Fig. 2b).

Several GPCRs are important for the maturation of the gastric mucosa [24] and for the ghrelin-secreting cells [25]. In the list of upregulated genes of these nodes, the highest rank was for the gene for 5-hydroxytryptamine (Serotonin) receptor 1B, G protein-coupled (*HTR1B*). Serotonin is produced in the gastric mucosa by the enterochromaffin cells and the variation of this gene demonstrates the interacting role of early priming by the microbiota and the diet which controls the serotonergic signaling in the oxyntic mucosa. Some metabolites derived from the gut microbiota can modulate serotonin expression in the colon [26]. Furthermore, the gene for another serotonin receptor (number 4) was co-associated with the serotonin transporter gene across different states of microbial association colonization stages in the mouse colon [27].

The CA treatment enriched the complex of nodes related to neurotransmitter binding, to neurotransmitter receptor activity and to transmembrane receptor activity. The gastric parietal cells are innervated by secretomotor neurons and stimulated to release acid, and the same is also true for the chief cells which release pepsinogen. The gastric acid secretomotor neurons are cholinergic [28]. Thus, the CA could have better activated neuronal control of the gastric secretions, as indicated by its first rank in the list of the genes of the cholinergic receptor, nicotinic, alpha 2 (neuronal) (*CHRNA2*). Another high-ranking gene was the gamma-aminobutyric acid (GABA) A receptor, gamma 2 (*GABRG2*). In the stomach, the cholinergic enteric neurons present GABAergic neuron immunoreactivity; GABA may also be sensed as hormonal and paracrine signaling [29]. Interestingly, GABA is produced by several common gut commensals, including some lactic acid bacteria [30]. In fact, our observation could be associated with the presence of a more developed and complex gastric microbiota.

Induction and regulation of the gastric immune system and local inflammation by both the microbiota and the diet

In general, it has been well demonstrated that, in mammals, the first week of life are extremely relevant for the induction of proper maturation of all the key actors of innate and acquired defenses in the gut, in functional dependence on the host antigenic experience. Specifically, in piglets, in the passage from the 2nd to the 4th wk of age, the intestinal mucosa becomes colonized by CD4 + T cells and the number of intraepithelial lymphocytes, and IgM + B cells start to appear in small numbers [31].

The degree of activation of different pathways related to the induction and the regulation of the gastric immune system and local inflammation by the microbiota changed with the age of the animals according to the type of previous association with microbiota and diet. With the control diet, the piglets sampled at three weeks of age which had been associated with a complex microbiota had overlapping enriched nodes related to interferon α and γ, cellular defense response and inflammatory response as compared to the SA piglets (Fig. 1a). This activation was also linked to more activity and the receptor binding activation of the cytokines (Fig. 1b). Of the genes in these groups, there was gene coding for chemokine (C-X-C motif) ligand 11 (CXCL11) which ranked 2nd in the list of all genes. The chemotactic protein CXCL11 is involved in mature T-cell recruitment and is induced by interferon γ [32]. All the piglets were apparently healthy and, thus, this could be interpreted in the general situation of the higher functional maturity of the CA piglets. The presence of a remodeling phase in the older CA piglets versus the SA piglets could also be supported by the enrichment of gene sets COAGULATION and EPITHELIAL_MESENCHIMAL_ TRANSITION where the first ranking gene was matrix metalloproteinase 3 (MMP3) responsible for tissue remodeling [33]. The differential maturation of the immune cells associated with the gastric tissue of the CA piglets as compared to the SA piglets is also indicated by the higher expression of a gene not considered in the two GSEA catalogues, JCHAIN (Joining chain of multimeric IgA and IgM). The product of the JCHAIN is essential for the fusion of two monomer units of either IgM or IgA, also cooperating with their binding to the secretory component. This gene is induced by long-term bacterial colonization [34].

With MCTs, the interferon α and γ gene sets were stimulated in the SA piglets (Fig. 2a), but not in the CA piglets, indicating that the presence of MCTs in the diet could condition these piglets to a maturation such as that of the CTRL-fed CA piglets. These observations relating to the CTRL-fed piglets are, in some ways, in contrast with those obtained in the 1st trial with the 2-week-old piglets. In this case, in addition to the already mentioned general mitotic stimulation, the sets of genes for cytokine/chemokine activity and chemokine receptor binding were enriched with the SA diet. Several chemokines characterized by the CXC motif were represented at the top of the core genes characterizing the enriched gene sets, including CXCL11. This led us to conclude that an early association of newborn piglets with a complex microbiota favorably prevents the activation of

pathways related to cell division and inflammatory development in the gastric mucosa at an age of two weeks as compared to the association with a simple microbiota.

The diet containing an oil source of medium chain fatty acids in each group of piglets assigned to different early priming programs (3 wk of age, trial 2), in general, activated the inflammatory response, with the involvement of T cell cycling (indicated by the enrichment of IL2_STAT5_sSIGNALING together with TNFA_SIGNALING_VIA_NFKB) (Fig. 3a). This was also particularly associated with the activation of CCL2 (Chemokine (C-C Motif) Ligand 2) and CXCL10 (Chemokine (C-X-C Motif) Ligand 10) involved in monocyte chemotaxis and T-cell migration [35]. However, this was particularly evident when MCTs were given to the SA piglets, with a general enrichment of the nodes related to chemokine activity and cytokine binding (Fig. 3b). Interestingly, feeding piglets a diet supplemented with coconut oil increased the intestinal translocation of the bacterial enterotoxin as compared to the control diet or a polyunsaturated oil [36]. The Authors of this study explained this observation with the increased endocytosis of the enterotoxin favored by the innate immune receptor complex based on CD 14/Toll-like receptor 4 (TLR4)/ Lymphocyte Antigen 96. The Authors detected consistent expression of the TLR4 gene in our data set and a similar action regarding the MCT diet cannot be excluded, with the consequent induction of the inflammatory response by the enterotoxin eventually present in the stomach and derived from resident or transient bacteria.

Conclusions

The quality of early microbial colonization is important in promoting the functional maturation of the oxyntic mucosa in an age-dependent manner, as indicated by the variation of several genes typical of the oxyntic mucosa and of the gastric tissue in general. Furthermore, this was relevant for establishing a strong balance between the regulation of the immune system and the inflammatory response. The partial dietary substitution of unsaturated fatty acids with a medium chain fatty acid source may have affected the activation of several pathways addressed to the recruitment and the functional maturation of immune cells, especially when the piglets were early associated with a simplified starter microbiota. On the whole, these observations are important for the feed industry when designing strategies to stimulate the gut and immune system maturation of the piglet. Furthermore, the presence of an interaction of the complexity of early microbial colonization with the diet can explain the frequent variable response to early supplemented probiotics to the piglets and be relevant also as a general model for young human studies.

Abbreviations

ATP4A: ATPase, H$^+$/K$^+$ exchanging, α polypeptide; ATP4B: ATPase, H$^+$/K$^+$ exchanging, α polypeptide; CA: Treatment with association with complex microbiota inoculum; CCL2: Chemokine (C-C Motif) Ligand 2; CHRNA2: Cholinergic receptor, nicotinic, alpha 2 (neuronal); CLCA1: For chloride channel accessory 1; CTRL: Control diet; CXCL10: Chemokine (C-X-C Motif) Ligand 10; CXCL10: Chemokine (C-X-C Motif) Ligand 10; EGF: Epidermal growth factor; FDR: False discovery rate; GABRG2: Gamma-aminobutyric acid (GABA) A receptor, gamma 2; GO: Gene Ontology; HTR1B: for 5-hydroxytryptamine (Serotonin) receptor 1B, G protein-coupled; JCHAIN: Joining chain of multimeric IgA and IgM; KCNE2: Potassium voltage-gated channel, Isk-related family, member 2; KCNJ15: Potassium inwardly rectifying channel, subfamily J, member 15; LPL: Lipoprotein lipase; MCT: Diet with medium chain tryglicerides; MMP3: Matrix metalloproteinase 3 matrix metalloproteinase 3; NES: Normalized enrichment score; SA: Treatment with simplified microbiota; SLC2A4: Solute carrier family 2 (facilitated glucose transporter), member 4; SLC4A2: Anion exchanger 2; T1-T2: Trial 1 and Trial 2; TGFB: Transforming growth factor β; TLR4: Toll like receptor 4

Acknowledgements

Not applicable.

Funding

This study was financially supported by the European Union (contract No. 227549) through the Interplay project. The authors are solely responsible for the study described in this paper, and their opinions are not necessarily those of the European Union.

Authors' contributions

AJMJ, SJK and PB conceived and designed the experiments; DP and PT performed the experiments; PB and VM analyzed the data; SJK, PT, DP and DL contributed reagents/materials/analysis tools; PT, DP, AJMJ and PB wrote the manuscript. All authors read and approved the final manuscript.

Competing interests

The authors declare that they have no competing interests.

Author details

[1]DISTAL, University of Bologna, Viale Fanin 46, 40127 Bologna, Italy.
[2]Wageningen University and Research, Wageningen, The Netherlands.

References

1. Trevisi P, Gandolfi G, Priori D, Messori S, Colombo M, Mazzoni M, et al. Age-related expression of the polymeric immunoglobulin receptor (pIgR) in the gastric mucosa of young pigs. PLoS One. 2013;8:e81473.
2. Mazzoni M, Bosi P, De Sordi N, Lalatta-Costerbosa G. Distribution, organization and innervation of gastric MALT in conventional piglet. J Anat. 2011;219:611–21.
3. De Bruyne E, Flahou B, Chiers K, Meyns T, Kumar S, Vermoote M, et al. An experimental Helicobacter suis infection causes gastritis and reduced daily weight gain in pigs. Vet Microbiol. 2012;160:449–54.
4. Hooper LV. Bacterial contributions to mammalian gut development. Trends Microbiol. 2000;12:129–34.
5. Stappenbeck TS. Accounting for reciprocal host–microbiome interactions in experimental science. Nature. 2016;534:191–9.
6. Laycock G, Sait L, Inman C, Lewis M, Smidt H, van Diemen P, et al. A defined intestinal colonization microbiota for gnotobiotic pigs. Vet Immunol Immunopathol. 2012;49:216–24.
7. Dierick NA, Decuypere JA, Molly K, Van Beek E, Vanderbeke E. The combined use of triacylglycerols (TAGs) containing medium chain fatty acids (MCFAs) and exogenous lipolytic enzymes as an alternative to nutritional antibiotics in piglet nutrition: II. In vivo release of MCFAs in gastric cannulated and slaughtered piglets by endogenous and exogenous lipases; effects on the luminal gut flora and growth performance. Liv Prod Sci. 2002;76:1–16.
8. Zentek J, Ferrara F, Pieper R, Tedin L, Meyer W, Vahjen W. Effects of dietary combinations of organic acids and medium chain fatty acids on the gastrointestinal microbial ecology and bacterial metabolites in the digestive tract of weaning piglets. J Anim Sci. 2013;91:3200–10.
9. Odle J. New insights into the utilization of medium-chain triglycerides by the neonate: observations from a piglet model. J Nutr. 1997;127:1061–7.
10. Bosi P, Mazzoni M, De Filippi S, Casini L, Trevisi P, Petrosino G, et al. Continuous dietary supply of free calcium formate negatively affects parietal cell population and gastric RNA expression for H$^+$/K$^+$–ATPase in weaning pigs. J Nutr. 2006;136:1229–35.
11. Mazzoni M, Le Gall M, De Filippi S, Minieri L, Trevisi P, Wolinski J, et al. Supplemental sodium butyrate stimulates different gastric cells in weaned pigs. J Nutr. 2008;138:1426–31.
12. Friis-Hansen L. Gastric functions in gastrin gene knock-out mice. Pharmacol Toxicol. 2002;91:363–7.
13. Godden S, McMartin S, Feirtag J, Stabel J, Bey R, Goyal S, et al. Heat treatment of bovine colostrum. II. Effects of heating duration on pathogen viability and IgG. J Dairy Sci. 2006;89:3476–83.
14. Elizondo-Salazar JA, Heinrichs AJ. Feeding heat-treated colostrum or unheated colostrum with two different bacterial concentrations to neonatal dairy calves. J Dairy Sci. 2009;92:4565–71.
15. Goebel M, Stengel A, Lambrecht NW, Sachs G. Selective gene expression by rat gastric corpus epithelium. Physiol Genomics. 2011;43:237–54.
16. Colombo M, Priori D, Trevisi P, Bosi P. Differential gene expression in the oxyntic and pyloric mucosa of the young pig. PLoS One. 2014;9(10):e111447.
17. Varro A, Noble PJ, Pritchard DM, Kennedy S, Hart CA, Dimaline R, et al. Helicobacter pylori induces plasminogen activator inhibitor 2 (PAI-2) in gastric epithelial cells through NF-κB and RhoA: implications for invasion and apoptosis. Cancer Res. 2004;64:1695–702.
18. Romano M, Ricci V, Zarrilli R. Mechanisms of disease: Helicobacter pylori-related gastric carcinogenesis–implications for chemoprevention. Nat Clin Pract Gastroenterol Hepatol. 2006;3:622–32.
19. Lam EK, Yu L, Wong HP, WK W, Shin VY, Tai EK, et al. Probiotic Lactobacillus rhamnosus GG enhances gastric ulcer healing in rats. Eur J Pharmacol. 2007;565:171–9.
20. Pegg AE, McCann PP. Polyamine metabolism and function. Am J Phys. 1982;243:C212–21.
21. Jansman AJM, Zhang J, Koopmans SJ, Dekker RA, Smidt H. Effects of a simple or a complex starter microbiota on intestinal microbiota composition in caesarean derived piglets. J Anim Sci. 2012;90(Supplement 4):433–5.
22. Merchant JL, Shiotani A, Mortensen ER, Shumaker DK, Abraczinskas DR. Epidermal growth factor stimulation of the human gastrin promoter requires Sp1. J Biol Chem. 1995;270:6314–9.
23. Roussa E, Wittschen P, Wolff NA, Torchalski B, Gruber AD, Thévenod F. Cellular distribution and subcellular localization of mCLCA1/2 in murine gastrointestinal epithelia. J Histochem Cytochem. 2010;58:653–8.
24. Nagata A, Ito M, Iwata N, Kuno J, Takano H, Minowa O, Chihara K, et al. G. protein-coupled cholecystokinin-B/gastrin receptors are responsible for physiological cell growth of the stomach mucosa in vivo. Proc Natl Acad Sci U S A. 1996;93:11825–30.
25. Engelstoft MS, Park WM, Sakata I, Kristensen LV, Husted AS, Osborne-Lawrence S, et al. Seven transmembrane G protein-coupled receptor repertoire of gastric ghrelin cells. Mol Metab. 2013;2:376–92.
26. Yano JM, Yu K, Donaldson GP, Shastri GG, Ann P, Ma L, et al. Indigenous bacteria from the gut microbiota regulate host serotonin biosynthesis. Cell. 2015;161:264–76.
27. Reigstad CS, Linden DR, Szurszewski JH, Sonnenburg JL, Farrugia G, Kashyap PC. Correlated gene expression encoding serotonin (5-HT) receptor 4 and 5-HT transporter in proximal colonic segments of mice across different colonization states and sexes. Neurogastroenterol Motil. 2016;28:1443–8.
28. Furness JB. Types of neurons in the enteric nervous system. J Auton Nerv Syst. 2000;81:87–96.
29. Tsai LH. Function of GABAergic and glutamatergic neurons in the stomach. J Biomed Sci. 2005;12:255–66.

The effects of starter microbiota and the early life feeding of medium chain triglycerides on the gastric...

191

30. Barrett E, Ross RP, O'Toole PW, Fitzgerald GF, Stanton C. γ-Aminobutyric acid production by culturable bacteria from the human intestine. J Appl Microb. 2012;113:411–7.

31. Rothkötter HJ, Mollhoff S, Pabst R. The influence of age and breeding conditions on the number and proliferation of intraepithelial lymphocytes in pigs. Scand J Immunol. 1999;50:31–8.

32. Cole KE, Strick CA, Paradis TJ, Ogborne KT, Loetscher M, Gladue RP, et al. Interferon–inducible T cell alpha chemoattractant (I-TAC): a novel Non-ELR CXC Chemokine with potent activity on activated T cells through selective high affinity binding to CXCR3. J Exp Med. 1998;187:2009–21.

33. Page-McCaw A, Ewald AJ, Werb Z. Matrix metalloproteinases and the regulation of tissue remodelling. Nat Rev Mol Cell Bio. 2007;8:221–33.

34. Johansson ME, Jakobsson HE, Holmén-Larsson J, Schütte A, Ermund A, Rodríguez-Piñeiro AM, et al. Normalization of host intestinal mucus layers requires long-term microbial colonization. Cell Host Microbe. 2015;18:582–92.

35. Zlotnik A, Yoshie O. Chemokines: a new classification system and their role in immunity. Immunity. 2000;12:121–7.

36. Mani V, Hollis JH, Gabler NK. Dietary oil composition differentially modulates intestinal endotoxin transport and postprandial endotoxemia. Nutri Metab. 2013;10:6.

21

Over-expression of Toll-like receptor 2 up-regulates heme oxygenase-1 expression and decreases oxidative injury in dairy goats

Shoulong Deng[1†], Kun Yu[2,3†], Wuqi Jiang[2], Yan Li[2], Shuotian Wang[2], Zhuo Deng[4], Yuchang Yao[5], Baolu Zhang[6], Guoshi Liu[2], Yixun Liu[1*] and Zhengxing Lian[2,3*] ⓘ

Abstract

Background: Mastitis, an infection caused by Gram-positive bacteria, produces udder inflammation and oxidative injury in milk-producing mammals. Toll-like receptor 2 (TLR2) is important for host recognition of invading Gram-positive microbes. Over-expression of *TLR2* in transgenic dairy goats is a useful model for studying various aspects of infection with Gram-positive bacteria, in vivo.

Methods: We over-expressed *TLR2* in transgenic dairy goats. Pam3CSK4, a component of Gram-positive bacteria, triggered the TLR2 signal pathway by stimulating the monocytes-macrophages from the *TLR2*-positive transgenic goats, and induced over-expression of activator protein-1 (AP-1), phosphatidylinositol 3-kinase (PI3K) and transcription factor nuclear factor kappa B (NF-κB) and inflammation factors downstream of the signal pathway.

Results: Compared with wild-type controls, measurements of various oxidative stress-related molecules showed that *TLR2*, when over-expressed in transgenic goat monocytes-macrophages, resulted in weak lipid damage, high level expression of anti-oxidative stress proteins, and significantly increased mRNA levels of transcription factor NF-E2-related factor-2 (Nrf2) and the downstream gene, heme oxygenase-1 (HO-1). When Pam3CSK4 was used to stimulate ear tissue in vivo the HO-1 protein of the transgenic goats had a relatively high expression level.

Conclusions: The results indicate that the oxidative injury in goats over-expressing *TLR2* was reduced following Pam3CSK4 stimulation. The underlying mechanism for this reduction was increased expression of the anti-oxidation gene *HO-1* by activation of the Nrf2 signal pathway.

Keywords: Haem oxygenase, Nrf2 signal pathway, Toll-like receptor 2, Transgenic goats

Background

Mastitis, an inflammatory and oxidative stress disease of udders, is caused by pathogenic bacteria. The disease is common in goats and other dairy animals. Mastitis reduces the quantity and quality of milk. Gram-positive bacteria such as staphylococcus, streptococcus and corynebacterium cause mastitis. *Staphylococcus aureus* is the main zoonotic agent of mastitis [1]. During infection, the innate immune system of the host animal provides the first line of defense in pathogen recognition, and pattern recognition receptors (PRRs) are necessary for this process to occur. Toll-like receptor (TLR) 2 plays a critical role in recognition of the conserved portion of Gram-positive bacteria [2], and it is known that mastitis can increase expression of the *TLR2* gene in mammary glands [3]. Genetic diversity in the *TLR2* gene is a risk factor for staphylococcus infection [4], and *TLR2*−/− mice are susceptible to *S. aureus* [5]. The *S. aureus* cell wall components (e.g., peptidoglycan, lipoteichoic

* Correspondence: liuyx@ioz.ac.cn; lianzhx@cau.edu.cn
†Equal contributors
[1]State Key Laboratory of Stem Cell and Reproductive Biology, Institute of Zoology, Chinese Academy of Sciences, Beijing 100101, China
[2]Laboratory of Animal Genetics and Breeding, College of Animal Science and Technology, China Agricultural University, Beijing 100193, People's Republic of China
Full list of author information is available at the end of the article

acid and lipoproteins) are recognizable by TLR2 [6]. Pam3CSK4, a synthetic bacterial lipoprotein, is an agonist for TLR2 [7].

TLR2 is widely expressed in all types of goat immune cells and epithelial cells, and is highly expressed in peripheral blood monocytes. TLR2 activation relies mainly on the MyD88 signaling pathway, which activates the NF-κB pathway and mitogen-activated protein kinase (MAPK) pathway. Phosphorylated MAPK activates the transcription factor activator protein-1 (AP-1) and activation of the PI3K/Akt signaling pathway. Expression of inflammatory cytokines such as tumor necrosis factor α (TNF-α), interleukin (IL)-1β, IL-6, IL-8, chemokines and nitrous oxide (NO) can all induce immune responses.

In mastitis, oxidative stress results from infection with pathogenic bacteria. Such infections in udders trigger host immune responses that induce production of large amounts of reactive oxygen species (ROS) in host cells [8]. Transcription factor NF-E2-related factor-2 (Nrf2) plays a critical role in anti-inflammation and oxidation stress, induces anti-oxidation proteins and expression of the type II detoxification enzyme, and triggers expression of the downstream target genes catalase (CAT), superoxide dismutase (SOD), glutathione S-transferase alpha 1 (GSTα1), quinone oxidoreductase 1 (NQO1), γ-glutamylcysteine synthetase (γ-GCS) and heme oxygenase-1 (HO-1). Some cellular signaling pathways such as PI3K and MAPK regulate activation of the Nrf2 antioxidative response element pathway [9]. Oxidative stress activates NF-κB, triggers

inflammatory responses, up-regulates expression of target genes such as cyclooxygenase-2 (COX-2), inducible nitric oxide synthase (iNOS) and the oxidation enzyme NADPH, and facilitates prostaglandin, ROS and NO production, while intensifying the oxidative stress response. Crosstalk between Nrf2 and NF-κB pathways cooperatively regulates the oxidative stress response [10]. HO-1 and its catalytic products enhance the innate protection system of the host [11]. The HO-1 gene contains AP-1, NF-κB, Nrf2, which are regulatory structural sites. IL-10 and TLR pathways are involved in HO-1 regulation [12].

The expression level of TLR2 is closely associated with lipid oxidation and inflammation reactions [13], and also plays an important role in triggering innate and adaptive immune responses against infection, thereby protecting the host from disease. In this study, we over-expressed the TLR2 receptor in laso-shan dairy goats. Through stimulation with the Pam3CSK4 ligand from Gram-positive bacteria, we observed the effect of TLR2 on the immune response. We discuss the mechanism whereby TLR2 regulates anti-oxidative stress.

Methods
Production of transgenic goats over-expressing toll-like receptor 2
Transgenic goats were produced by microinjecting plasmid constructs into fertilized goat eggs (Fig. 1a). The transformed exogenous TLR2 genes in the offspring were

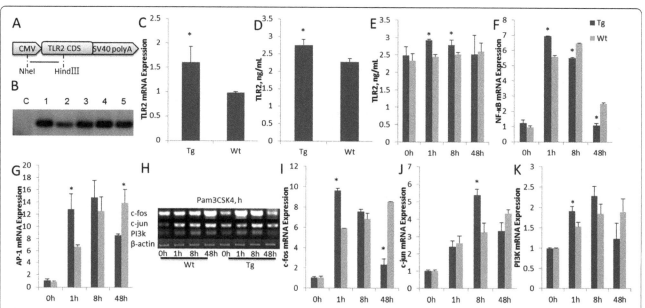

Fig. 1 Overexpression of TLR2 in goats. **a** Structure of TLR2 plasmids. **b** TLR2-positive goats were detected by Southern blot analysis. Lanes: **c**, control (Wt); 1–5, genomic DNA. **c** TLR2 expression in monocytes-macrophages was detected by qRT-PCR. **d** ELISA analysis of TLR2 expression in goat sera. **e** ELISA analysis of TLR2 expression in monocytes-macrophages. **f** Expression of NF-κB and **g** AP-1 was detected by qRT-PCR. (**h**), (**i**), (**j**) and (**k**). The gene expression levels of c-Fos, c-Jun and PI3K were quantified by RT-PCR. Tg: Transgenic goat, Wt: wild type. The results are expressed as the means ± SEM. *P < 0.05 in Tg versus Wt groups

analyzed by Southern blotting. Genomic DNA was extracted from ear tissue, and a gene-specific digoxigenin-labeled probe (Roche Diagnostics, Mannheim, Germany) was generated by PCR amplification with the following primer pair, oTLR2, forward: 5′-TTC TCC CAC TTC CGT CTC-3′; reverse: 5′-CCC TAT CTC GGT CTA TTC TT-3′, resulting in a 618-bp fragment. Genomic DNA was isolated from ear tissue and digested with *Nhe*I and *Hin*dIII (NEB, Beverly, MA, USA) [14].

Isolation purification and culturing of goat monocytes-macrophages

Each group had 5 transgenic goats. Peripheral blood mononuclear cells (PBMCs) were isolated from the peripheral blood of 6-month-old transgenic goats using goat lymphocyte separation medium (TBD, Tianjin, China). The PBMCs containing thrombocytes, lymphocytes, and monocytes were washed twice with Hanks' buffer (Invitrogen, Beijing, China). The harvested cells at a concentration of 2×10^7 cells/well were seeded in 6-well plate. The plates were incubated at 37 °C in an incubator with a 5% CO_2 for 2 h and then the non-adherent cells (most of lymphocytes) were washed out with PBS. The adherent cells were cultured with RPMI1640 (Gibco, Grand Island, NY, USA) medium containing 10% FBS (Gibco). After 24 h of incubation and 3 times washing, the cells were digested with 0.02% EDTA for 5 min, and then the cells were harvested. The cells (1×10^5) were seeded in each well of a 6-well plate with three repeats for each group. RPMI1640 medium containing 10% FBS was changed every 24 h. After 48 h of incubating and washing away non-adherent cells, the adherent cells were mainly monocytes-macrophages [15, 16]. Monocytes-macrophages were stimulated by Pam3CSK4 (1 µg/mL) (InvivoGen, San Diego, CA, USA) [17]. Cell suspensions were collected at 0, 1, 8 and 48 h post-stimulation. RNA was extracted from adherent cells using TRIzol plus RNA purification kit (Invitrogen, Carlsbad, CA, USA) and treated with RNase-free DNase (Promega, Madison, WI, U.S.) per the manufacturer's instructions. cDNA was synthesized with a Reverse Transcription kit (Promega, Madison, WI, USA).

Reverse transcription (RT) and quantitative real-time (qRT)-PCR

cDNA from Pam3CSK4-stimulated monocytes-macrophages was obtained at different time points. The abundance of *TLR2*, *AP-1*, *NF-κB*, *Nrf2*, *SOD1*, *CAT*, *GSTα1*, *NQO1* and *HO-1* mRNA molecules was measured by qRT-PCR. The gene expression levels of *c-Fos*, *c-Jun* and *PI3K* were quantified by RT-PCR. *β-actin* was used as an internal standard. PCR products were analyzed by agarose gel electrophoresis. Primer sequences are shown in Table 1. qRT-

Table 1 The primer sequences

Primer	Forward (5'→3')	Reverse (5'→3')
TLR2	TGCTGTGCCCTCTTCCTGTT	GGGACGAAGTCTCGCTTATGAA
AP-1	GAAGGAAGAGCCGCAGACG	CCACCTGTTCCCTGAGCATA
NF-κB	TTCTCCAAATGGCTGAAGGTA	TTGTTTGAGGGCCATAAGGAT
Nrf2	CCAACTACTCCCAGGTAGCCC	AGCAGTGGCAACCTGAACG
c-Fos	GAGGACCTTATCTGTGCGTGAA	GGTTAATTCCAATAACGAACCCAAT
c-Jun	AAGATGGAAACGACCTTCTACGA	TTGATCCGCTCCTGGGACT
PI3K	CCGACAGTGAGCAACAAGC	AGGAGGCGGCATCACAATC
CAT	TTTTCTGCTGAAGCCCTATGA	GTCCTTTCAGGGAGAATGGTG
GSTα1	TGAGCCAAGTGGGAGACAGAC	ATCCAGGTCTTCTGGTTGTTCTAT
SOD1	TCTGCGTGCTGAAGGGTGA	CTTTGGCCCACCGTGTTTT
HO-1	ACACCCAGGCGGAGAATG	CTCCTGGAGTCGCTGAACATAG
NQO1	AACTTCAATCCCGTCATCTCC	CCTTTCAGGATGGCAGGGACT
β-actin	AGATGTGGATCAGCAAGCAG	CCAATCTCATCTCGTTTTCTG

PCR was performed using a Realtime Master Mix SYBR Green kit (Tiangen, China) using MX300P (Stratagene).

Enzyme-linked immunosorbent assays (ELISAs)

During the TLR2 receptor agonist experiments, cell culture supernatants from Pam3CSK4-stimulated monocytes-macrophages were collected at different time points (0, 1, 8, and 48 h). The concentrations of TLR2, TNF-α, IL-1β, IL-8, IL-12p40, IL-4 and monocyte chemoattractant protein-1 (MCP-1) in the supernatants were measured using ELISA kits (CUSABIO, Hubei, China). Sera were obtained to detect TLR2 concentration using ELISA kit. Blood samples were collected from goats and allowed to clot at 37 °C for 30 min. Sera were separated immediately by centrifugation at 3,000 r/min for 15 min. All experimental operations were performed according to the kit instructions.

Oxidative stress measurements

Pam3CSK4-stimulated monocytes-macrophages suspensions were collected at different time points (0, 1, 8, and 48 h). The activities of iNOS, SOD, CAT, GSH, COX-2, NADPH oxidase and malondialdehyde (MDA) contents were examined by spectrophotometry in accordance with the manual supplied with the detection kit (Nanjing Jiancheng, China).

Immunocytochemical staining

The ears of three 8-month-old transgene-positive goats were injected intradermally with 100 µL of 3 mg/mL Pam3CSK4, after which the tissues were collected at 1, 8, and 48 h [18, 19]. Samples were fixed with 4% paraformaldehyde and embedded in paraffin. Hematoxylin and eosin staining was used to investigate inflammatory responses and immunohistochemistry was used to detect HO-1 protein expression (Abcam, ab13248, Cambridge,

UK). Six fields from each slide were randomly selected. Optical densities were quantified by scanning densitometry and expressed in arbitrary units determined by Image J software (NIH, USA).

Statistical analyses

Individual experiments were repeated three times. Statistical tests to determine differences of *TLR2* mRNA expression between two groups were performed with one-way ANOVA followed by Tukey's HSD using Statistical Analysis System (SAS Institute, Cary, NC, USA). A repeated measures ANOVA with SAS proc GLM followed by Tukey's HSD post-hoc tests were performed to determine the statistical significance between the relative groups in different time points. All data were expressed as mean ± SEM. Differences were considered to be significant when $P < 0.05$.

Results

Toll-like receptor 2 over-expression triggers activation of down-stream transcription factors

The microinjection technique successfully generated *TLR2* over-expression in goats. Transfected goats positive for exogenous *TLR2* were identified by Southern blot analysis of genomic DNA (Fig. 1b), the positive rate was 9.82% (Table 2). Figure 1c shows that *TLR2* expression in the transgenic goats was higher than that of the wild-type (Wt) animals ($P < 0.05$). The expression of TLR2 protein in sera in the transgenic group was significantly higher than that in the Wt group ($P < 0.05$) (Fig. 1d).

ELISA was performed to measure the TLR2 expression levels in the Pam3CSK4-stimulated monocytes-macrophages. TLR2 expression levels in the transgenic group were significantly higher than those of the Wt group at 1 and 8 h post-stimulation ($P < 0.05$) (Fig. 1e). These results confirmed that TLR2 expression of the monocytes-macrophages of the transgenic group was up-regulated after Pam3CSK4 stimulation. Next, qRT-PCR was conducted to determine *NF-κB* and *AP-1* mRNA expression levels. In this experiment, *NF-κB* expression in the transgenic group was significantly higher than that in the Wt group at 1 h, and reduced at 8 h post-stimulation compared with Wt (Fig. 1f). These results show that TLR2 triggered activation of the NF-κB signal pathway, with *NF-κB* expression levels decreasing at an earlier stage than those of Wt. *AP-1* expression in the transgenic group was significantly higher than that of the Wt group at 1 h post-stimulation, reached a peak at 8 h, while its expression in

Wt group continued to increase over time (Fig. 1g). The expression levels of *c-Fos*, *c-Jun* and *PI3K* were similar as *AP-1* (Fig. 1h, i, j and k).

Toll-like receptor 2 over-expression improves expression of the anti-inflammatory factor

After Pam3CSK4 stimulation of goat monocytes-macrophages, expression of the pro-inflammatory factors IL-1β and TNF-α were similar (Fig. 2a and b). At 1 h post-stimulation, the expression level of the transgenic groups was significantly higher than that of the Wt group ($P < 0.05$). Furthermore, expression in the transgenic group both IL-1β and TNF-α peaked at 1 h, while in the Wt group the peak occurred at 8 h, IL-8 expression in the transgenic and Wt groups was not significantly different at the different time points (Fig. 2c). Expression of the type Th1 cytokine IL-12p40 in the Wt group peaked at 1 h and was significantly higher than that of the transgenic group; it subsequently decreased and then increased at 48 h post-stimulation. The expression level of the transgenic group peaked at 8 h, which was slower than that of the Wt cells (Fig. 2d). This result shows that over-expressed TLR2 retarded the expression of type Th1 cytokines. IL-4 expression continuously increased in transgenic group during the experiment (Fig. 2e). For MCP-1 expression (Fig. 2f), the transgenic group was significantly higher than the Wt group at 1 h ($P < 0.05$), but significantly lower at 8 h post-stimulation.

Weakens oxidative lipid injury in monocytes-macrophages

In this experiment, after Pam3CSK4 stimulation of goat monocytes-macrophages, the NADPH oxidase expression level in the transgenic group was higher than that of the Wt group at 1 h and 8 h post-stimulation ($P < 0.05$) (Fig. 3a). iNOS expression in the transgenic and Wt groups showed similar trends (Fig. 3b), though the level in the transgenic group was significantly lower than that of the Wt group at 1 h and 8 h post-stimulation ($P < 0.05$). COX-2 is involved in the stress reaction against pathological conditions. In this experiment, the COX-2 expression level reached a peak at 8 h in both group (Fig. 3c). At 1 h, the level in the Wt group was significantly higher than that of the transgenic group ($P < 0.05$). The expression level of the over-oxidative lipid product (MDA) in the transgenic group was relatively stable over time unlike that of the Wt group (Fig. 3d), and was also significantly lower than that of the Wt group at 1 and 8 h post-stimulation ($P < 0.05$).

Over-expressed toll-like receptor 2 retains anti-oxidative protein activity in goat monocytes-macrophages

In this experiment, after Pam3CSK4 stimulation of goat monocytes-macrophages, the GSH content was maintained

Table 2 Microinjection production of over-expression of *TLR2* gene in goats

No. of donors	No. of microinjections	No. of embryo transfer recipients	No. of lambs	Southern positive rate, %
11	127	30	22	22.73%

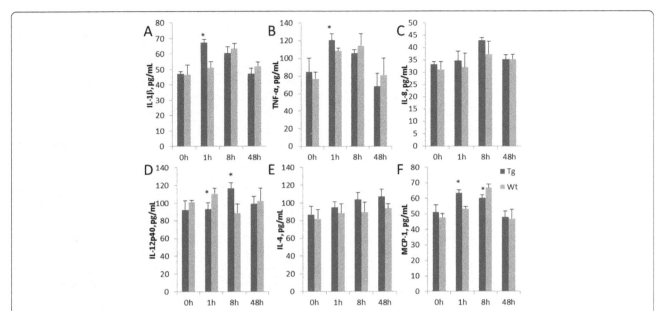

Fig. 2 ELISA testing of inflammatory cytokines changes in TLR2 monocytes-macrophages under Pam3CSK4 stimulation. **a**, **b**, **c**, **d** and **e** show the effects of IL-1β, TNF-α, IL-8, IL-12p40 and IL-4 immune factors, respectively. **f** shows MCP-1 expression. Tg: Transgenic goat, Wt: wild type. The results are expressed as the means ± SEM. * $P < 0.05$ in Tg versus Wt groups

at a stable level in monocytes-macrophages the transgenic group (Fig. 4a). In contrast, the GSH content in the Wt group reduced over time, was significantly different from the transgenic group at 8 h and at 48 h post-stimulation ($P < 0.05$). The activity expression trends of SOD and CAT were similar to each other in that both showed decreasing trends (Fig. 4b and c). The enzyme activities in the transgenic group was higher than those of the Wt group, and there were significant differences between the two groups at 1 h and 8 h post-stimulation ($P < 0.05$).

Over-expressed toll-like receptor 2 up-regulates expression of the anti-oxidative gene heme oxygenase-1

Following Pam3CSK4 stimulation TLR2 over-expressed in goat monocytes-macrophages, *Nrf2* expression was significantly up-regulated at 1 and 8 h ($P < 0.05$), after which the expression level decreased, while *Nrf2* expression in the Wt group continuously increased over time

(Fig. 5a). The *HO-1* expression level reached a peak at 8 h in both groups. The level of HO-1 expression in the transgenic group was significantly higher than that of the Wt at 1, 8 and 48 h post stimulation ($P < 0.05$) (Fig. 5b). Expression of the *NQO1* and *GSTα1* genes showed similar decreases over time, but were higher in Tg group than in Wt group (Fig. 5c and d). The expression levels of *SOD1* and *CAT* were not significantly different (Fig. 5e and f).

Toll-like receptor 2 overexpression triggers infiltration of neutrophils and high-level expression of heme oxygenase-1 protein

Pam3CSK4 was injected into the ear of each transgenic goat and the resulting inflammatory infiltrate was observed by light microscopy after hematoxylin and eosin staining (Fig. 6a). In the transgenic group, a number of segmented neutrophils infiltrated the dermis at 1 h. Few

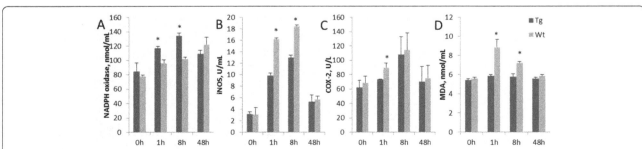

Fig. 3 Over-expressed TLR2 confers weak lipid oxidative injury. The activities of NADPH oxidase, iNOS, COX-2 and MDA content (**a**, **b**, **c** and **d**, respectively). Tg: Transgenic goat, Wt: wild type. The results are expressed as the means ± SEM. * $P < 0.05$ in Tg versus Wt groups

Over-expression of Toll-like receptor 2 up-regulates heme oxygenase-1 expression...

197

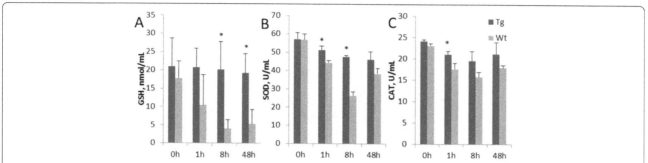

Fig. 4 Oxidative stress protein expressions in goat monocytes-macrophages. The activities of GSH, SOD and CAT are shown in **a**, **b** and **c**, respectively. Tg: Transgenic goat, Wt: wild type. The results are expressed as the means ± SEM. $^*P < 0.05$ in Tg versus Wt groups

infiltrating inflammatory cells were evident after 8 h. However, no significant lesions were observed after 48 h. In contrast, in the Wt group, dermal bleeding occurred in conjunction with inflammatory cell infiltration at 1 h. After 8 h, many erythrocytes were present on the skin surface and between the connective tissues and inflammatory cells, including many neutrophils that had infiltrated around the blood vessels. Few infiltrating inflammatory cells were evident after 48 h. The transgenic group displayed a strong inflammatory response at 1 h, which is consistent with the result we observed for TLR2 expression. Immunohistochemistry was used to observe HO-1 protein expression. HO-1-positive tissues showed claybank (Fig. 6b). The expression of HO-1 was maintained at a high level in transgenic group, and was

significantly higher than that in the Wt group at 1 h and 8 h ($P < 0.05$).

Discussion

Activated TLR2 triggers MyD88 protein, and I-κB kinase phosphorylation activates transcription factor NF-κB or alternatively, the MAPK pathway activates AP-1 to induce the expression of inflammatory cytokines. AP-1 is a pro-inflammation mediated transcription factor, the core components of which are c-Jun and c-Fos. Overexpression of c-Jun has been shown to induce production of inflammation mediators [20], while c-Jun/AP-1 interacts with the transcription factor, NF-κB. In cells under oxidation stress, TLR2 inhibition stopped transportation of NF-κB and AP-1 [21]. We have obtained transgenic

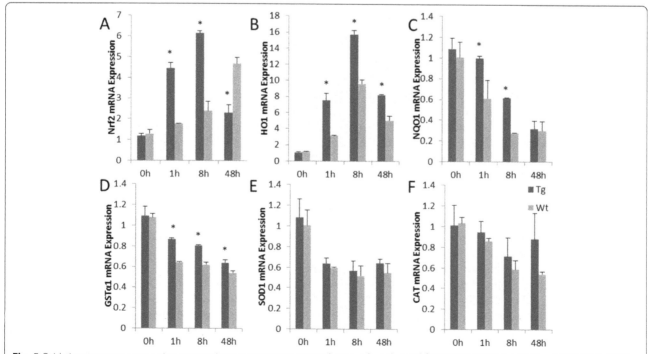

Fig. 5 Oxidative stress gene expression patterns in goat monocytes-macrophages. **a**, **b**, **c**, **d**, **e** and **f** represent mRNA expression of *Nrf2*, *HO-1*, *NQO1*, *GSTα1*, *SOD1* and *CAT*, respectively. Tg: Transgenic goat, Wt: wild type. The results are expressed as the means ± SEM. $^*P < 0.05$ in Tg versus Wt groups

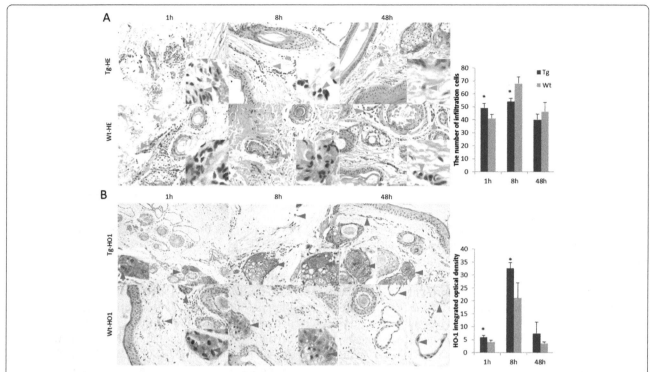

Fig. 6 Neutrophil infiltrations in sections of dermis. **a** Pathological changes were examined microscopically (hematoxylin and eosin staining, ×400), *Green* triangle pointing to infiltration cells. **b** HO-1 expression was detected by immunohistochemical staining (×200), *Blue* triangle pointing to HO-1 positive cells. Tg: Transgenic goat, Wt: wild type. The results are expressed as the means ± SEM. *P < 0.05 in Tg versus Wt groups

goats with TLR2 over-expression. Excessive expression of *TLR2* mRNA was detected in monocytes-macrophages of the transgenic group. Transgenic group expressed more TLR2 protein than the Wt group in the ear tissue section [14]. In goat sera, TLR2 protein levels of the transgenic group were higher too. However without stimulation, there was no significant difference in the expression of TLR2 protein between the transgenic group and the Wt group in monocytes-macrophages. This may be caused by the negative regulation of TLRs receptors [22]. Previous study indicated, TLRs levels were kept in relative stable levels by intracellular modulation, such as alternative splice, degradation by ubiquitination and deubiquitination [23]. After Pam3CSK4-stimulation of transgenic goat monocytes-macrophages, the expression of TLR2 protein was up-regulated in the transgenic group and triggered activation of downstream transcription factors, promoted the expression of AP-1 and inhibited the activity of NF-κB. Protein levels of TLR2 in different tissues were different, and further research on the different oxidative stress response caused by it will be studied.

TNF-α and IL-1β are both able to induce the production of MCP-1. In animal experiments, it has been shown that an increase in oxidation stress can induce the expression of MCP-1 [24]. Inflammation was accompanied with an anti-inflammation reaction, and IL-4

played an important role in it. The Th1/Th2 balance is determined by two different types of cytokine secretion. The Th1 response produces IL-12 to promote the production of IFN-γ, whereas the Th2 response is distinguished by low IFN-γ levels and high IL-4 levels. TLR2 activated by different ligands generates different immune effects, and the expression of TLR2 shows time-dependence [25, 26]. It has been shown that stimulation by the TLR2 agonist Pam3CSK4 reduces the infiltration of chemokine and inflammatory cells within mouse tissues, in vivo [19]. In the present study, Pam3CSK4 could activate the monocytes-macrophages, and over-expressed TLR2 caused early expression of the pro-inflammatory cytokines TNF-α and IL-1β, and a continuous increase in expression of the anti-inflammatory factor IL-4, as well as inducing expression of Th1 type cytokines. The result of our skin inflammation experiment indicated that over-expression of the *TLR2* gene accelerated the inflammation process.

Under inflammatory conditions, free radicals are necessary for host defense. Inflammatory cells produce ROS and NO to expose the area of inflammation under oxidative stress. The transcription factors AP-1, NF-κB and Nrf2 all exhibit sensitivity to reductant-oxidant stressors. TLR2 activation of mouse macrophages produces NO resulting in damage to bacteria within cells

[27], while NADPH oxidase regulates the production of ROS, over-expresses pro-inflammatory cytokines, and positively facilitates oxidative stress. Within cells, there is both elevation of ROS and activation of NF-κB. The production of NO is regulated by iNOS, and the activation of Nrf2 can reduce induction of iNOS expression by IL-1β and inhibit the activity of NF-κB [28]. The results of the present study showed that Pam3CSK4 stimulated goat monocytes-macrophages, and over-expressed TLR2 up-regulated the activity of NADPH oxidase. Furthermore, the low activity of iNOS in the TLR2 over-expressed group showed that NO was synthesized in small amounts, and that the oxidative stress injury to cells was weak. Over-expressed *TLR2* caused the over-expression of Nrf2 and inhibited iNOS activity. It has been shown that an Nrf2 agonist inhibited TNF-α and IL-1β and induced COX-2 expression by blocking the activation of NF-κB [29]. High expression of COX-2 inhibited PI3K activity and the anti-oxidation reaction mediated by Nrf2 [30]. In addition, activation of Nrf2 can regulate c-Jun signal activity directly, and COX-2 expression negatively. Our results of showed that over-expressed TLR2 induced the up-regulation of Nrf2, increased the expression of the *c-Jun* gene, and inhibited the expression of COX-2 in monocytes-macrophages under Pam3CSK4 stimulation.

Too many free radicles can disturb the balance between the oxidation and anti-oxidation systems of a host. GSH, SOD and CAT are important anti-oxidative substances in a host organism. Over-expression of anti-oxidants can block NF-κB activation. Also, activated NF-κB can induce secretion of pro-inflammatory factors, reduce SOD activity and increase the MDA content of a cell [31]. Our results indicate that the consumption of anti-oxidation stress enzymes in TLR2 over-expressed cells was lower than that of the Wt cells, thereby effectively maintaining the oxidative stress balance of the cells. The result of this study also shows that TLR2 over-expression in monocytes-macrophages stimulated by Pam3CSK4 improved GSH activity and that the GSH consumed could be synthesized rapidly.

We also found that oxidative stress activated NF-κB and AP-1 leading to the production of pro-inflammatory cytokines, while excessive oxidative stress activated the Nrf2 signal pathway [32]. Nrf2-induced HO-1 expression was found to inhibit the activation of NF-κB and the secretion of MCP-1 in endothelial cells stimulated by TNF-α [33]. Additionally, expression of iNOS in Nrf2 knockout mice was significantly higher than that seen in Wt mice [34]. Here, at 8 h post-stimulation with Pam3CSK4, TLR2 over-expression in transgenic goat monocytes-macrophages was found to up-regulate expression of the Nrf2 gene. Expression of the pro-inflammatory factors TNF-α decreased. The activity of

iNOS was lower than that of the control. Importantly, the transgenic group experienced a weaker inflammatory reaction than that of the control group by the ear skin regional inflammation test. It indicates that TLR2 up-regulated expression of the *Nrf2* gene, thereby reducing the level of oxidative injury in the host.

AP-1 family members include Jun and Fos. Jun subclasses include c-jun and JunB. Fos subclasses include c-fos and FosB. Different types of AP-1 transcription factor dimer combinations have different functions in gene expression regulation. AP-1 positively regulates the expression of most genes, but increased levels of its c-Fos and c-Jun subunits negatively regulate expression of some genes [35]. The NQO1 gene ARE contains an AP-1 binding site. NQO1 induced by ARE is positively regulated by Nrf2 and negatively regulated by c-Fos [36]. NQO1 and GST are highly expressed in c-Fos knockout mice [37], and our results are consistent with these findings. Over-expressed TLR2 up-regulated the expression of *AP-1* and its *c-Fos* and *c-Jun* gene subunits; indeed, their expression trends were similar. Expression of the anti-oxidation genes *NQO1*, *GSTα1*, *SOD1* and *CAT* were inhibited.

HO-1 is a rate-limiting enzyme in heme catalysis and the protection system comprising HO-1–bilirubin–carbon monoxide (CO) widely participate in anti-inflammatory reactions and oxidative stress injuries. Similar to NO, CO was also found to reduce the production of inflammatory factors [38]. Researchers have discovered that activation of the *HO-1* gene inhibited the activity of the AP-1 protein [39]. However, many studies have shown that AP-1 is involved in the induction of HO-1 expression [40, 41]. In Nrf2 knockout cells, nitrite improved the activity of AP-1 through the JNK/c-Jun pathway and up-regulated HO-1 expression. Lipoteichoic acid induced expression of the *HO-1* gene through the TLR2/MyD88/c-Src/NADPH oxidase pathway in tracheal smooth muscle cells [42]. The results of our study showed that, in Pam3CSK4 stimulated goat monocytes-macrophages, the expression of AP-1 and PI3K were up-regulated in the TLR2 over-expression group; this facilitated expression of the *Nrf2* gene and induced an increase in *HO-1* gene expression.

Conclusions

Our study showed that, stimulated by the synthetic bacterial lipoprotein Pam3CSK4, TLR2 over-expressed in goat monocytes-macrophages triggered the expression of the anti-oxidation gene *Nrf2*, up-regulated expression of the *HO-1* gene, inhibited an excessive immuno-response, and enhanced the anti-oxidative stress injury in the host through activating the AP-1 and PI3K signal pathways. TLR2 over-expressed in goats might reduce the inflammatory and oxidative stress damage.

Acknowledgments
This work was supported by grants from National Transgenic Creature Breeding Grand Project (2014ZX08008-005), Chinese Universities Scientific Fund (2014BH032), and Natural Science Foundation of China (31501953, 31471352, 31471400 and 31171380).

Funding
Not applicable.

Authors' contributions
Conceived and designed the experiments: SLD and ZXL. Performed the experiments: SLD, KY and WQJ. Analyzed the data: SLD and ZXL. Contributed reagents/materials/analysis tools: YL, STW, GSL and YCY. Wrote the paper: SLD, BLZ, ZD and YXL. All read and approved the final manuscript.

Competing interests
The authors declare that they have no competing interests.

Author details
[1]State Key Laboratory of Stem Cell and Reproductive Biology, Institute of Zoology, Chinese Academy of Sciences, Beijing 100101, China. [2]Laboratory of Animal Genetics and Breeding, College of Animal Science and Technology, China Agricultural University, Beijing 100193, People's Republic of China. [3]National key Lab of Agrobiotechnology, College of Biological Sciences, China Agricultural University, Beijing 100193, People's Republic of China. [4]Department of Animal Science, Oklahoma State University, Stillwater, OK 74078, USA. [5]College of Animal Science and Technology, Northeast Agricultural University, Harbin 150030, People's Republic of China. [6]State Oceanic Administration, Beijing 100860, People's Republic of China.

References
1. Weese JS, Dick H, Willey BM, McGeer A, Kreiswirth BN, Innis B, et al. Suspected transmission of methicillin-resistant Staphylococcus aureus between domestic pets and humans in veterinary clinics and in the household. Vet Microbiol. 2006;115(1–3):148–55.
2. Du E, Wang W, Gan L, Li Z, Guo S, Guo Y. Effects of thymol and carvacrol supplementation on intestinal integrity and immune responses of broiler chickens challenged with Clostridium perfringens. J Anim Sci Biotechnol. 2016;7:19.
3. Goldammer T, Zerbe H, Molenaar A, Schuberth HJ, Brunner RM, Kata SR, et al. Mastitis increases mammary mRNA abundance of beta-defensin 5, toll-like-receptor 2 (TLR2), and TLR4 but not TLR9 in cattle. Clin Diagn Lab Immunol. 2004;11(1):174–85.
4. Ogus AC, Yoldas B, Ozdemir T, Uguz A, Olcen S, Keser I, et al. The Arg753GLn polymorphism of the human toll-like receptor 2 gene in tuberculosis disease. Eur Respir J. 2004;23(2):219–23.
5. Echchannaoui H, Frei K, Schnell C, Leib SL, Zimmerli W, Landmann R. Toll-like receptor 2-deficient mice are highly susceptible to Streptococcus pneumoniae meningitis because of reduced bacterial clearing and enhanced inflammation. J Infect Dis. 2002;186(6):798–806.
6. Schmaler M, Jann NJ, Götz F, Landmann R. Staphylococcal lipoproteins and their role in bacterial survival in mice. Int J Med Microbiol. 2010;300(2–3):155–60.
7. Fournier B. The function of TLR2 during staphylococcal diseases. Front Cell Infect Microbiol. 2013;2:167.
8. Giacco F, Brownlee M. Oxidative stress and diabetic complications. Circ Res. 2010;107(9):1058–70.
9. Tkachev VO, Menshchikova EB, Zenkov NK. Mechanism of the Nrf2/Keap1/ARE signaling system. Biochemistry (Mosc). 2011;76(4):407–22.
10. Bellezza I, Mierla AL, Minelli A. Nrf2 and NF-κB and their concerted modulation in cancer pathogenesis and progression. Cancers (Basel). 2010;2(2):483–97.
11. Piantadosi CA, Withers CM, Bartz RR, MacGarvey NC, Fu P, Sweeney TE, et al. Heme oxygenase-1 couples activation of mitochondrial biogenesis to anti-inflammatory cytokine expression. J Biol Chem. 2011;286(18):16374–85.
12. Paine A, Eiz-Vesper B, Blasczyk R, Immenschuh S. Signaling to heme oxygenase-1 and its anti-inflammatory therapeutic potential. Biochem Pharmacol. 2010;80(12):1895–903.
13. Frantz S, Vincent KA, Feron O, Kelly RA. Innate immunity and angiogenesis. Circ Res. 2005;96(1):15–26.
14. Deng S, Yu K, Zhang B, Yao Y, Liu Y, He H, et al. Effects of over-expression of TLR2 in transgenic goats on pathogen clearance and role of up-regulation of lysozyme secretion and infiltration of inflammatory cells. BMC Vet Res. 2012;8:196.
15. Ma H, Ning ZH, Lu Y, Han HB, Wang SH, Mu JF, et al. Monocytes-macrophages phagocytosis as a potential marker for disease resistance in generation 1 of dwarf chickens. Poult Sci. 2010;89(9):2022–9.
16. Satoh M, Shimoda Y, Akatsu T, Ishikawa Y, Minami Y, Nakamura M. Elevated circulating levels of heat shock protein 70 are related to systemic inflammatory reaction through monocyte Toll signal in patients with heart failure after acute myocardial infarction. Eur J Heart Fail. 2006;8(8):810–5.
17. Song Z, Deng X, Chen W, Xu J, Chen S, Zhong H, et al. Toll-like receptor 2 agonist Pam3CSK4 up-regulates FcεRI receptor expression on monocytes from patients with severe extrinsic atopic dermatitis. J Eur Acad Dermatol Venereol. 2015;29(11):2169–76.
18. Supajatura V, Ushio H, Nakao A, Akira S, Okumura K, Ra C, et al. Differential responses of mast cell Toll-like receptors 2 and 4 in allergy and innate immunity. J Clin Invest. 2002;109(10):1351–9.
19. Mersmann J, Berkels R, Zacharowski P, Tran N, Koch A, Iekushi K, et al. Preconditioning by toll-like receptor 2 agonist Pam3CSK4 reduces CXCL1-dependent leukocyte recruitment in murine myocardial ischemia/reperfusion injury. Crit Care Med. 2010;38(3):903–9.
20. Shi Q, Le X, Abbruzzese JL, Wang B, Mujaida N, Matsushima K, et al. Cooperation between transcription factor AP-1 and NF-kappaB in the induction of interleukin-8 in human pancreatic adenocarcinoma cells by hypoxia. J Interferon Cytokine Res. 1999;19(12):1363–71.
21. Abarbanell AM, Wang Y, Herrmann JL, Weil BR, Poynter JA, Manukyan MC, et al. Toll-like receptor 2 mediates mesenchymal stem cell-associated myocardial recovery and VEGF production following acute ischemia-reperfusion injury. Am J Physiol Heart Circ Physiol. 2010;298(5):H1529–36.
22. Zhong J, Shi QQ, Zhu MM, Shen J, Wang HH, Ma D, et al. MFHAS1 is associated with sepsis and stimulates TLR2/NF-κB signaling pathway following negative regulation. PLoS One. 2015;10(11), e0143662.
23. Wang J, Hu Y, Deng WW, Sun B. Negative regulation of Toll-like receptor signaling pathway. Microbes Infect. 2009;11(3):321–7.
24. Ruth MR, Field CJ. The immune modifying effects of amino acids on gut-associated lymphoid tissue. J Anim Sci Biotechnol. 2013;4(1):27.
25. Blumenthal A, Kobayashi T, Pierini LM, Banaei N, Ernst JD, Miyake K, et al. RP105 facilitates macrophage activation by Mycobacterium tuberculosis lipoproteins. Cell Host Microbe. 2009;5(1):35–46.
26. Agrawal S, Agrawal A, Doughty B, Gerwitz A, Blenis J, Van Dyke T, et al. Cutting edge: different Toll-like receptor agonists instruct dendritic cells to induce distinct Th responses via differential modulation of extracellular signal-regulated kinase-mitogen-activated protein kinase and c-Fos. J Immunol. 2003;171(10):4984–9.
27. Kawai T, Akira S. The role of pattern-recognition receptors in innate immunity: update on Toll-like receptors. Nat Immunol. 2010;11(5):373–84.
28. Ho FM, Kang HC, Lee ST, Chao Y, Chen YC, Huang LJ, et al. The anti-inflammatory actions of LCY-2-CHO, a carbazole analogue, in vascular smooth muscle cells. Biochem Pharmacol. 2007;74(2):298–308.
29. Lee SH, Sohn DH, Jin XY, Kim SW, Choi SC, Seo GS. 2', 4', 6'-tris (methoxymethoxy) chalcone protects against trinitrobenzene sulfonic acid-induced colitis and blocks tumor necrosis factor-alpha-induced intestinal epithelial inflammation via heme oxygenase 1-dependent and independent pathways. Biochem Pharmacol. 2007;74(6):870–80.
30. Healy ZR, Lee NH, Gao X, Goldring MB, Talalay P, Kensler TW, et al. Divergent responses of chondrocytes and endothelial cells to shear stress: cross-talk among COX-2, the phase 2 response, and apoptosis. Proc Natl Acad Sci U S A. 2005;102(39):14010–5.
31. Yao XM, Chen H, Li Y. Protective effect of bicyclol on liver injury induced by hepatic warm ischemia/reperfusion in rats. Hepatol Res. 2009;39(8):833–42.
32. Pedersen TX, Leethanakul C, Patel V, Mitola D, Lund LR, Danø K, et al. Laser capture microdissection-based in vivo genomic profiling of wound keratinocytes identifies similarities and differences to squamous cell carcinoma. Oncogene. 2003;22(25):3964–76.

33. Malhotra D, Thimmulappa R, Navas-Acien A, Sandford A, Elliott M, Singh A, et al. Decline in NRF2-regulated antioxidants in chronic obstructive pulmonary disease lungs due to loss of its positive regulator, DJ-1. Am J Respir Crit Care Med. 2008;178(6):592–604.

34. Innamorato NG, Rojo AI, García-Yagüe AJ, Yamamoto M, De Ceballos ML, Cuadrado A. The transcription factor Nrf2 is a therapeutic target against brain inflammation. J Immunol. 2008;181(1):680–9.

35. Ogawa S, Lozach J, Jepsen K, Sawka-Verhelle D, Perissi V, Sasik R, et al. A nuclear receptor corepressor transcriptional checkpoint controlling activator protein 1-dependent gene networks required for macrophage activation. Proc Natl Acad Sci U S A. 2004;101(40):14461–6.

36. Jaiswal AK. Regulation of genes encoding NAD (P) H: quinone oxidoreductases. Free Radic Biol Med. 2000;29(3–4):254–62.

37. Wilkinson J, Radjendirane V, Pfeiffer GR, Jaiswal AK, Clapper ML. Disruption of c-Fos leads to increased expression of NAD (P) H: quinone oxidoreductase1 and glutathione S-transferase. Biochem Biophys Res Commun. 1998;253(3):855–8.

38. MacGarvey NC, Suliman HB, Bartz RR, Fu P, Withers CM, Welty-Wolf KE, et al. Activation of mitochondrial biogenesis by heme oxygenase-1-mediated NF-E2-related factor-2 induction rescues mice from lethal Staphylococcus aureus sepsis. Am J Respir Crit Care Med. 2012;185(8):851–61.

39. Calkins MJ, Johnson DA, Townsend JA, Vargas MR, Dowell JA, Williamson TP, et al. The Nrf2/ARE pathway as a potential therapeutic target in neurodegenerative disease. Antioxid Redox Signal. 2009;11(3):497–508.

40. Terry CM, Clikeman JA, Hoidal JR, Callahan KS. Effect of tumor necrosis factor-alpha and interleukin-1 alpha on heme oxygenase-1 expression in human endothelial cells. Am J Physiol. 1998;274(3 Pt 2):H883–91.

41. Harada H, Sugimoto R, Watanabe A, Taketani S, Okada K, Warabi E, et al. Differential roles for Nrf2 and AP-1 in upregulation of HO-1 expression by arsenite in murine embryonic fibroblasts. Free Radic Res. 2008;42(4):297–304.

42. Lee IT, Wang SW, Lee CW, Chang CC, Lin CC, Luo SF, et al. Lipoteichoic acid induces HO-1 expression via the TLR2/MyD88/c-Src/NADPH oxidase pathway and Nrf2 in human tracheal smooth muscle cells. J Immunol. 2008;181(7):5098–110.

Permissions

All chapters in this book were first published in JASB, by BioMed Central; hereby published with permission under the Creative Commons Attribution License or equivalent. Every chapter published in this book has been scrutinized by our experts. Their significance has been extensively debated. The topics covered herein carry significant findings which will fuel the growth of the discipline. They may even be implemented as practical applications or may be referred to as a beginning point for another development.

The contributors of this book come from diverse backgrounds, making this book a truly international effort. This book will bring forth new frontiers with its revolutionizing research information and detailed analysis of the nascent developments around the world.

We would like to thank all the contributing authors for lending their expertise to make the book truly unique. They have played a crucial role in the development of this book. Without their invaluable contributions this book wouldn't have been possible. They have made vital efforts to compile up to date information on the varied aspects of this subject to make this book a valuable addition to the collection of many professionals and students.

This book was conceptualized with the vision of imparting up-to-date information and advanced data in this field. To ensure the same, a matchless editorial board was set up. Every individual on the board went through rigorous rounds of assessment to prove their worth. After which they invested a large part of their time researching and compiling the most relevant data for our readers.

The editorial board has been involved in producing this book since its inception. They have spent rigorous hours researching and exploring the diverse topics which have resulted in the successful publishing of this book. They have passed on their knowledge of decades through this book. To expedite this challenging task, the publisher supported the team at every step. A small team of assistant editors was also appointed to further simplify the editing procedure and attain best results for the readers.

Apart from the editorial board, the designing team has also invested a significant amount of their time in understanding the subject and creating the most relevant covers. They scrutinized every image to scout for the most suitable representation of the subject and create an appropriate cover for the book.

The publishing team has been an ardent support to the editorial, designing and production team. Their endless efforts to recruit the best for this project, has resulted in the accomplishment of this book. They are a veteran in the field of academics and their pool of knowledge is as vast as their experience in printing. Their expertise and guidance has proved useful at every step. Their uncompromising quality standards have made this book an exceptional effort. Their encouragement from time to time has been an inspiration for everyone.

The publisher and the editorial board hope that this book will prove to be a valuable piece of knowledge for researchers, students, practitioners and scholars across the globe.

List of Contributors

Ju Kyoung Oh, Edward Alain B. Pajarillo, Jong Pyo Chae, In Ho Kim and Dae-Kyung Kang
Department of Animal Resources Science, Dankook University, 119 Dandae-ro, Cheonan 31116, Republic of Korea

Dong Soo Yang
Abson BioChem, Inc, 10-1 Yangjimaeul-gil, Sangrok-gu, Ansan 15524, Republic of Korea

Bożena Króliczewska, Dorota Miśta and Wojciech Zawadzki
Department of Animal Physiology and Biostructure, Faculty of Veterinary Medicine, Wroclaw University of Environmental and Life Sciences, C.K Norwida 31, 50-375 Wrocław, Poland

Stanisław Graczyk and Aleksandra Pliszczak-Król
Department of Immunology, Pathophysiology and Veterinary Preventive Medicine, Faculty of Veterinary Medicine, Wroclaw University of Environmental and Life Sciences, Wrocław, Poland

Jarosław Króliczewski
Department of Chemical Biology, Faculty of Biotechnology, University of Wrocław, Fryderyka Joliot-Curie 14a, 50-383 Wrocław, Poland

Xiaoyong Chen
College of Animal Science and Technology, China Agricultural University, Beijing 100193, China
Institute of Animal Science and Veterinary of Hebei Province, Baoding 071000, China

Dan Wang, Hai Xiang, Dave O. H. Brahi, Tao Yin and Xingbo Zhao
College of Animal Science and Technology, China Agricultural University, Beijing 100193, China

Weitao Dun
Institute of Animal Science and Veterinary of Hebei Province, Baoding 071000, China

Omar Bulgari
Department of Animal Sciences and Division of Nutritional Sciences, University of Illinois at Urbana-Champaign, Urbana, IL 61801, USA
Department of Molecular and Translational Medicine, University of Brescia, Brescia 25123, Italy

Xianwen Dong
Department of Animal Sciences and Division of Nutritional Sciences, University of Illinois at Urbana-Champaign, Urbana, IL 61801, USA
Institute of Animal Nutrition, Sichuan Agricultural University, Chengdu 611130, China

Alfred L. Roca
Department of Animal Sciences and Division of Nutritional Sciences, University of Illinois at Urbana-Champaign, Urbana, IL 61801, USA

Anna M. Caroli
Department of Molecular and Translational Medicine, University of Brescia, Brescia 25123, Italy

Juan J. Loor
Department of Animal Sciences and Division of Nutritional Sciences, University of Illinois at Urbana-Champaign, Urbana, IL 61801, USA
Division of Nutritional Sciences, University of Illinois at Urbana-Champaign, Urbana, IL 61801, USA

Yingping Xiao, Kaifeng Li, Guohong Gui and Hua Yang
Institute of Quality and Standards for Agro-products, Zhejiang Academy of Agricultural Sciences, Hangzhou 310021, China

Choufei Wu
College of Life Sciences, Huzhou University, Huzhou 313000, China

Guolong Zhang
Department of Animal Science, Oklahoma State University, Stillwater, Oklahoma 74078, USA

Tao Ma, Yan Tu, Naifeng Zhang, Bingwen Si and Qiyu Diao
Feed Research Institute/Key laboratory of Feed Biotechnology of the Ministry of Agriculture, Chinese Academy of Agricultural Sciences, Beijing, China

Kaidong Deng
College of Animal Science, Jinling Institute of Technology, Nanjing, Jiangsu, China

Guishan Xu
Feed Research Institute/Key laboratory of Feed Biotechnology of the Ministry of Agriculture, Chinese Academy of Agricultural Sciences, Beijing, China
College of Animal Science, Tarim University, Alar, Xinjiang, China

Thiago L. A. C. de Araújo, Elzânia S. Pereira, Ana C. N. Campos, Marília W. F. Pereira, Eduardo L. Heinzen and Luciano P da Silva
Department of Animal Science, Federal University of Ceara, Fortaleza 60356001, Ceara, Brazil

Ivone Y. Mizubuti
Department of Animal Science, State University of Londrina, Londrina 86051990, Paraná, Brazil

Hilton C. R. Magalhães
Laboratory of Sensory Analysis, Agency for Agricultural Research (EMBRAPA - Tropical Agroindustry), Fortaleza 60511110, Ceará, Brazil

Leilson R. Bezerra
Department of Animal Science, Campus Professora Cinobelina Elvas, Federal University of Piauí, Bom Jesus 64900000, Piaui, Brazil

Ronaldo L. Oliveira
Department of Animal Science, School of Veterinary Medicine and Animal Science/Federal University of Bahia, Salvador City, Bahia State 40.170-110, Brazil

Adrianna Sobolewska, Joanna Bogucka, Agata Dankowiakowska, Anna Kułakowska, Agata Szczerba, Katarzyna Stadnicka and Marek Bednarczyk
Department of Animal Biochemistry and Biotechnology, University of Science and Technology in Bydgoszcz, Mazowiecka 28 Street, 85-084 Bydgoszcz, Poland

Gabriela Elminowska-Wenda and Michał Szpinda
Department of Normal Anatomy, the Ludwik Rydygier Collegium Medicum in Bydgoszcz, the Nicolaus Copernicus University in Torun, 24 Karłowicza Street, Bydgoszcz 85-092, Poland

Wei Jin, Yanfen Cheng and Weiyun Zhu
Jiangsu Province Key Laboratory of Gastrointestinal Nutrition and Animal Health; Laboratory of Gastrointestinal Microbiology, College of Animal Science and Technology, Nanjing Agricultural University, Nanjing 210095, China

Achille Schiavone
Department of Veterinary Sciences, University of Turin, Largo Paolo Braccini 2, 10095 Grugliasco, TO, Italy
Institute of Science of Food Production, National Research Council, Largo Paolo Braccini 2, 10095 Grugliasco, TO, Italy

Michele De Marco and Pierluca Costa
Department of Veterinary Sciences, University of Turin, Largo Paolo Braccini 2, 10095 Grugliasco, TO, Italy

Silvia Martínez, Josefa Madrid and Fuensanta Hernandez
Department of Animal Production, University of Murcia, Campus de Espinardo, 30071 Murcia, Spain

Sihem Dabbou
Department of Veterinary Sciences, University of Turin, Largo Paolo Braccini 2, 10095 Grugliasco, TO, Italy
Department of Agricultural, Forest and Food Sciences, University of Turin, Largo Paolo Braccini 2, 10095 Grugliasco, TO, Italy

Manuela Renna and Luca Rotolo
Department of Agricultural, Forest and Food Sciences, University of Turin, Largo Paolo Braccini 2, 10095 Grugliasco, TO, Italy

Francesco Gai
Institute of Science of Food Production, National Research Council, Largo Paolo Braccini 2, 10095 Grugliasco, TO, Italy

Laura Gasco
Department of Agricultural, Forest and Food Sciences, University of Turin, Largo Paolo Braccini 2, 10095 Grugliasco, TO, Italy
Institute of Science of Food Production, National Research Council, Largo Paolo Braccini 2, 10095 Grugliasco, TO, Italy

Moana Rodrigues França, Maressa Izabel Santos da Silva, Guilherme Pugliesi and Mario Binelli
Department of Animal Reproduction, School of Veterinary Medicine and Animal Science, University of São Paulo, 225, Duque de Caxias Norte Ave. Jd. Elite, 13635-900 Pirassununga, SP, Brazil

Veerle Van Hoeck
Gamete Research Centre, University of Antwerp, Antwerp, Belgium

Kaiji Sun
State Key Laboratory of Animal Nutrition, College of Animal Science and Technology, China Agricultural University, Beijing 100193, China

Yan Lei
DadHank (Chengdu) Biotech Corp, Sichuan, China

Renjie Wang
State Key Laboratory of Animal Nutrition, College of Animal Science and Technology, China Agricultural University, Beijing 100193, China
DadHank (Chengdu) Biotech Corp, Sichuan, China

Zhenlong Wu
State Key Laboratory of Animal Nutrition, College of Animal Science and Technology, China Agricultural University, Beijing 100193, China
Department of Animal Nutrition and Feed Science, China Agricultural University, Beijing 100193, China

Guoyao Wu
State Key Laboratory of Animal Nutrition, College of Animal Science and Technology, China Agricultural University, Beijing 100193, China
Department of Animal Science, Texas A&M University, College Station, TX 77843, USA

A. Sobolewskak, J. Bogucka, A. Dankowiakowska, K. Stadnicka and M. Bednarczyk
Department of Animal Biochemistry and Biotechnology, University of Science and Technology in Bydgoszcz, Mazowiecka 28, 85-084 Bydgoszcz, Poland

So Dam Jin, Bo Ram Lee, Young Sun Hwang, Hong Jo Lee and Jong Seop Rim
Department of Agricultural Biotechnology, Research Institute of Agriculture and Life Sciences, College of Agriculture and Life Sciences, Seoul National University, Seoul 08826, South Korea

Jae Yong Han
Department of Agricultural Biotechnology, Research Institute of Agriculture and Life Sciences, College of Agriculture and Life Sciences, Seoul National University, Seoul 08826, South Korea
Institute for Biomedical Sciences, Shinshu University, Minamiminowa, Nagano 399-4598, Japan

Lu Liu, Huanxian Cui, Ruiqi Fu, Maiqing Zheng, Ranran Liu, Guiping Zhao and Jie Wen
Institute of Animal Sciences, Chinese Academy of Agricultural Sciences, Beijing 100193, China
State Key Laboratory of Animal Nutrition, Beijing 100193, China

Feng Zhu, Yahui Gao, Fangbin Lin and Zhuocheng Hou
National Engineering Laboratory for Animal Breeding and MOA Key Laboratory of Animal Genetics and Breeding, Department of Animal Genetics and Breeding, China Agricultural University, Beijing 100193, China

Jinping Hao and Fangxi Yang
Beijing Golden Star Duck Inc., Beijing 100076, China

Katerina Adamkova and Pavla Jelinkova
Department of Veterinary Sciences, Faculty of Agriculture, Food and Natural Resources, Czech University of Life Sciences Prague, 6-Suchdol, Prague, Czech Republic

Young-Joo Yi
Division of Biotechnology, Safety, Environment and Life Science Institute, College of Environmental and Bioresource Sciences, Chonbuk National University, Iksan 54596, South Korea

Jaroslav Petr
Institute of Animal Science, 10-Uhrineves, Prague, Czech Republic

Tereza Zalmanova and Kristyna Hoskova
Department of Veterinary Sciences, Faculty of Agriculture, Food and Natural Resources, Czech University of Life Sciences Prague, 6-Suchdol, Prague, Czech Republic
Institute of Animal Science, 10-Uhrineves, Prague, Czech Republic

Jiri Moravec
Proteomic Laboratory, Biomedical Center of Faculty of Medicine in Pilsen, Charles University, Pilsen, Czech Republic

Milena Kralickova
Laboratory of Reproductive Medicine of Biomedical Center, Charles University, Pilsen, Czech Republic
Department of Histology and Embryology, Faculty of Medicine in Pilsen, Charles University, Pilsen, Czech Republic

Jan Nevoral
Department of Veterinary Sciences, Faculty of Agriculture, Food and Natural Resources, Czech University of Life Sciences Prague, 6-Suchdol, Prague, Czech Republic
Laboratory of Reproductive Medicine of Biomedical Center, Charles University, Pilsen, Czech Republic
Department of Histology and Embryology, Faculty of Medicine in Pilsen, Charles University, Pilsen, Czech Republic

Miriam Sutovsky
Division of Animal Science, University of Missouri, Columbia, MO, USA

Peter Sutovsky
Division of Animal Science, University of Missouri, Columbia, MO, USA
Departments of Obstetrics, Gynecology and Women's Health, University of Missouri, Columbia, MO, USA

A. K. Kelly
School of Agriculture and Food Science, College of Health and Agricultural Sciences, University College Dublin, Belfield, Dublin 4, Ireland

P. Lawrence and M. McGee
Livestock Systems Research Department Teagasc, Animal & Grassland Research and Innovation Centre, Grange, Dunsany, Co. Meath, Ireland

Mette Olaf Nielsen
Department of Veterinary and Animal Sciences, Faculty of Health and Medical Sciences, University of Copenhagen, Grønnegårdsvej 3, 1st floor, DK-1870 Frederiksberg C, Denmark

Prabhat Khanal
Department of Veterinary and Animal Sciences, Faculty of Health and Medical Sciences, University of Copenhagen, Grønnegårdsvej 3, 1st floor, DK-1870 Frederiksberg C, Denmark
Department of Nutrition, Faculty of Medicine, Transgenic Animal and Lipid Storage, Norwegian Transgenic Centre (NTS), University of Oslo, Oslo, Norway

Paolo Trevisi, Davide Priori, Vincenzo Motta, Diana Luise and Paolo Bosi
DISTAL, University of Bologna, Viale Fanin 46, 40127 Bologna, Italy

Alfons J. M. Jansman and Sietse-Jan Koopmans
Wageningen University and Research, Wageningen, The Netherlands

Shoulong Deng and Yixun Liu
State Key Laboratory of Stem Cell and Reproductive Biology, Institute of Zoology, Chinese Academy of Sciences, Beijing 100101, China

Kun Yu and Zhengxing Lian
Laboratory of Animal Genetics and Breeding, College of Animal Science and Technology, China Agricultural University, Beijing 100193, People's Republic of China
National key Lab of Agrobiotechnology, College of Biological Sciences, China Agricultural University, Beijing 100193, People's Republic of China

Wuqi Jiang, Yan Li, Shuotian Wang and Guoshi Liu
Laboratory of Animal Genetics and Breeding, College of Animal Science and Technology, China Agricultural University, Beijing 100193, People's Republic of China

Zhuo Deng
Department of Animal Science, Oklahoma State University, Stillwater, OK 74078, USA

Yuchang Yao
College of Animal Science and Technology, Northeast Agricultural University, Harbin 150030, People's Republic of China

Baolu Zhang
State Oceanic Administration, Beijing 100860, People's Republic of China

Index